T0181733

An Introduction to Ocean Turbulence

This textbook provides an introduction to turbulent motion occurring naturally in the ocean on scales ranging from millimetres to hundreds of kilometres. It describes how turbulence is created and varies from one part of the ocean to another, what its properties are (particularly those relating to energy flux and the dispersal of pollutants) and how it is measured. Examples are given of real data and the instruments that are commonly used to measure turbulence. Chapters describe turbulence in the mixed boundary layers at the sea surface and seabed, turbulent motion in the density-stratified water between, and the energy sources that support and sustain ocean mixing.

Little prior knowledge of physical oceanography is assumed and the book is written at an introductory level that avoids mathematical complexity. The text is supported by numerous figures illustrating the methods used to measure and analyse turbulence, and by more than 50 exercises, which are graded in difficulty, that will allow readers to expand and monitor their understanding and to develop analytical techniques. Detailed solutions to the exercises are available to instructors online at www.cambridge.org/9780521676809. Further reading lists give direction to additional information on the background and historical development of the subject, while suggestions for further study encourage readers to probe further into more advanced aspects.

An Introduction to Ocean Turbulence is intended for undergraduate courses in physical oceanography, but will also form a useful guide for graduate students and researchers interested in multidisciplinary aspects of how the ocean works, from the surface to the seabed and from the shoreline to the deep abyssal plains. It complements the graduate-level text *The Turbulent Ocean*, also written by Professor Thorpe (Cambridge University Press, 2005).

STEVE THORPE was a Senior Scholar at Trinity College, Cambridge, where he studied mathematics and fluid mechanics, his PhD being awarded in 1966. He then spent 20 years at the UK Institute of Oceanographic Sciences, before being appointed Professor of Oceanography at Southampton University in 1986. He has carried out laboratory experiments on internal waves and turbulent mixing, and has measured and developed instrumental and analytical methods for studying waves and mixing in lakes, as well as making seagoing studies of turbulence in the boundary layers of the deep ocean and shelf seas. Professor Thorpe was awarded the Walter Munk Award by the US Office of Naval Research and the Oceanography Society, for his work using underwater acoustics, The Fridtjof Nansen Medal of the European Geophysical Society, for his fundamental experimental and theoretical contributions to the study of mixing and internal waves, and the Society's Golden Badge for introducing a scheme to assist young scientists. He became a Fellow of the Royal Society in 1991 and is now an Emeritus Professor at the University of Southampton and an Honorary Professor at the School of Ocean Sciences, Bangor.

An Introduction to Ocean Turbulence

S. A. Thorpe

CAMBRIDGE
UNIVERSITY PRESS

CAMBRIDGE
UNIVERSITY PRESS

University Printing House, Cambridge CB2 8BS, United Kingdom

One Liberty Plaza, 20th Floor, New York, NY 10006, USA

477 Williamstown Road, Port Melbourne, VIC 3207, Australia

314-321, 3rd Floor, Plot 3, Splendor Forum, Jasola District Centre, New Delhi - 110025, India

79 Anson Road, #06-04/06, Singapore 079906

Cambridge University Press is part of the University of Cambridge.

It furthers the University's mission by disseminating knowledge in the pursuit of education, learning and research at the highest international levels of excellence.

www.cambridge.org
Information on this title: www.cambridge.org/9780521676809

First published 2007

A catalogue record for this publication is available from the British Library

ISBN 978-0-521-85948-6 Hardback
ISBN 978-0-521-67680-9 Paperback

Contents

Preface

My book entitled *The Turbulent Ocean* (referred to later as TTO) was written in 2003. It provides an account of much of the knowledge that there was then of the processes leading to turbulence in the ocean, but it was not written as a course that might be followed and used to introduce students to turbulent flow. Rather, it is a text useful for those beginning or already involved in research. It might form the basis of a number of advanced courses about ocean physics, teachers selecting material according to their needs or specialities.

I was asked to write a shorter book, an introductory course on turbulence in the ocean. Although believing that the best undergraduate and postgraduate courses are based and modelled on a teacher's own experience and enthusiasms, and that to follow a 'set text' may be less enjoyable for students, I became convinced that a simplified text, more directly usable in teaching students unfamiliar with fluid motion, might be of value. Turbulence is a subject of which at least a basic understanding is essential in engineering and in many of the natural sciences, but particularly for students of oceanography. Moreover, many students, whose main interests are not in oceanography and who will not later address their talents to the study of the ocean, find interest in the sea and are motivated by aspects of their studies that are related or have application to matters of public and international concern, for example those of pollution and climate change that are at present being addressed by ocean scientists. A study of turbulent motion set in an oceanographic context can be attractive, satisfying and stimulating.

The purpose of the present book is consequently to provide a text that might be used in constructing and teaching an introductory course to students with a variety of academic abilities but who know little of ocean physics or turbulence. Much of the content has developed from a second-year course on ocean physics given in the Department of Oceanography at Southampton University, UK, in some 16 hours of lectures over a period of 4 weeks, supplemented by problems and additional reading undertaken by the students, a course attended by students whose main interests were in mathematics, physics, geology, biological oceanography or, generally, in marine science.

As in *The Turbulent Ocean*, the intricacy of turbulence theory is omitted. I recall, when an undergraduate, being totally mystified, if not frightened, by introductory lectures on turbulent motion that dealt with the subject in a largely statistical and analytical way, giving little or no insight into the dynamical processes of how it works. Unless students have a relatively high degree of ability in mathematics, the theoretical background is better faced after the basic concepts and ideas underlying the processes relating to turbulent motion have been absorbed and understood, and perhaps even after students have some understanding of the methods used to observe and measure turbulence. Neither is the numerical modelling of turbulence discussed here. That is best introduced to students in a separate and probably more mathematically demanding course once the processes involved in turbulence are firmly understood.

Unlike the earlier text, the material is almost entirely (but not quite!) restricted to what is well established and known, but I have also tried to explain the present limits of knowledge. I have taken the opportunity to include information that has been published since TTO was written, and to draw attention to errors that have come to my attention (specifically in footnotes 6 and 13 of Chapter 1, and footnote 13 of Chapter 6). I should be glad to be informed of any further errors found by readers in either that or this book.

S. A. Thorpe
'Bodfryn', Glanrafon, Llangoed, Anglesey LL58 8PH, UK

Notes on the text

The symbol • denotes important points or summary statements.

There are six chapters with substantial cross-referencing between them. The first is intended as a general introduction, before means of quantifying and measuring turbulence are introduced in Chapter 2. Chapter 3 deals with the turbulent boundary layers near the sea surface and seabed. Chapter 4 describes the relatively weak and patchy turbulent motion that is found in the density-stratified water between these two boundary layers. Chapter 5 is about turbulent dispersion, whilst Chapter 6 is a discussion of the present (and rapidly developing) knowledge of the sources and rates of supply of turbulence energy required to support mixing in the deep ocean.

The **illustrations** are a very important supplement to the text. It is through pictures that information is carried most readily, and often in the most pleasurable form, to the mind and memory of a reader. 'Cartoons' (or sketches) conveying new ideas or concepts, photographs and data presented in graphical form are often an output of research, to which they provide a useful introduction or overview. The figure captions add substantial information that is not always included within the text.

Lists of **Suggested further reading** are provided at the end of each chapter. These are of literature that students might be expected to peruse, if not read in detail, in the course of their study of the contents of the chapter, e.g., to appreciate better the historical derivation of knowledge. Also listed are reference works that will provide information about basic fluid dynamics or ocean physics, should it be required.

Papers referred to under **Further study** are guides to encourage more extensive in-depth study of the material of the chapter, possibly leading to new research. In many cases another pathway into such further study is through the sources of figures referred to in the figure captions.

Problems are listed at the end of each chapter and are denoted at a point in the text where they might be attempted by [**Pm.n**], where **m** is a chapter number and **n** the problem number within the chapter. Each problem number in this list is followed by a letter that denotes the problem's degree of difficulty: E = easy, M = mild, D = difficult and F = fiendish. The problems allow students to re-discover for themselves some of the now-accepted relationships, and provide experience in calculation and problem solving. These problems are *essential* elements in developing the ideas introduced in the text, and provide much additional information. They should preferably be read (if not solved) as students or readers advance through the course. The solutions to the problems are not given in this book but password-protected solutions to the problems are available online at www.cambridge.org/9780521859486. Quantitatively correct solutions are less important than the concepts introduced by the problems.

Lists of abbreviations, useful values etc. are provided on pages xv–xx for easy reference, and a map showing locations of places to which reference is made in the text is included on page xxii.

Scientific papers and books mentioned in text are all listed in the **References**, together with the numbers of pages on which they are mentioned.

The **Index** provides an entry to subjects that students may wish to locate or pursue, including 'dimensional arguments'.

Acknowledgements

Mrs Kate Davis has kindly helped in the preparation of figures and I am most grateful for her care and professional attention to detail. I am also grateful to Drs Larry Armi and Jim Moum for providing very helpful information about their teaching courses.

I am particularly grateful to the many individuals who have allowed me to reproduce their figures and photographs, or who have helped me to find suitable material, including M. H. Alford, L. Armi, E. D'Asaro, O. Brown, B. Brügge, J. Bryan, D. R. Caldwell, M. Carle, R. E. Davis, A. J. Elliott, R. Evans, D. M. Farmer, B. Ferron, O. Fringer, M. C. Gregg, A. J. Hall, A. D. Heathershaw, B. Hickey, H. E. Huppert, B. C. Kenney, B. King, J. Larsen, J. R. Ledwell, J. E. Lupton, W. K. Melville, J. Miles, J. N. Moum, A. Nimmo Smith, T. R. Osborn, G. Ostlund, P. E. Oswald, K. Polzin, B. S. Rarity, R. D. Ray, A. Roshko, P. L. Richardson, L. St. Laurent, R. W. Schmitt, R. Scorer, J. H. Simpson, W. D. Smyth, A. Souza, P. Stegmann, J. C. Stephens J. R. Toggweiler, C. Troy, F. Veron, H. W. Wijesekera, P. J. Wiles, J. D. Woods and V. Zhurbas, and to the publishers and organizations that have granted permission, including The American Association for the Advancement of Science, The American Geophysical Union, The American Institute of Physics, The American Meteorological Society, Blackwell Publishing Co., Cambridge University Press, CEFAS, Elsevier Ltd, The Journal of Marine Research/Yale University, MacMillian Publishers Ltd, NASA/GSFC and ORBIMAGE, The Royal Society and the Scientific Commission on Oceanic Research.

I also much appreciate the friendly help provided by the staff of Cambridge University Press in the preparation of this book and their care in its publication.

Abbreviations

AABW	Antarctic Bottom Water
abl	atmospheric boundary layer
ADCP	acoustic Doppler current profiler
ALACE	Autonomous Lagrangian Circulation Explorer
AMP	Advanced Microstructure Profiler
AUV	autonomous underwater vehicle
bbl	benthic or bottom boundary layer
CTD	conductivity–temperature–depth probe
FLIP	Floating Instrument Platform
HAB	harmful algal bloom
HOME	Hawaiian Ocean Mixing Experiment
HRP	High Resolution Profiler
LES	large eddy simulation
lhs	left-hand side
MSP	Multi-Scale Profiler
PIV	particle image velocimetry
pd	potential difference
pdf	probability distribution function (or histogram)
rhs	right-hand side
rms	root mean square
RFZ	Romanche Fracture Zone
SOFAR	SOund Fixing And Ranging
STABLE	Sediment Transport And Boundary Layer Equipment
TTO	*The Turbulent Ocean* by S. A.Thorpe, Cambridge University Press, 2005
VACM	vector-averaging current meter

Standard parameters and symbols

(with the section and, where appropriate, equation in which they are introduced)

C	Cox number (Section 4.4.2; (4.7)–(4.8))
C_D	the drag coefficient on the seabed (Section 3.4.1). (C_{Da} is used in this text to denote the drag coefficient of the wind on the water surface, with subscript a – standing for air – to emphasize that its value is different from C_D; it is defined in Section 3.4.1.)
I	isotropy parameter (Section 2.3.5; (2.14))
K_H	eddy dispersion coefficient (Section 5.2.1; (5.5))
$K_{H\infty}$	eddy dispersion coefficient at times $\gg T_L$ (Section 5.2.1; (5.6))
K_S	eddy diffusion coefficient of salinity (Section 2.2.2)
K_T	eddy diffusion coefficient of heat or eddy diffusivity of heat (Section 2.2.2; (2.5))
K_ρ	eddy diffusion coefficient of density (Section 2.2.2)
K_ν	eddy viscosity (Section 2.2.1; (2.2))
L_L	Lagrangian integral length scale (Section 5.2.1; (5.3))
L_{MO}	Monin–Obukov length scale (Section 3.4.1; (3.6)–(3.7))
L_O	Ozmidov length scale (Section 4.4.1; (4.4))
L_{Ro}	Rossby radius (of deformation) (Section 1.8.2)
l_K	Kolmogorov length scale (Section 2.3.4)
N	buoyancy frequency (Section 1.7.2; (1.5))
Ra	Rayleigh number (Section 3.2.1, footnote 4)
Re	Reynolds number (Section 1.2; (1.1))
R_f	flux Richardson number (Section 4.4.2; (4.11))
Ri	gradient Richardson number (Section 4.2; (4.1))
Ri_B	bulk Richardson number (Section 4.5)
R_ρ	density gradient ratio (Section 4.8; (4.16))

T_L	Lagrangian integral time scale (Section 5.2.1; (5.2))
Γ	efficiency factor (Section 4.4.2; (4.9))
ε	rate of dissipation of turbulence kinetic energy per unit mass (Section 2.3.1; (2.9)–(2.11))
σ_T	sigma-T (temperature) (Section 1.7.1)
σ_θ	sigma-theta (potential temperature) (Section 1.7.1, footnote 15)
χ_S	rate of loss of salinity variance (Section 2.3.3)
χ_T	rate of loss of temperature variance (Section 2.3.3; (2.12)–(2.13))

Units and their symbols

Unit	SI symbol (name)	kg–m–s equivalent
Force	N (Newton)	kg m s^{-2}
Pressure (force per unit area)	Pa (Pascal, 1 Pa = 10^{-5} bar)	kg m^{-1} s^{-2}
Energy	J (Joule)	kg m^2 s^{-2}
Power, energy flux	W (Watt, 1W = 1 J s^{-1})	kg m^2 s^{-3}
Energy dissipation rate per unit mass	W kg^{-1}	m^2 s^{-3}
Volume flux	Sv (Sverdrup)	10^6 m^3 s^{-1}

SI prefixes

Symbol	Name	Factor
E	exa	10^{18}
P	peta	10^{15}
T	tera	10^{12}
G	giga	10^{9}
M	mega	10^{6}
k	kilo	10^{3}
d	deci	10^{-1}
m	milli	10^{-3}
μ	micro	10^{-6}
n	nano	10^{-9}
p	pico	10^{-12}
f	femto	10^{-15}

Approximate values of commonly used measures

Radius of a sphere with the same volume as the Earth $= 6371$ km
Rotation rate of the Earth, $\Omega = 7.292 \times 10^{-5}\ \mathrm{s}^{-1}$
Mean depth of the ocean $= 3.795$ km
Area of the ocean surface $= 3.61 \times 10^{14}\ \mathrm{m}^2$
Mean area of sea ice $\approx 1.7 \times 10^{13}\ \mathrm{m}^2$ in March and $2.8 \times 10^{13}\ \mathrm{m}^2$ in September
Volume of the ocean $= 1.37 \times 10^{18}\ \mathrm{m}^3$
Mass of the atmosphere $= 5.3 \times 10^{18}$ kg
Mass of the ocean $= 1.4 \times 10^{21}$ kg
Mass of water in lakes and rivers $\approx 5 \times 10^{17}$ kg
Speed of sound $\approx 1500\ \mathrm{m\ s}^{-1}$
von Kármán's constant, $k = 0.40$–0.41
1 knot $= 0.5148\ \mathrm{m\ s}^{-1}$

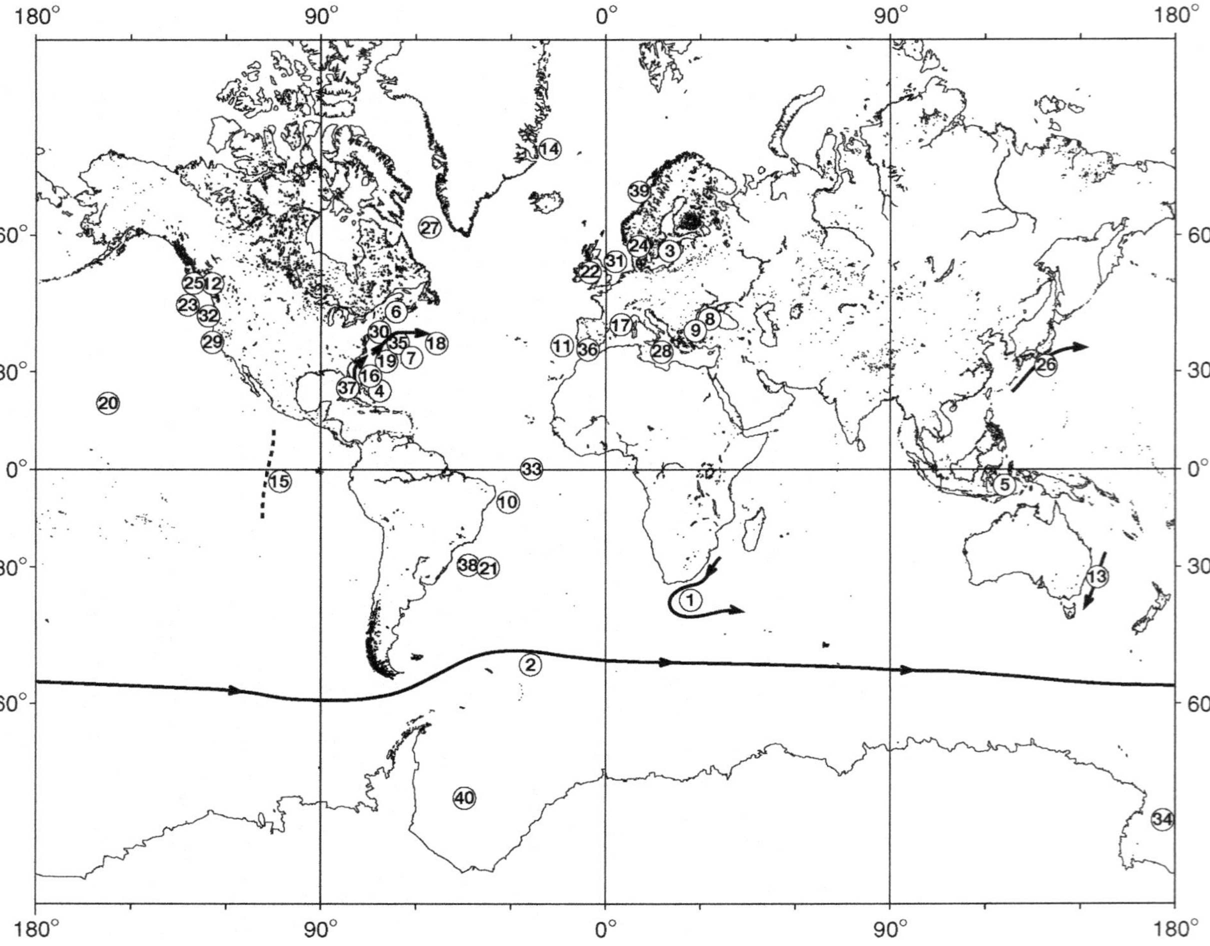
180°
90°
0°
90°
180°
60°
30°
0°
30°
60°

The general positions of some of the locations and currents referred to in the text.

1. Agulhas Retroflection Zone
2. Antarctic Circumpolar Current
3. Baltic Sea
4. Bahamas
5. Banda Sea
6. Bay of Fundy
7. Bermuda
8. Black Sea
9. Bosphorus
10. Brazil Basin
11. Discovery Gap
12. Discovery Passage
13. East Australia Current
14. East Greenland Sea
15. East Pacific Rise
16. Florida Current
17. Gulf of Lions
18. Gulf Stream
19. Hatteras Abyssal Plain
20. Hawaiian Ridge
21. Hunter Channel
22. Irish Sea
23. Juan de Fuca Ridge
24. The Kattegat
25. Knight Inlet, British Columbia
26. Kuroshio
27. Labrador Sea
28. Mediterranean Sea
29. Monterey Bay, California
30. New England Continental Shelf
31. North Sea
32. Oregon Continental Shelf
33. Romanche Fracture Zone (RFZ)
34. Ross Sea
35. Sargasso Sea
36. Strait of Gibraltar
37. Straits of Florida
38. Vema Channel
39. Vøring Plateau
40. Weddell Sea

Chapter 1

Turbulence, heat and waves

1.1 Introduction

Turbulence is the dominant physical process in the transfer of momentum and heat, and in dispersing solutes and small organic or inorganic particles, in the lakes, reservoirs, seas, oceans and fluid mantles of this and other planets. Oceanic turbulence has properties that are shared by turbulence in other naturally occurring fluids and in flows generated in civil, hydraulic and chemical engineering installations and in buildings. The study of turbulence consequently has applications well beyond the particular examples in the ocean[1] that are selected for description below.

Figure 1.1 shows the sea surface in a wind of about 26 m s^{-1}. It is covered by waves, many of them breaking and injecting their momentum and bubbles of air from the overlying atmosphere into the underlying seawater. Immediately below the surface, and even at great depths, the water is generally in the state of irregular and variable motion that is referred to as 'turbulence', although there is no simple and unambiguous definition of the term. Turbulence has, however, characteristics that, as will be explained, can be quantified and which make it of vital importance. Many of the figures in this book illustrate the nature of turbulent motion, the processes that drive turbulence, or the measurements that can be made to determine its effects.

•[2] Turbulence is generally accepted to be an energetic, rotational and eddying state of motion that results in the dispersion of material and the transfer of momentum, heat and solutes at rates far higher than those of molecular processes alone. It disperses,

1 By the 'ocean' is meant, here and later, the sum of the major oceans and their connected seas, including the continental shelf seas and those seas, such as the Mediterranean, Black Sea and Baltic, connected by straits to the larger ocean basins.

2 The symbol • is used to draw attention to paragraphs of particularly important information or summaries of the earlier text.

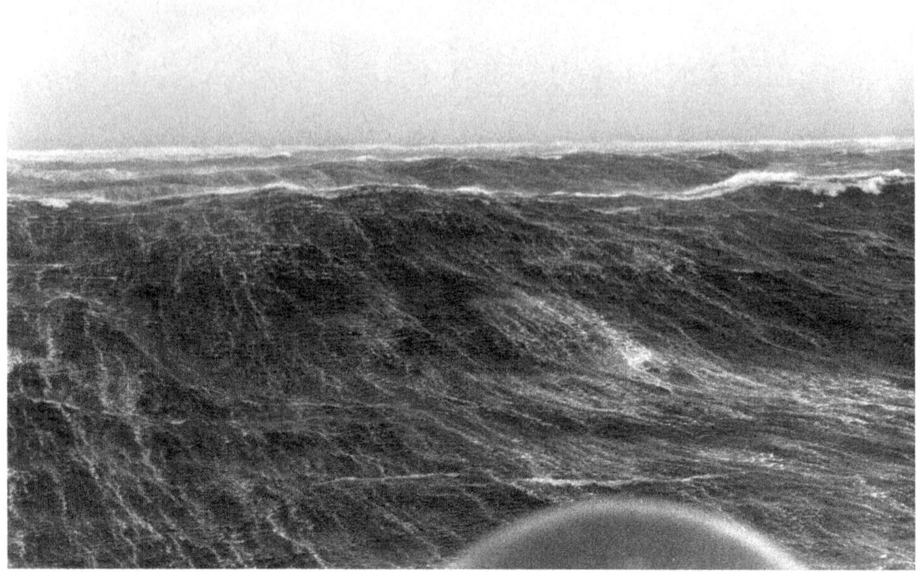

Figure 1.1. The sea surface in the Bay of Biscay looking upwind in a wind speed of about 26 m s^{-1}. Breaking is occurring at the crests of the larger waves, separated by over 100 m, and there are numerous bands of foam aligned in the downwind direction, as well as evidence of short waves with a typical wavelength of 0.2 m. (Photograph taken by Mr J. Bryan from RRS Charles Darwin off northwest Spain and reproduced with his kind permission.)

stresses and strains clusters (or flocs) of sediment or atmospheric dust particles and living organisms within the ocean, and it stirs, spreads and dilutes the chemicals that are dissolved in the seawater or released into the ocean from natural and anthropogenic sources. Perhaps its most important property, and one that is generally used to characterize it, is that by generating relatively large gradients of velocity at small scales, typically 1 mm to 1 cm, turbulence promotes conditions in which, relatively rapidly, viscous dissipation transfers the kinetic energy of turbulent motion into heat, a process of energy transfer and 'dissipation'.

Since the natural state of the ocean is one of turbulent motion, knowledge of turbulence and its effects is crucial in understanding how the ocean works and in the construction of numerical models to predict how, in the future, the ocean will adjust as the forcing by the atmosphere is altered by changes in the world's climate. Although estimates of the rate of dissipation of the energy of the tides through turbulence in shallow seas were made as early as 1919, direct measurement of turbulence in the ocean dates back only to the observations of near-bed turbulent stress made in the 1950s and to studies of the spectra of small-scale motions in the tidal Discovery Passage off the west coast of Canada in the early 1960s. In spite of the developments of ingenious techniques to measure turbulent motions, the geographical variation of turbulence in the ocean is still poorly known, its range of variability is often grossly under-sampled and, in comparison with the atmosphere, there are few sets of data against which to test those models of the ocean that include representation of its turbulent nature. Much is still to be discovered and quantified.

This chapter describes some of the ideas and discoveries that form a basis for understanding of the part played by turbulence in the ocean. Much of this background is derived from studies of turbulence and heat transfer in laboratory experiments, some of which were made well before the first measurements of turbulence in the ocean itself.

1.2 Reynolds' experiment

The scientific study of turbulence did not begin until late in the nineteenth century. The first substantial step was the publication in 1883 of a paper by Osborne Reynolds. He described how a smooth flow of water through long circular tubes with diameters ranging from about 0.6 to 2.5 cm is disrupted and cannot be sustained when the mean speed of the flow, U, exceeds a value that is related to the tube diameter, d, and to the viscosity of water. In his laboratory experiments Reynolds introduced a thin line of dye into the water entering through one end of a horizontal tube from a large tank of stationary water, dye that made the flow visible (Fig. 1.2). He described his observations as follows:

> *When the velocities were sufficiently low, the streak of colour extended in a beautiful straight line through the tube*

but later

> *As the velocity was increased by small stages, at some point in the tube, always at a considerable distance from . . . the intake, the colour band would all at once mix up with the surrounding water, and fill the tube with a mass of coloured water. On viewing the tube by light of an electric spark, the mass of colour resolved itself into a mass of more or less distinct curls, showing eddies.*

• Reynolds' remarkable experiments show that the 'laminar flow', the smooth flow through the tube at low flow speeds, breaks down into a random eddying 'turbulent' motion at higher speeds when a non-dimensional parameter, now known as the Reynolds number,

$$Re = Ud/\nu, \tag{1.1}$$

exceeds a value of about 1.3×10^4.[3] Here ν is the kinematic viscosity, which for water has a value of about 10^{-6} m^2 s^{-1}.

3 The critical value of Re is now known to depend on the level of the background disturbances to the flow (sometimes described as 'noise'), particularly near the entry of flow to the tube, the critical Re consequently ranging from about 1×10^3 for relatively substantial disturbances to about 4.5×10^4 in very carefully controlled, low-disturbance, tube flows. (See Further study.)

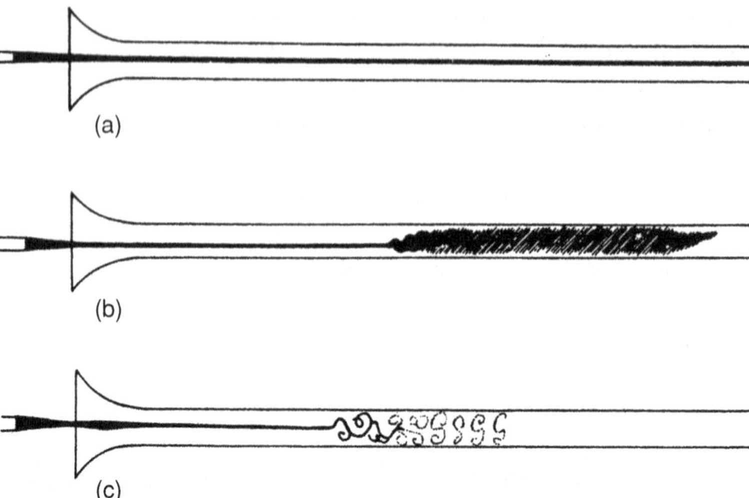

Figure 1.2. Reynolds' experiment, described in his paper published in 1883. Flow through the tube is from left to right. The shape of the entry to the tube within a large tank of still water on the left is to ensure a smooth flow. (a) A band of dye passes down the tube when the flow is relatively slow, or at a low Reynolds number, *Re*. (b) When $Re > 1.3 \times 10^4$, the flow becomes turbulent. As observed by eye, the band of dye is dispersed across the width of the tube. (c) An image obtained with a very brief electric spark, showing that the onset of turbulence, and its later form, is associated with eddies of size comparable to the tube diameter.

This provides the first example of a relationship pertaining to turbulence that can be determined on dimensional grounds. A condition for a transition to turbulence can depend only on the independent dimensional quantities that characterize or determine the state of the flow in the tube. These are its mean velocity, U (with dimensions LT^{-1}), its density, ρ (dimensions ML^{-3}), the diameter of the tube, d (dimension L), and the kinematic viscosity, ν (dimensions $\mathrm{L}^2\mathrm{T}^{-1}$), where L stands for length, T for time and M for mass. The velocity varies with distance, r, from the axis of the tube but in a way determined by d and ν, and the pressure gradient along the tube is directly related to U, ν and ρ, and so is not a quantity independent of the four chosen. The tube walls are smooth and so do not introduce a further length scale. If also the tube is 'very long', or of very much greater length than either of the two possible length scales, d and ν/U, and so does not introduce a further relevant length scale, the only non-dimensional parameter which can characterize whether the flow may become turbulent is Ud/ν, the Reynolds number, *Re*. There is no other parameter possible. Although the value of *Re* at which a transition from laminar to turbulent flow occurs cannot be determined from the dimensional argument, it serves to identify in a logical way the combination of dimensional quantities that characterize the onset of turbulence. The power of dimensional arguments in characterizing the nature of turbulent flows is demonstrated later by other examples. (For easy reference and comparison they are listed together under 'dimensional arguments' in the index.)

Some of the eddies in the turbulent flow in Reynolds' experiment are of size comparable to the tube radius, but many are smaller, smaller therefore than the distance (about the radius of the tube) over which the mean flow itself varies. As well as dispersing dye, the eddies carry fluid and momentum from the tube walls, where the presence of the boundaries and associated viscosity effects are important constraints, into the interior of the tube.

Reynolds' experiment underpins many of the concepts that we now have of turbulence. It shows that turbulence may occur as a *transition* from one state of flow to another: even the way in which the mean (time-averaged) flow speed varies with radius in the tube is changed at the onset of turbulence. Turbulence involves eddying motions, some of which are small relative to the characteristic length scale of the flow, namely the tube diameter, d, in Reynolds' experiments. It disperses dissolved matter in an irreversible way – the mixed dye cannot be unmixed [**P1.1**].

• An irregular state of fluid motion, referred to as turbulence, occurs when a *critical* value of the parameter Re, characterizing the flow, is exceeded, replacing the smooth laminar flow found at lower values of Re.

The precise value of Re at which turbulence sets in depends on the particular geometry of the flow and the nature of disturbances to which it is subjected. Geophysical flows in which a value of Re characterizing the flow exceeds about 10^4 are generally found to be turbulent unless affected by other factors (e.g., density stratification, described in Section 1.7) that suppress or delay the onset of turbulence until higher values of Re are reached. In the ocean, the values of the speed U and length d that appear in Re (1.1) are usually taken to be those characterizing the flow, for example the mean speed of the local flow and water depth (or, in mid-water, a change in mean speed over a vertical distance, d). The characteristic Re commonly exceeds 10^4 in the ocean [**P1.2**].

1.3 Joule's experiment

Heat is a form of energy contained at a molecular level within a fluid, and temperature is a measure of heat content. Heat is transferred (as a heat flux) by turbulent motion in the ocean, and turbulent energy is largely dissipated by viscosity into heat. The density of seawater depends on temperature (and usually to a lesser extent on salinity and pressure – see Section 1.7.1), and the variation of density with depth in the ocean, normally an increase in density as depth increases, strongly affects or regulates the processes leading to turbulence, as we shall see later. Heat and temperature[4] are consequently important factors in oceanic turbulence.[5]

4 The temperature is measured in degrees Kelvin, K, or °C. Both units are used, selection being made on the basis of which appears most appropriate.

5 Although variations in heat content or a flux of heat can, through the production of buoyancy forces (Section 1.7.2), lead to convective motions and therefore kinetic energy, the heat energy at molecular level is not available for transfer back to kinetic energy. This is why, although the geothermal energy flux into the ocean may exceed, for example, the flux of tidal energy, it is not as effective in producing mixing, as is explained further in Section 6.9.

• The relation between heat and temperature can be expressed in terms of a change, ΔH, in the heat per unit mass (measured in $\mathrm{J\,kg^{-1}}$) over a period of time and the corresponding change, ΔT, in temperature (in °C or K) by

$$\Delta H = c_p \, \Delta T, \tag{1.2}$$

where c_p is the specific heat at constant pressure which, for seawater, has a value now known to be about $3.99 \times 10^3\ \mathrm{J\,kg^{-1}\,K^{-1}}$.[6]

The unit in which energy is measured is named after J. P. Joule.[7] Joule's most celebrated experiment, an account of which was published in 1850, is that in which he found the relation between changes of energy in its most commonly known, well-described and easily quantified forms (potential or kinetic energy) and the temperature change resulting from the dissipation of that energy. The experiment contains at its heart the substantial effects of turbulent motion, in this case turbulence artificially induced by rotating paddles.

Joule's apparatus is sketched in Fig. 1.3 and described in the caption. The essence of the experiment is that the falling weights lose a measured amount of potential energy in driving paddles, which churn a fluid (Joule used water and mercury) in the cylinder, leading to its heating. The lost potential energy can be related to gains in two different forms of energy, that of heat and kinetic energy. In the experiment the weights descended through a distance of 1.6 m, reaching a speed of about $6.1\ \mathrm{cm\,s^{-1}}$. They were repeatedly lifted and, over a period of some 35 min in which the weights descended 20 times, the temperature of the water in the cylinder increased by about 0.31 °C. This temperature change was carefully measured, the accuracy attained being about 3 mK. (Temperature is now routinely measured at sea to an accuracy of 1 mK and often, with specially designed equipment, with a resolution of 0.1 mK or better, e.g., in studies of mixing in boundary layers where the temperature is relatively uniform.) Joule took great care to minimize heat loss during the period of the experiment by insulating the cylinder, and a wooden screen was erected to avoid effects of radiant heat from the observer. Joule calculated the total potential energy lost by the weights in descending, and, by subtracting their kinetic energy at the end of their descent and accounting for a small unavoidable heat loss to and from the cylinder during the experiment, was able to relate the mechanical energy imparted to the fluid per unit volume through the

6 The specific heat varies with temperature, salinity and pressure. For more precise values of c_p see Gill (1982, Section A3.4 and Table A3.7). Note that the heat per unit volume is $\rho\,\Delta H = \rho c_p\,\Delta T$, where ρ is the density. Equation (1.2) is given incorrectly (ρ should not be included) in TTO (Thorpe, 2005).

7 James Prescott Joule (1818–1889) was given private lessons in chemistry in his home city of Manchester by John Dalton (1766–1844), now best known as the discoverer of the law of partial pressures of gases. As a young man, Joule observed the aurora borealis and sounded the depth of Lake Windermere in northwest England with his elder brother, Benjamin. The Joule family owned and managed a brewery but to what extent James Joule was actively engaged in its running is unclear; Osborne Reynolds (see Section 1.2), a friend and biographer, asserts that Joule had little to do with the brewery, although he did do experiments within its premises as part of an extensive study of the relationship between different forms of energy. Cardwell's (1989) biography of Joule provides informative details of his early years and of his contacts with other scientists of the time, including Michael Faraday, who communicated Joule's paper describing his experiments to the Royal Society, which published his results.

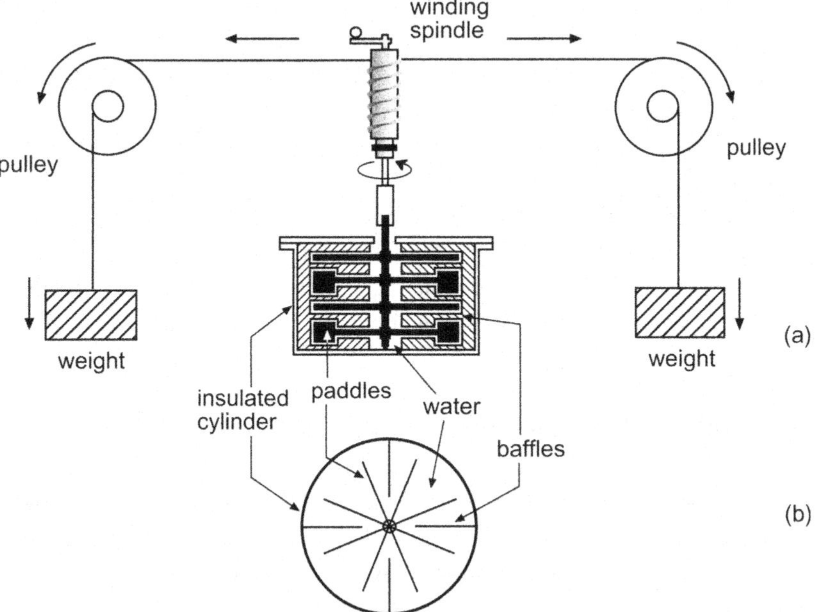

Figure 1.3. A side view of Joule's experiment, as described in his 1850 paper. The insulated cylinder (shown also in plan view) is filled with water or mercury and stirred with paddles driven by falling weights through the linkage pulley system. By calculating the energy lost by the weights, Joule was able to estimate the energy dissipated by stirring the fluid within the cylinder. Joule carefully measured the rise of the temperature of the fluid, and was then able to determine the relationship between the mechanical energy dissipated and the gain in heat energy of the fluid, proportional to its temperature rise, the constant of proportionality giving the specific heat.

paddles (equal to its change in heat energy) to its rise in temperature and so, from (1.2), to calculate c_p.

This experiment was later refined to obtain greater accuracy, but, as it is, it contains a major subtlety that involves the motion of the fluid within the cylinder. As Fig. 1.3 shows, there are baffles fixed to the inside of the cylinder. They are important for two reasons. The first is that without them a circulatory flow would be set up, which, containing kinetic energy, would have to be accounted for in the energy balance. (Alternatively, Joule could have waited until the circulation had died out before measuring the temperature, but that would have required a means to ensure that there was no substantial residual motion and would have taken time, during which heat would have been lost from the cylinder to the air.) The second reason is perhaps more important. The rotating paddles drive fluid past the stationary fixed baffles, and this promotes a transfer of kinetic energy from the mean flow to irregular and interacting small-scale eddies, characteristic of turbulence, that are shed by flow separation from the edges of the baffles and paddles. These eddies or 'turbulence' enhance the shear within the fluid and greatly increase the rate at which molecular viscosity dissipates the kinetic energy imparted to the fluid, transferring mechanical energy into heat much more rapidly than can a mean circulation gradually spun down through viscous drag at the cylinder walls. The potential energy of the falling weights that is not transferred to their kinetic energy consequently passes into turbulent energy that, in dissipating through viscosity, results in heating.

• The important factor is that turbulence transfers the energy involved in motion – the kinetic energy – to heat.

The heating caused by turbulent dissipation in the ocean turns out to be generally insignificant (e.g., see P1.6, P2.4 and P6.2), but the energy lost in turbulent motion is very important in the budget of the ocean's energy, and the effect of turbulent motion in mixing the ocean is a vital element in ocean circulation and in climate change.

1.4 The surf zone: waves and turbulence

The surf zone on a gently shelving beach provides an example of a turbulent region of the ocean that is familiar to many, and one that has some properties that relate to, and some that contrast with, those of Reynolds' experiment. It is also a region of the ocean in which turbulence is most energetic, and one in which (as in Reynolds' and Joule's experiments) the principal source of turbulent energy – in this case waves – can be identified. It provides an opportunity to introduce several ideas about energy, dispersion and structure relating to turbulent motion.

Figure 1.4 shows the water surface within a surf zone, the shallow and gradually shoaling region at the edge of the sea. It is partly covered by floating bubbles of foam that have been produced by waves as they approach the beach and break, carrying air into the water and producing clouds of subsurface bubbles. The bubbles rise buoyantly, some reaching the surface either to burst, producing tiny droplets in the air, or to float, contributing to the visible foam layer. A few bubbles may completely dissolve before reaching the surface, transporting all their component atmospheric gases into the seawater. The breaking waves also generate motions that disperse both the subsurface bubble clouds and the floating foam. In a casual viewing, the foam layer appears random, without any structure, rather like Reynolds' experiment when viewed without the advantage of the instantaneous spark image to make eddies visible. On more careful inspection some larger, repetitive and regular features with coherent structures can, however, be seen in a foam layer within the surf zone, notably bands or filaments and near-circular holes. How these are produced is described later.

Several distinct 'processes' are associated with the waves. In deep water (before waves break at the edge of the surf zone) the wave-induced motions are relatively regular, benign and, except in high winds when wave breaking becomes frequent, quiescent in comparison with the violent motion within the surf zone. The deep-water waves cause water particles to move in nearly circular orbits. (It may appear at first sight that these motions in the water column beneath surface waves in deep water are like eddies in a turbulent flow, and will result in the overturning and mixing of the water. This is not the case, as explained in Fig. 1.5(a).)

Figure 1.4. Plunging breakers, and foam within the surf zone. The plunging breakers are visible at the outer edge of the surf zone. Nearer shore, these become bores (or hydraulic jumps, abrupt changes in water level) advancing towards shore and producing a surface mat of white foam that is broken up by motions within the underlying water.

The waves also produce a mean movement of water called 'Stokes drift' in the direction of wave propagation. The drift at the sea surface is typically about 2% of the wind speed and much less than the speed of wave propagation, and it decreases with depth. It carries surface or near-surface floating particles towards shore. Although at this stage the waves contribute very little to the spreading or dispersion of such particles, because of the Stokes drift the particle paths are not exactly circles. This acts like a mean shear, slowly tilting and stretching fluid columns as shown in Fig. 1.5(b).

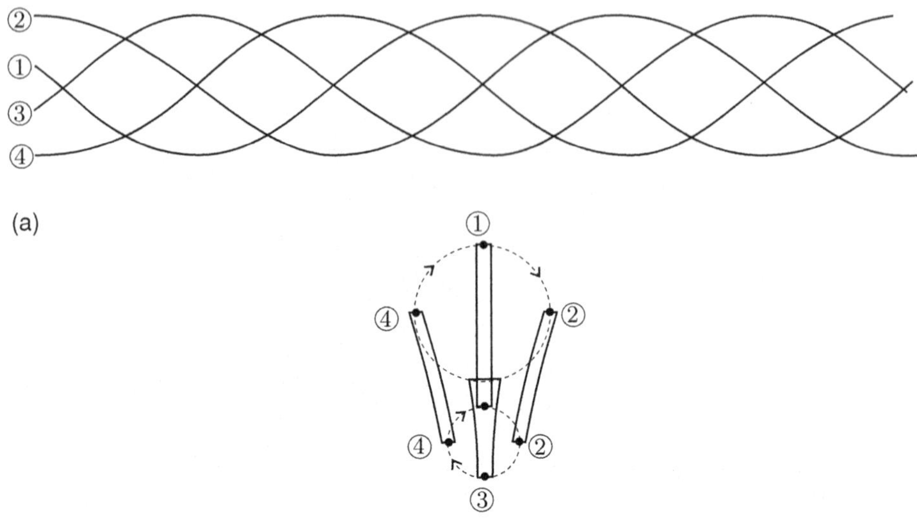

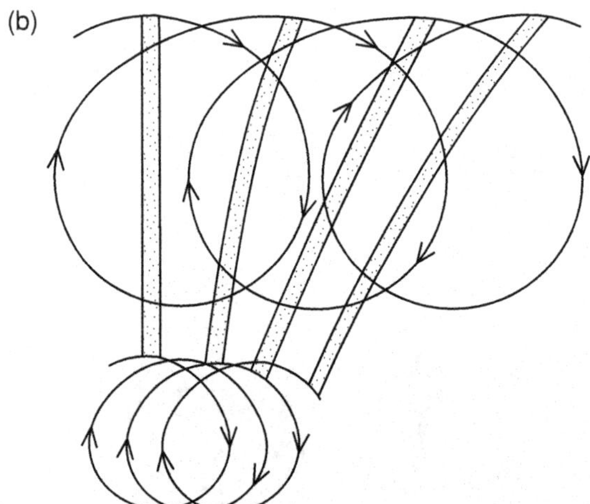

Figure 1.5. A sketch showing motion induced by non-breaking surface gravity waves. (a) Top – the surface of waves propagating to the right at increasing times, 1–4, and, below, the corresponding approximately circular motions of particles at two depths under waves in deep water. For clarity the size of the orbital motions is exaggerated. The radius of the particle orbits decreases approximately as $\exp(-kz)$, where z is their mean depth below the level of the mean water surface and $k = 2\pi/\text{wavelength}$ is the wavenumber of the waves. Columns of water joining the particles in the two orbits at times 1–4 are shown. These do not overturn, but are periodically slightly stretched, tilted and thinned. In reality, the orbits of particles at the two levels are not exactly closed circles, however, but drift slowly in the direction of the waves as sketched in (b). The Stokes drift decreases with depth below the surface. The stretching of an initially vertical column of water particles (stippled) by the drift is shown as three successive wave crests pass by. The Stokes drift at depth z is approximately $a^2k\sigma\exp(-2kz)$, provided that the amplitude of the waves, a (half their height, i.e., half the crest-to-trough distance), is small (so that $ak \ll 1$), where their frequency is σ, equal to $2\pi/(\text{wave period})$.

That deep-water waves impart a mean drift, the Stokes drift, to the water implies that they carry momentum (or at least a form referred to as pseudo-momentum). The flux of momentum (sometimes called the wave radiation stress) carried by waves entering the surf zone leads to a rise in water level at the beach (typically only a few centimetres in height and hardly noticeable); the resulting pressure force balances the waves' momentum flux. When waves enter the surf zone from a direction that is not normal to the beach, their momentum flux drives along-shore currents that are partly balanced by frictional drag forces on the seabed. The movement of water into the surf zone and instability of the along-shore drift currents sometimes leads to localized and dangerous rip-currents returning water – and carrying swimmers unfortunate enough to be caught in the currents – from the surf zone into the deeper water beyond.

The waves approaching the beach also carry energy: there is a shoreward flux of energy maintaining the continual turbulent motion that, like the motion within Joule's cylinder, leads to a relatively rapid dissipation of energy. [**P1.3**] When the waves break at the outer edge of the surf zone, some of their energy is lost, and the waves change from having a rather smooth and almost sinusoidal shape outside the surf zone to become advancing walls of foaming water, or hydraulic jumps. (An example of a steep, but stationary, hydraulic jump is given in Fig. 6.5(b) later.) Much of the energy lost by the waves in breaking contributes to the kinetic energy of the dispersive turbulent motions, but some of their energy provides the potential and surface energy of bubbles within the bubble clouds and foam.[8] The water in the overturning and plunging breakers at the edge of the surf zone may not be turbulent or generate turbulence until the falling jet of water from the wave crest impacts with the water ahead of the breaking wave, locally creating high shear and producing bubbles. The process of trapping air within the overturn, and of carrying the air into the water, provides an example of entrainment or engulfment, characteristic of the way in which turbulent motions spread as described in Section 1.5.2.

As in Reynolds' experiment, there is a transition from a relatively smooth flow of water, that outside the surf zone, to one that is turbulent within it. In this case turbulence at the outer edge of the zone does not begin at a critical value of a characteristic Reynolds number, *Re*, but is initiated by wave breaking, a transition that is determined by parameters of the incident wave field and the slope and nature of the seabed. Within the surf zone, turbulence is partly sustained by the hydraulic jumps. There is a decreasing shoreward flux of wave energy (i.e., a divergence of the energy flux towards shore).

• The important thing of note here is that breaking waves generate turbulence. We shall find that there are types of waves other than those at the sea surface, which lead to turbulence in the ocean.

8 In deep water, the efficiency of a plunging breaker, namely the energy lost divided by the original wave energy, can be as large as 0.25 (i.e., through breaking, a wave may lose 25% of the energy it carries). Since – as mentioned in P1.3 – wave energy is proportional to the square of the wave amplitude, the visible effect of breaking on the form of a wave is less readily apparent than is the production of foam. An efficiency factor of 0.1 is more typical of a spilling breaker. As much as 40% of the energy lost by a wave in breaking may go into the production of bubbles rather than directly into turbulence.

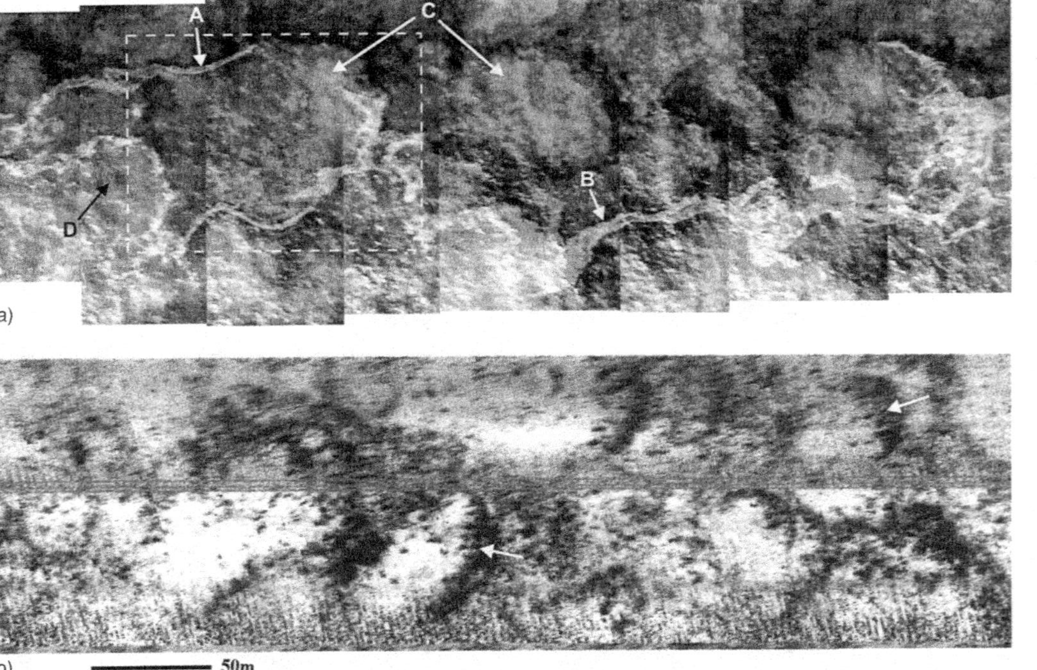

Figure 1.6. Boils affecting the dispersion of buoyant oil in the southern North Sea. (a) A composite image constructed from video taken from an aircraft. Parts of a slick of floating oil carried by the tidal current from a continuous fixed source to the right of the image are visible, e.g. at A and B. Boils are made visible, e.g., at C, through the sediment they carry to the surface from the seabed, 45 m below the water surface. Some (e.g., as shown in the dashed box) make holes in the oil slick, breaking it into filaments, one being that marked A. (b) An acoustic image of the sea surface obtained using an upward-pointing side-scan sonar mounted on the seabed, with the same scale and orientation but not coincident with the image in (a). The upwind edges of the boils appear as dark crescent-shaped bands where steeper and possibly breaking waves reflect sound to the sonar. (The steepening of waves results from the currents induced by the boils that oppose the waves.) The near-vertical bands, seen best at the lower edge of the image, are reflections from individual waves. Bands (arrowed) in the crescents are aligned with the wind and are possibly caused by Langmuir circulation (Section 3.4.3). The horizontal line halfway up the image is caused by the multiple reflection of vertically propagating sound from the seabed and water surface. The water in this southern North Sea location (south of the front labelled 1 in Fig. 3.17) is strongly mixed and of uniform density. Boils are observed when, or soon after, the tidal current reaches its greatest value of about 1 m s^{-1}. Their diameter is comparable to the water depth, here about 45 m. (From Nimmo Smith *et al.*, 1999.)

As well as being produced by wave breaking, turbulence is generated in the surf zone by the rapid flows periodically induced by the waves over the underlying beach sediment,[9] a process of generation like that caused by flow past the immobile tube walls in Reynolds' experiment. The image of the water surface (Fig. 1.4) provides no immediate indication of this process or that of the *vertical* dispersion of subsurface bubbles. The floating foam does, however, indicate the areas where clouds of bubbles produced by breakers reach the water surface and how, once at the surface, the floating bubbles are moved about by the variable currents caused by the breaking waves, by the water motions induced by the clouds of bubbles themselves in rising to the surface and by the rapid surging wave-induced flows over the seabed. These flows over the seabed produce 'boils', eddies rising to and visible on the surface, much as they are seen in rapidly flowing rivers. At a larger scale, boils are found in strong tidal flows in shallow seas, as illustrated in Fig. 1.6. The floating foam in the surf zone is affected by these three-dimensional motions, but its dispersion and spread, confined to the water surface, is (unlike the dye in Reynolds' experiment) primarily two-dimensional. [**P1.4**]
• Turbulent dispersion consequently depends on the nature of the dispersant, e.g., whether it floats, sinks or is, like a dye, soluble in water. This is discussed further in Chapter 5.

1.5 The nature of turbulent flow

In this section we describe some of the processes involved in turbulent motion.

1.5.1 Stirring + diffusion = mixing

In the 1940s, Eckart drew attention to two distinct mechanisms by which turbulence operates in promoting mixing and dispersion. The first is one of stirring. Turbulent eddies stretch fluid elements to produce narrow 'streaky' or 'filamentary' distributions of water properties, like those apparent when cream is stirred in a cup of coffee or the 'curls' observed by Reynolds (Fig. 1.1(c)). Welander illustrates the process graphically, as shown in Fig. 1.7. Particles of fluid, initially close together, become separated or dispersed by the turbulence; the stirring extends the surfaces of contact between fluid volumes or, in the case of the figure, the length of the lines between the black and white areas, increasing the area across which diffusive transfers of fluid properties such as temperature, or dissolved dye, may occur. (In looking at the figure or at eddies on the surface of a cup of coffee, remember, however, that small-scale turbulent motions in the sea are commonly three-dimensional, and the eddies in the body of a fluid should not be thought of as 'flat'.)

But the same process of stirring enhances the gradients, filaments becoming narrower and interfaces between marked and unmarked fluid becoming sharper. This

9 The rate of production of turbulence by flows over solid boundaries is discussed further in Sections 2.5.3 and 6.7.

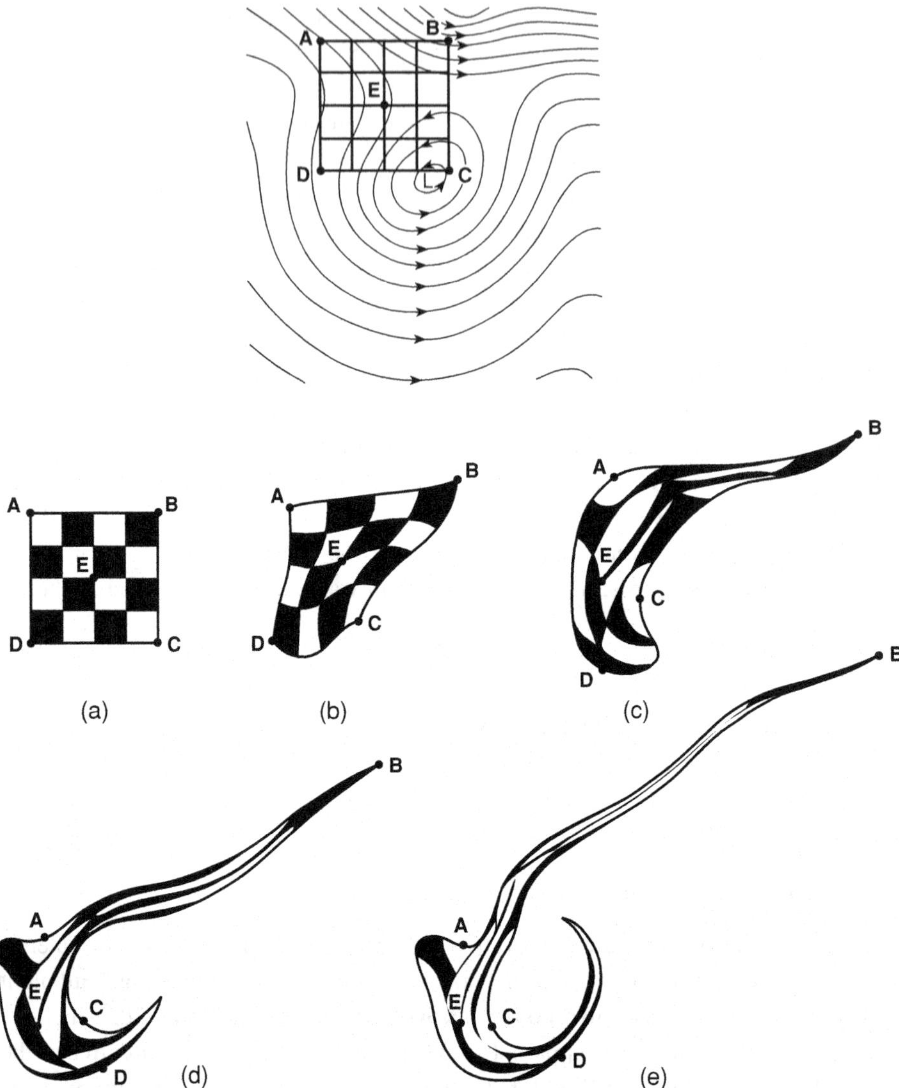

Figure 1.7. The distortion of a regular checkerboard pattern by an eddying motion. Successive stages are shown, (a)–(e), and in each the locations of points A–E carried with the flow are marked to indicate their dispersion by the eddying flow field. The sketch correctly conveys the notion of a spreading patch, but is slightly incorrect because area is not conserved: e.g., the black region on the left in (e) has an area greater than the corresponding area in (a). This is impossible in a flow that is strictly two-dimensional. (After Welander, 1955.)

eventually leads to the second process that operates in turbulent flows, one that is effective in transporting solutes or water properties between fluid particles, rather than simply *moving* particles as does stirring. In this process, molecular diffusion (or, for heat, thermal conduction) causes transfers of fluid properties that rapidly reduce both the gradients and the extremes of temperature or solute concentration. (The most intensely coloured dye patches become paler after this mixing process.) The flux of heat by thermal conduction, for example, in a direction denoted by z (i.e., the flux of heat per unit area of a surface normal to z, with units W m^{-2}) is given by

$$F = -\rho c_p \kappa_T \, \mathrm{d}T/\mathrm{d}z, \tag{1.3}$$

where ρ is the density of the seawater (kg m^{-3}) and κ_T is the molecular conductivity of heat (about 1.4×10^{-7} m^2 s^{-1} in seawater). The sign is negative because the flux is in a direction opposite to that in which the temperature increases. The flux increases in proportion to the temperature gradient, $\mathrm{d}T/\mathrm{d}z$; processes that enhance the gradient also increase the magnitude of the flux.

• The action of turbulence is consequently one of dispersion of material particles by stirring whilst homogenizing fluid properties by diffusion. Together the processes lead to mixing. This is irreversible: the dye mixed in the transition from laminar to turbulent flow in Reynolds' experiment cannot be unmixed without the operation of an additional and artificial process, for example a machine that removes the dye from the water and reintroduces it at the same location in a re-concentrated form.

The terms diffusion and dispersion are commonly interchanged and confused. Diffusion, as used here, involves the transfer of fluid properties at molecular scale and should not be applied to solid particles that retain their properties as they are moved by the surrounding turbulent motion. Dispersion involves spreading, the moving apart of solid particles and particles of fluid – and their contents, which might include heat or dye – by the turbulent motion.

1.5.2 Entrainment and detrainment

Another important process is the enlargement of a region of turbulence by the entrainment of fluid from beyond the original turbulent region. This process is seen in the broadening of a plume of smoke as it rises from a chimney in calm weather (Fig. 1.8); turbulent eddies at its edge engulf the surrounding air. J. S. Turner (to whose work reference will be made later) describes this 'entrainment' as follows: there is '*a sharp boundary separating nearly uniform turbulent buoyant fluid from the surroundings. This boundary is indented by large eddies and the mixing process takes place in two stages, the engulfment of external fluid by the large eddies, followed by a rapid smaller scale mixing across the central core*' of the rising plume. There is no formal, commonly accepted definition of the term engulfment, but here it involves the movement of fluid in the plume around the surrounding or ambient fluid, in a manner similar to the way in which a plunging breaker at the edge of a surf zone (Section 1.4) captures air and carries it into the water.

A further process by which a region of turbulence may spread is that of 'detrainment'. In this, pairs of relatively large, coherent vortices escape a turbulent region, driven by their mutual interaction,[10] carrying small-scale turbulent fluid within their

10 As explained, for example, by Batchelor (1967, section 7.3) in his description of the motion of pairs of vortices through the surrounding fluid.

Figure 1.8. A plume rising over the nuclear power plant at Gosgen, Switzerland, in the Swiss Foothills. The plume rises, eddies at its edge entraining surrounding air, until it meets and spreads on an inversion (a stably stratified layer). (Photograph kindly provided by Paul E. Oswald.)

circulating cores into the ambient, non-turbulent water. The self-induced motion of pairs of vortices through their surroundings is illustrated in Fig. 1.9. The boils described in Fig. 1.6 may be viewed as the consequence of the movement of eddies or vortices from a turbulent boundary layer close to the seabed upwards to the water surface, carrying both the fluid – which may contain vigorous (although perhaps decaying)

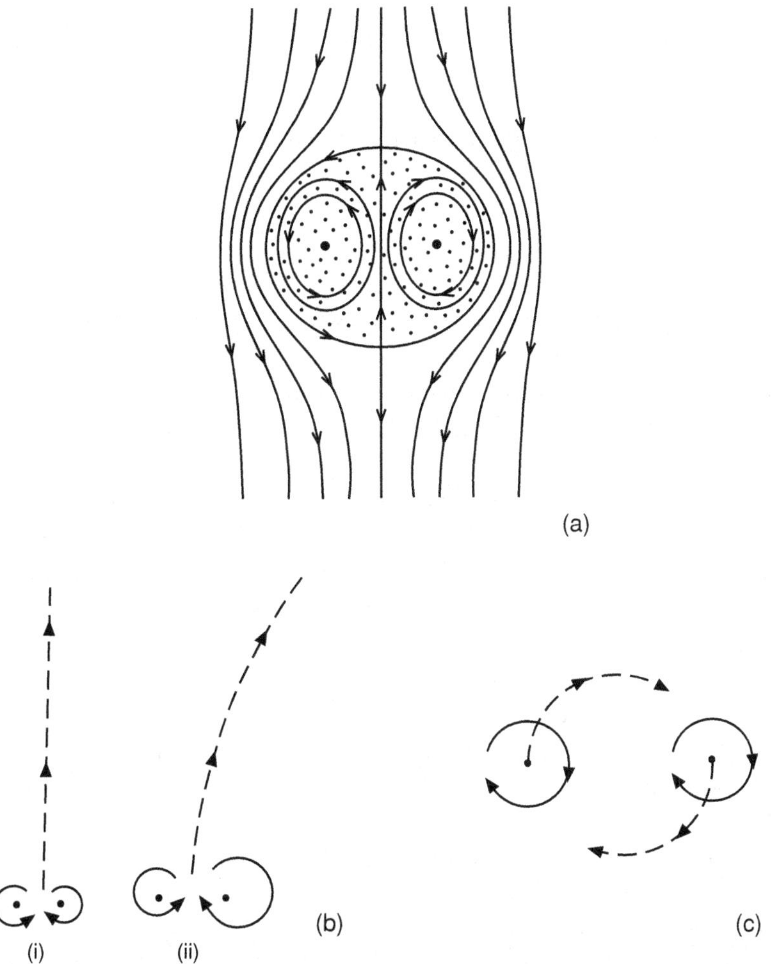

Figure 1.9. The motion of a pair of vortices (or eddies in a turbulent flow) under their mutual interaction. (a) Flow around a pair of stationary equal point vortices of opposite sign. If, instead of the vortices being stationary, there is no average flow in the fluid around them (or, more precisely, if the fluid at infinity is at rest), the vortices will move vertically, as shown in part b(i), carrying fluid in the stippled volume with them. (In a confined region the fluid beyond the region of influence of the vortices must, on average, move slightly downwards to replace the fluid carried upwards by the vortices, just as air between rising convective plumes forming cumulus clouds descends between them.) (b) The (dashed) paths of vortices of opposite sign but (i) of equal vortex strength, as in (a), and (ii) of unequal strength. In three dimensions, a smoke ring provides an example of self-induced vortex motion and the transport of material (smoke) through surrounding air. (The legs of the hairpin vortices illustrated in Fig. 3.15 may interact to drive one another away from the underlying plane boundary, although in this case the virtual images of the vortices in the boundary – required to ensure zero flow through the boundary – need also to be taken into account.) (c) The motion induced by a pair of equal vortices of the same sign. In practice, such vortices may merge or 'pair', leading to a single larger vortex, a process of energy transfer to a larger scale of motion.

small-scale turbulent motions – and sediment with them. This process is described further in Section 3.4.4.

1.6 Shear, convergence and strain

Both shear and convergence (or divergence) are inherent in the three-dimensional eddying motion of a turbulent flow. Shear is the spatial gradient of the speed of a current in a direction normal to the direction in which it flows. Shear is represented in Fig. 1.10(a) by a 'vertical shear', du/dz, where u is the horizontal current and z is the vertical, which results in an increase of the distance between particles separated by distances (z) normal to the flow direction (for example, of material fluid elements at points A and D in Fig. 1.10(a)). If the current, u, is steady, the distance between A and D increases in proportion to time, t. As well as stretching the area, ABCD, shear results in a thinning process: the thickness of marked patches of water (i.e., d) decreases in proportion to t^{-1} (keeping the area ABCD constant). These effects are evident in the sketch (Fig. 1.5(b)) of a column of water being stretched by Stokes drift, and are also present in the eddying motion sketched in Fig. 1.7, where points A and B get further apart but the width of the marked region becomes less, and points C and D get (temporarily) closer.

Shear can result in the production of vertical gradients of water properties of increasing magnitude when initially there is only a horizontal gradient. For example, if, in Fig. 1.10(a), the temperature is less on (and to the left of) the line AD than it is on (and to the right of) the line BC and if, during the time t, the temperatures remain unchanged, the effect of the shear is to produce a vertical temperature gradient of increasing magnitude over a surface, represented by the length of the lines AD and BC, the area of which also increases with time. The magnitude of the molecular flux of heat given by (1.3) will therefore increase with time. Moreover, if the parallelogram ABCD is a region of anomalous properties, for example a patch of plankton that is passively advected by the flow, the thickness of the patch will decrease with time, eventually resulting in a 'thin layer'.

A simple local, perhaps transient, region of convergent and divergent fields of motion is sometimes referred to as a strain,[11] and leads to similar effects of stretching and thinning. In the steady motion field represented in Fig. 1.10(b) by $u = qx$ and $v = -qy$, the x-length of a patch increases as e^{qt} and, to conserve the area of the patch, the y-scale decreases as e^{-qt}. [**P1.5**] The overall effect of stretching and thinning is like that of shear (Fig. 1.10(a)), but the thinning is exponential in t rather than proportional to t^{-1}, and therefore may develop more rapidly.

• Processes of shear and strain are inherent in turbulent stirring, acting to increase interfacial areas and enhance the concentration gradients of advected solutes or fluid properties across them, thereby increasing the rates of molecular transfer.

11 Strain is also used as a measure of the change in density gradient, see (4.15).

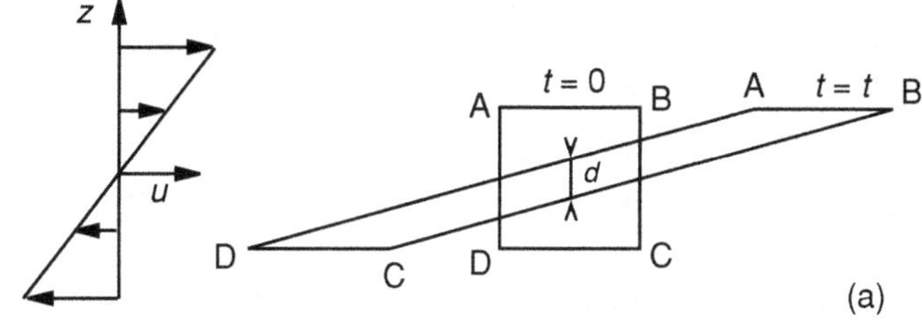

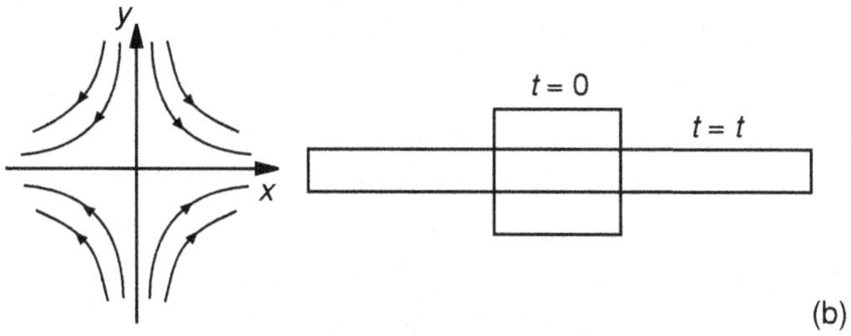

Figure 1.10. The effects of (a) shear and (b) convergence. (a) A vertical shear in which the speed of the horizontal flow varies with depth. Such shear flows are commonly encountered in the ocean, although the diagram illustrates an idealization of what might be a transient, but coherent, shear within a turbulent region. As time, t, increases, A and D become further apart; the horizontal width of the initially square patch (the distance A–B) remains unchanged, but the vertical thickness, d, becomes smaller. If AD and BC are isothermal surfaces (viewed 'end-on'), the gradient of temperature between these two surfaces increases with time. (b) The distortion of a square at time $t = 0$ by convergent–divergent flow, that reduces its height and increases its width. Gradients in the y direction are increased by the flow field, whilst those in the x direction are reduced.

1.7 Ocean stratification and buoyancy

1.7.1 Density

The mean depth of the ocean is 3795 m and there are differences in current speed typically exceeding 0.1 m s^{-1} through the ocean depth. The characteristic Reynolds number, Re, of the ocean consequently appears to be of order 4×10^8, far exceeding the critical Reynolds number of about 10^4 defined in Section 1.2. If it is valid to interpolate from Reynolds' experiment to the scale of the ocean, the ocean should therefore be turbulent with large overturning eddies (like those observed by Reynolds to be of size comparable to the tube radius) mixing through much of the water column. (See also P1.2.) Whilst the ocean is indeed turbulent, turbulent motions are patchy or 'intermittent' in the body of the ocean and (except in shallow water and in deep convection) eddies do *not* overturn on a scale comparable to the water depth, their maximum vertical scale being typically 1–100 m.

A factor not included in Reynolds' experiment has a vital role in determining the nature of turbulence in much of the ocean. This is the density variation or stratification resulting largely from atmospheric heating and the distribution of heat by the ocean's circulation. The introduction of stratification takes us briefly away from the subject of turbulence, but is essential because of the profound effects of density and buoyancy on ocean turbulence.

The mean density of the ocean increases with depth: it is 'stratified' in density.[12] The density of seawater depends on temperature, salinity and pressure through a relation known as the equation of state. The majority of liquids expand when heated and contract when cooled. Their density consequently increases as their temperature decreases. Freshwater is a well-known exception between its freezing point at 0 °C and a temperature of about 4 °C, becoming denser as temperature increases.

This freshwater behaviour is modified by the presence of the salts dissolved in seawater and quantified as 'salinity'. The main components contributing to salinity in the ocean are chlorine (55.0%), sodium (30.6%), sulphate (7.7%), magnesium (3.7%), calcium (1.2%) and potassium (1.1%). (The proportions, given as percentages, are nearly constant.) The unit of salinity is the practical salinity unit or psu, approximately equal to 1000 times the mass of dissolved salts per unit mass of seawater. An increase in salinity increases density because of the increase in the mass of salts, and may be a very substantial component of density in some circumstances. The freezing point of seawater differs from that of freshwater, and decreases as salinity increases. The density of seawater with salinity greater than about 24.7 psu (and most of the ocean has a higher salinity, typically about 35 psu) behaves like that of most liquids, increasing as temperature decreases until freezing occurs (at a temperature of −1.92 °C for seawater with a salinity of 35 psu at atmospheric pressure) as shown in Fig. 1.11. As a consequence, the dynamical properties of the ocean may differ from those of freshwater lakes at temperatures near 0 °C. (See, for example, P3.8.)

Although density depends in a complex and non-linear way on temperature, salinity and pressure, for small variations of temperature, T, and salinity, S, from reference values where the density is ρ_0, the equation of state for density, ρ, may be approximated by

$$\rho = \rho_0(1 - \alpha T + \beta S). \tag{1.4}$$

The coefficients, α and β, relate to the expansion of seawater and are specified at the reference values of T and S, but depend on depth. Seawater near the sea surface with a salinity of 35 psu has a thermal expansion coefficient, α (the increase in volume per unit volume per degree Kelvin), of $5.26 \times 10^{-5}\ \mathrm{K}^{-1}$ at 0 °C, $7.81 \times 10^{-5}\ \mathrm{K}^{-1}$ at 2 °C, 1.67×10^{-4} at 10 °C and $2.97 \times 10^{-4}\ \mathrm{K}^{-1}$ at 25 °C. The value of α increases with pressure or depth, at 1000 m being equal to $1.84 \times 10^{-4}\ \mathrm{K}^{-1}$ in water at 10 °C and 35 psu. At

12 Although layers of relatively uniform density do occur, 'stratification' here refers to the general increase of density downwards through the water column (although with relatively very much weaker horizontal changes) and not necessarily, as in geology, to a layered structure with abrupt discontinuities of properties.

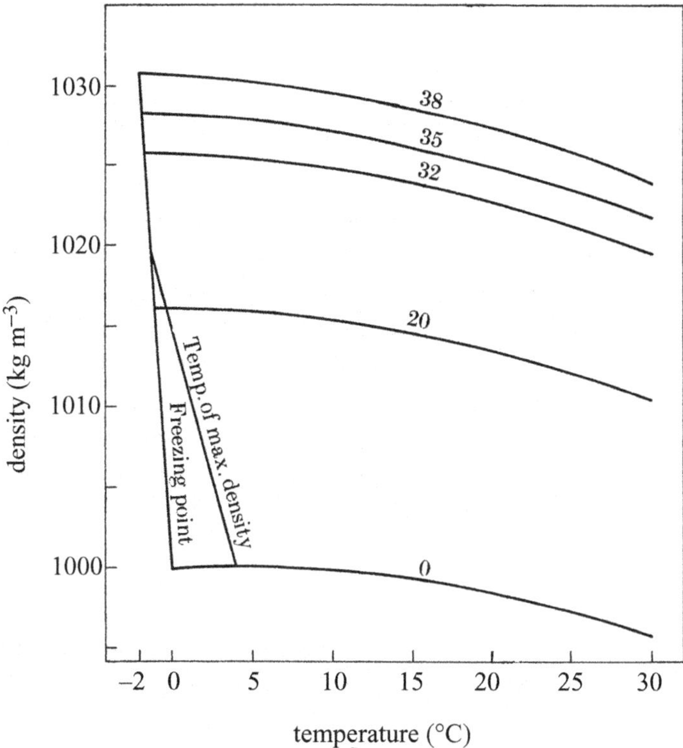

Figure 1.11. The variation of density with temperature for various values of the salinity but at atmospheric pressure. The curve on the left shows the temperature at which freezing occurs. The density of freshwater (salinity = 0 psu) is greatest at about 4 °C, but, in seawater of salinity typically ranging from 32 psu to 38 psu, the maximum density is reached at the freezing point.

a depth of 4000 m, with a temperature of 2 °C and salinity of 35 psu, fairly typical of the deep ocean bed, α is equal to $1.71 \times 10^{-4}\ \mathrm{K}^{-1}$. Near the sea surface, when the salinity is about 35 psu, the coefficient for the contribution of salinity to density, β (or the expansion coefficient for salinity), is about $7.9 \times 10^{-4}\ \mathrm{psu}^{-1}$ at 0 °C and $7.6 \times 10^{-4}\ \mathrm{psu}^{-1}$ at 10 °C.[13] An increase in temperature of $\Delta T = 0.1$ K in an upper layer of the ocean of thickness $h = 1000$ m at a temperature of 10 °C would result in expansion causing a 'steric' rise in sea level of $\alpha h\, \Delta T$, or about $1.76 \times 10^{-4} \times 0.1 \times 1000\ \mathrm{m} = 1.76$ cm.[14] Changes in sea level are a very important consequence of changing climate.

Temperature, salinity and pressure are measured routinely from research vessels to the full depth of the ocean basins using lowered conductivity–temperature–depth (CTD) probes, salinity being derived from the conductivity with a temperature correction. Rather than the density, ρ, measured in kg m^{-3}, values of a quantity referred to as sigma-T (σ_T) expressed without dimensions and equal to ($\rho - 1000$) are usually

13 Other values of α and β are tabulated by Gill (1982; see his table A3.1). Values of β given in TTO (Thorpe, 2005; p. 7) are too high by factors of about 10^3.

14 An average value of α between the surface and 1000-m values has been used in this estimate.

quoted.[15] The density of seawater in the upper 100 m of the ocean is typically 1.028×10^3 kg m^{-3} or $\sigma_T = 28$.

1.7.2 Buoyancy, and the buoyancy frequency, N

Archimedes showed that the upthrust on a body submerged in water is equal to the weight of water it displaces. A body of volume V and density ρ has a weight $g\rho V$, where g is the downward acceleration due to gravity, about 9.81 m s^{-2}. In water of density ρ_0 the body displaces a mass $\rho_0 V$ or a weight of water $g\rho_0 V$. The net (upward) buoyancy force on the submerged body is $g(\rho_0 - \rho)V$; therefore, using Newton's second law (force equals mass times acceleration), its vertical acceleration from rest is $g(\rho_0 - \rho)/\rho$. This is often referred to as the reduced acceleration due to gravity.

The body referred to may itself be a volume of water of density ρ surrounded by seawater of density ρ_0.[16] The buoyancy of such a volume is defined as the acceleration, $b = g(\rho_0 - \rho)/\rho_0$, which is positive (i.e., upwards) if the density of the volume is less than that of its surroundings.

If a small volume of water of density ρ is displaced upwards by a small distance, η, from its initial position in a uniform density gradient $\mathrm{d}\rho/\mathrm{d}z$, with the z coordinate in the vertically upward direction, the density difference between it and its new surroundings is $-\eta\, \mathrm{d}\rho/\mathrm{d}z$. If it is then released, its upward acceleration will be $(g\eta/\rho)\mathrm{d}\rho/\mathrm{d}z$. If $\mathrm{d}\rho/\mathrm{d}z > 0$ (density increasing upwards), the net force is upwards and the acceleration is positive, so the volume of water will move away from its initial position.

- The stratification is then 'unstable' or in a state of static instability in which (ignoring effects of viscosity and the conduction of the water properties, e.g., heat and salinity, that contribute to density) small disturbances grow and convection occurs. If, however, $\mathrm{d}\rho/\mathrm{d}z < 0$ (density increasing with depth, as is usual in the ocean), the motion following an upward (or downward) displacement of a small volume of fluid is oscillatory with simple harmonic motion of frequency $N = [-(g/\rho)\mathrm{d}\rho/\mathrm{d}z]^{1/2}$. The water is then said to be 'stable' or stably stratified (or, more precisely – ignoring processes such as viscosity that lead to damping – 'neutrally stable'). The value N characterizes the local density stratification.

A further dimensional scale, one of length, $[(1/\rho)\mathrm{d}\rho/\mathrm{d}z]^{-1}$, can be defined to characterize the density variation. This is known as the 'scale height' and is typically 40 times the ocean depth. The scale height is relatively large in the ocean,[17] the density

15 The potential temperature, usually denoted by θ (with corresponding potential density, σ_θ), is the temperature fluid would have if moved adiabatically (with no exchange of heat with its surroundings) to a given (stated) reference level. Figure 4.14, for example, shows σ_θ = (potential density (kg m^{-3}) – 1000), referenced to 2000 m and zonally (i.e., east–west) averaged across the Atlantic.

16 When in motion, a submerged body will experience other forces, including drag, viscosity and lift forces. A sphere that is small (small enough that a Reynolds number based on its speed and diameter and on kinematic viscosity is much less than unity) and of density, ρ_1, greater than that of the surrounding water, ρ_0, will fall at a uniform speed, $[2a^2/(9\nu)]g(\rho_1 - \rho_0)/\rho_0$, when the viscous drag forces and the upward buoyancy force balance its weight.

17 Relative to its thickness, the scale height of the atmosphere, about 7.4 km, is not large, and there care needs to be taken in formulating equations of large-scale motions – see Gill (1982).

varying by only a small amount over the ocean depth relative to the mean density, so that, to a good approximation, N can be written

$$N = [-(g/\rho_0)\mathrm{d}\rho/\mathrm{d}z]^{1/2}, \tag{1.5}$$

where ρ_0 is a mean or characteristic density.

• The frequency, N, is called the buoyancy frequency and characterizes the highest frequency of the small-amplitude free oscillations that occur naturally following a disturbance in a stably stratified fluid (i.e., with $\mathrm{d}\rho/\mathrm{d}z < 0$). The maximum value of N in a given region provides an upper limit to the frequency of oscillations known as internal waves and, in principle, provides a means of discriminating between waves and turbulence: fluctuations observed in a frame of reference following the mean motion which have a frequency greater than N cannot be internal waves and are usually associated with turbulence. (For example, see Fig. 2.6 later. Internal waves are described in Section 1.8.1.)

1.7.3 The oceanic density profile

The variation with depth of a quantity such as density or buoyancy frequency is commonly referred to as a 'profile' (see Fig. 1.12). The term 'velocity profile', for example, usually applies to the variation of the mean horizontal water speed with depth.

The region immediately below the sea surface is often of almost uniform density, typically to a depth of about 50 m, but with diurnal, seasonal and latitudinal variations in both density and depth (e.g., as illustrated later in Figs. 3.8–3.10). In regions of sufficiently large water depth, this 'mixed layer'[18] is stirred, and turbulence within it is generated and sustained, mainly by the processes arising from air–sea interaction described in Chapter 3. Even though it is actively mixed, small variations in temperature are usually detectable within the mixed layer. Diurnal changes in heating or cooling at the sea surface lead to temperature changes that are transmitted by turbulence through the layer; an eddying turbulent field of temperature with variations of typically 1 mK is commonly found. Although the temperature variations are small, they may exhibit coherent patterns that provide information about the structure of the turbulent motion within the mixed layer. The turbulence processes at the sea surface and within the mixed layer (a 'buffer zone' between the deep ocean and the atmosphere) are of considerable importance in transferring heat and momentum from the atmosphere into the body of the ocean. In some parts of the shallow shelf seas, the turbulence caused by the tidal flow over the seabed may have a substantial effect, mixing the entire water column, including water near the surface, to a nearly uniform density with no distinguishable surface mixed layer with properties different from those of the deeper water. (Section 3.4.5).

18 The term 'mixed layer' is ambiguous. The layer is mixed by the action of the atmosphere, but it should not be thought of as being mixed in the sense of being uniform in its properties. There are almost always variations in temperature within the layer and these, although only amounting to a few milli-degrees, can be good indicators of the processes active in the layer that are described in Chapter 3.

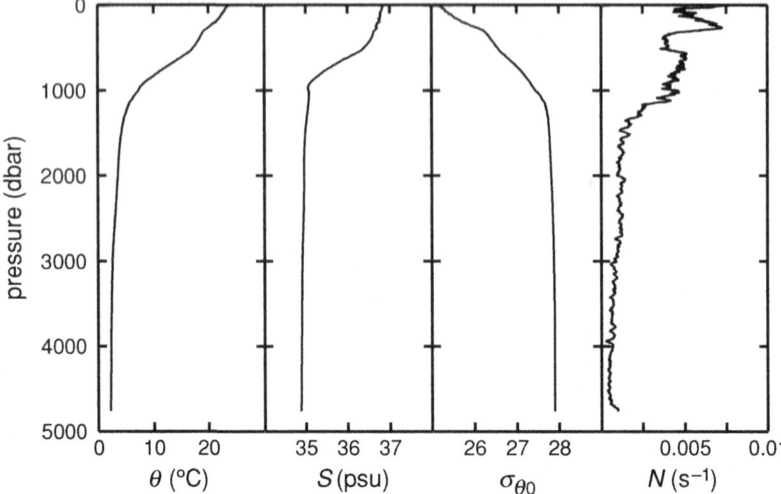

Figure 1.12. The variation of water properties with depth in the North Atlantic at 26° 29.50′ N, 75° 42.22′ W in April 2004. From left to right: potential temperature (θ), salinity (S), sigma-theta, a measure of density (here referred to the surface, $\sigma_{\theta 0}$), and the buoyancy frequency (N) change with pressure (100 dbar = 1 MPa $\approx$ 100 m). The profile of potential temperature exhibits a rapid decrease in temperature in the seasonal thermocline at about 200 m depth, followed by a further decrease down to about 1200 m in the main thermocline (or main pycnocline). These changes are matched in the $\sigma_{\theta 0}$ profile by a rapid increase of density in the seasonal thermocline and a more gradual increase in the main pycnocline. Relatively large values of buoyancy frequency, N, accompany these thermoclines. The small-scale structure in N possibly indicates the variations in the density gradient caused by internal waves or double diffusive convection. The near-surface and bottom mixed layers are not resolved in these profiles but may be seen in Fig. 2.4(a). (Figure kindly supplied by Dr B. A. King, National Oceanographic Centre, Southampton.)

• The mixed layer, together with the stratified region immediately below it that is affected by the dynamical processes within the mixed layer, is sometimes called the 'upper ocean boundary layer'. It is a region through which changes and the effects of the overlying atmosphere are transmitted to the underlying ocean. Its turbulence properties are discussed in Chapter 3 and Section 4.5.

Below the surface mixed layer, the density increases with depth in a layer known as the 'pycnocline'. The density increase is most commonly related to a decrease in temperature and, since it varies seasonally (e.g., see Fig. 3.10 later), the pycnocline is then usually called the seasonal 'thermocline'. Its maximum buoyancy frequency, N, is typically about 10^{-2} s^{-1}, corresponding to a period, $2\pi/N$, of about 5 min. In some regions, however, particularly near a source of freshwater, as in a river estuary or fjord or near melting ice, the density increase may be mainly a consequence of increased salinity (a 'halocline'). At all depths, the density generally increases monotonically with depth, but irregularly, exhibiting variable gradients with vertical scales of order 1–50 m, or density 'fine-structure'.

Below the seasonal thermocline, the mean temperature decreases further through the main thermocline or pycnocline at depth 500–1000 m. At greater depth still (the depth of water overlying the abyssal plains is typically 5000 m), the mean density continues to increase slowly, N reaching values of about 10^{-4} s^{-1} in the deep ocean below about 4000 m with corresponding periods of 12–20 h.[19] Turbulence generated at the seabed results in another mixed layer with little vertical gradient in temperature, salinity and density: the benthic (or bottom) boundary layer that extends to heights of typically 5–60 m above the sea floor.

• The density stratification in the various parts of the water column relates to the local processes that most commonly lead to turbulence. The stratification effectively divides the ocean into zones of different turbulence characteristics. This division is used to discriminate between the topics discussed in Chapter 3, namely turbulence in the upper ocean and benthic boundary layers, and Chapter 4, namely turbulence in the ocean interior or at 'mid-depth' when water is generally stratified.

1.8 Consequences of stratification

1.8.1 Internal waves and turbulent motion

As shown in Section 1.7.2, denser seawater moved upwards into less dense water in the stratified ocean is subjected to a net downward force, whereas an upwards force is experienced by water moved downwards. These forces lead to oscillatory motions or waves, called 'internal waves'. An example is shown in Fig. 1.13. Although these waves may, like surface waves, travel horizontally along density gradients or pycnoclines, in continuous stratification internal waves can propagate at an angle to the horizontal, transporting energy in a vertical direction through the stratified ocean and crossing surfaces of constant density (called isopycnals) in beams or rays, as illustrated in Fig. 1.14 and explained in its caption. They have both horizontal and vertical wavenumber components. Such beams, particularly those of internal waves of tidal period, can sometimes be detected near topographic sources. Distant from such sources, upward and downward beams combine to form horizontally propagating internal waves having distinct vertical structures or 'modal' properties, wave modes rather than wave rays.

The maximum frequency, $2\pi/T$, of small internal waves of period T, measured relative to the movement of the mean flow in a density-stratified region is equal to the buoyancy frequency, N. (By 'small' is meant that the waves move surfaces of constant density, or 'isopycnal surfaces', up and down through a distance small relative to their horizontal wavelength and that the slopes of the isopycnal surfaces are very much less than 90°. It also implies that the vertical wavelength of the waves is small

19 There is no formal definition of the 'abyssal ocean', but commonly depths below about 1500 m are implied, below the (ill-defined) base of the main thermocline, beyond the depth to which light penetrates from the sea surface and below that to which diurnal vertical migrations of zooplankton extend.

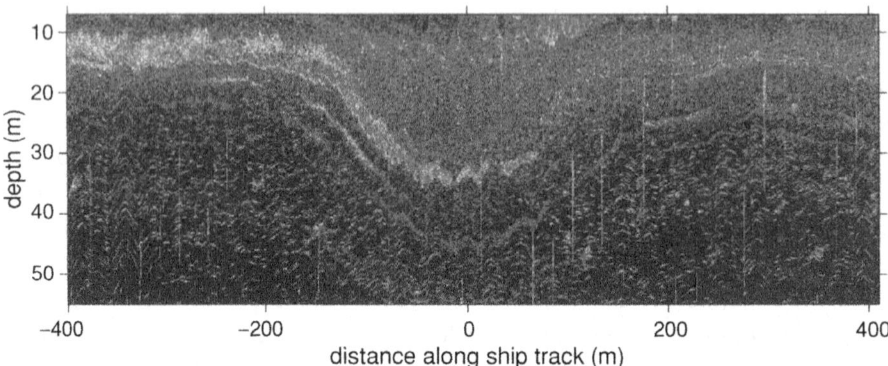

Figure 1.13. An internal wave on the Oregon continental shelf. The wave, one of a train of similar waves, is travelling from left to right. Although such waves are most commonly observed through the variations in temperature caused as they raise and lower thermally stratified water past fixed sensors on a mooring, this wave was detected acoustically by a downward-pointing sonar on a moving ship. Reflections from scatterers of sound such as zooplankton and fish with swim bladders in the thermocline make the wave 'visible'. It appears as a depression in the thermocline, some 20 m deep, advancing through the water at a speed of about 1 m s^{-1}. Within the wave trough, the internal wave causes currents of about 0.5 m s^{-1} in the direction of its propagation. The increased acoustic scattering from in front of the trough at about +70 m along the ship track, which extends to –400 m, appears to be the result of an increase in acoustic scattering from centimetre-scale temperature fluctuations caused by the wave's breaking. (From Moum *et al.*, 2003.)

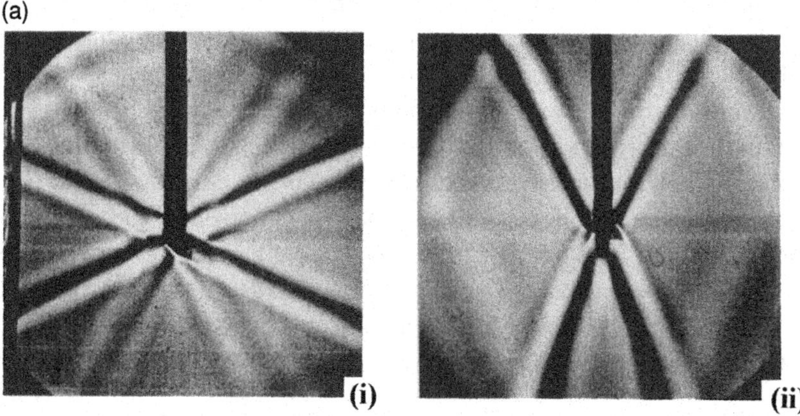

Figure 1.14. Internal-wave radiation. (a) A laboratory experiment showing internal waves generated in brine with a uniform density gradient (N constant) by a cylinder oscillating horizontally at frequency $\sigma < N$, which is seen 'end-on'. The ratio σ/N is 0.419 in (i) (where harmonic waves of frequency 2σ can be faintly detected) and 0.900 in (ii). Waves have radiated from the cylinder, crossing the horizontal isopycnal surfaces of constant density, along four beams or rays at angles of $\beta = \sin^{-1}(\sigma/N)$ to the horizontal and therefore at a greater angle in (ii) than in (i). These directions are those in which energy is transmitted by the waves from the cylinder, and are therefore the directions of the waves' group velocity, $\boldsymbol{c}_g$ (see part (b)). The black and white bands aligned in these directions correspond to the wave crests (*continued opposite*)

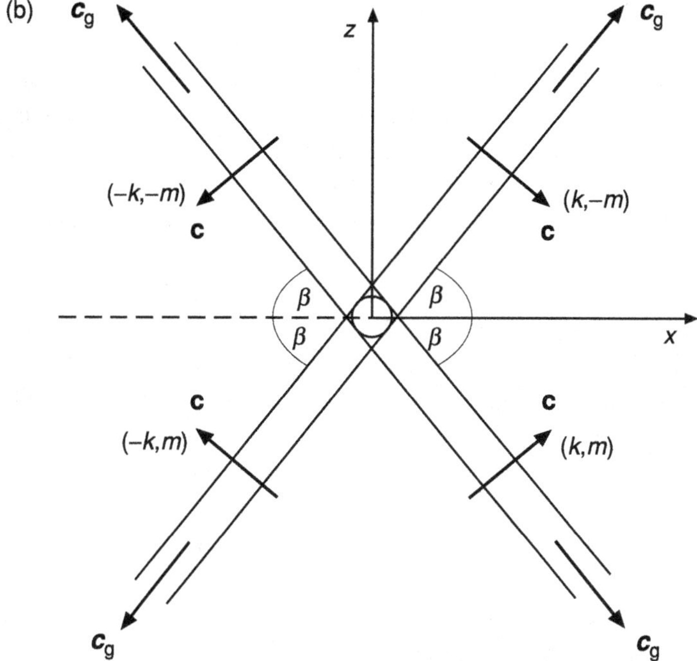

Figure 1.14. (*cont.*) or troughs, lines of constant phase, made visible using a Schlieren technique (sensitive to variations in the gradient in the refractive index within the brine that are caused by the waves). The lengths of the four beams increase with time after the cylinder has first been set into motion. If it is stopped, the beams detach from the cylinder and move away along their four directions. A brief period of cylinder motion results in four 'packets' of waves moving with the four directions of the group velocity, $\boldsymbol{c}_g$. The lines of constant phase move in directions normal to the group velocity (see part (b)). If the cylinder is oscillated at a frequency greater than N, an initial disturbance radiates away in all directions and further disturbances are confined close to the cylinder; no internal waves with the frequency of the cylinder can be formed or radiated. (From Mowbray and Rarity, 1967.) (b) The group ($\boldsymbol{c}_g$) and phase ($\boldsymbol{c}$) velocities of the waves. The latter advance in the direction of the wavenumber vectors with horizontal and vertical wavenumber components ($\pm k$, $\pm m$), where $k = m \tan(\sigma/N)$. As the arrows marking the group and phase velocities indicate, the phase and group velocities are at right angles and downward-propagating energy (indicated by the beams below the level of the oscillating cylinder) is accompanied by upward-propagating wave phase. The phase speed of the near-inertial-frequency waves in the Banda Sea shown in Fig. 4.11 is directed upwards – the dark bands of higher shear becoming shallower as time increases – indicating their probable generation near the surface and downward energy propagation.

in comparison with the scale over which N varies.) Care is required in determining or measuring frequency, however, because of the possible effects of Doppler shifts induced by mean flows. A wave with a short horizontal wavelength carried rapidly by flow past a fixed point may appear to have a relatively high frequency fluctuation in time. If this frequency is greater than N, the motion might therefore be interpreted as being caused by turbulence rather than by internal waves unless the effects of mean flow are properly accounted for. Internal waves break in a variety of ways, leading to turbulence within the body of the ocean. The nature of the turbulence they produce is described in Chapter 4.

• The links between waves and turbulence are central to the study of ocean physics. The properties of propagating waves are constrained by a wave dispersion relation. The dispersion relation is an equation relating wave frequency and wavenumber. (It may be expressed, e.g., as in P1.3, as a relation between the speed of advance of wave crests and the wave period.) No such relation between frequency and wavenumber applies to the fluctuations in turbulence. Turbulent eddies may propagate through the ocean as a result of their mutual interactions as suggested in Fig. 1.9, and larger eddies generally last longer than small. Unlike waves, however, other than through their advection by a mean flow turbulent eddies or turbulent motions do not conform to a relation that specifies a connection between their frequency and their size or scale. Whilst the wave group velocity defines the speed at which waves transport energy, there is no equivalent 'energy-transfer velocity' for turbulent motions. The absence of a dispersion relation and an energy-propagation velocity makes turbulent motion fundamentally different from wave motion. We return to this in Section 2.3.6.

A distinction between waves (surface or internal) and turbulence is not, however, always very clear or well-defined. Waves produce turbulence, for example by breaking, and turbulence may generate waves, e.g., internal waves may radiate from a turbulent patch of water, in particular downwards into the pycnocline from the bottom of the near-surface mixed layer. A breaking wave continues to have regular wavelike characteristics over much of the fluid space in which it is present, and generally the breaking is confined to relatively small regions in which, consequently, the motion is highly irregular. For illustration, a surface wave can break at the sea surface, entraining air and producing violent turbulent motions near the surface, but the motions induced by the wave well below the surface may continue to remain periodic and wavelike. Regions of mixing caused by breaking internal waves can be similarly localized within the larger field of oscillatory motion induced by the waves.

1.8.2 Isopycnal and diapycnal mixing

Mixing that involves the transfer of fluid across the isopycnal surfaces of constant density is called diapycnal mixing, whilst mixing involving the transfer of fluid properties[20] parallel to isopycnal surfaces, so involving no density change, is isopycnal mixing. (Diapycnal dispersion and isopycnal dispersion are similarly across and along isopycnals, respectively.)

Turbulent eddies in stratified water must raise dense fluid above less dense fluid (and carry less dense fluid below dense fluid) to 'overturn' in a stratified ocean (Fig. 1.15). This process requires an increase in potential energy [**P1.6**, **1.7**], work being done

20 Properties of, for example, temperature and salinity or some marker such as dye, but – being isopycnal – not density.

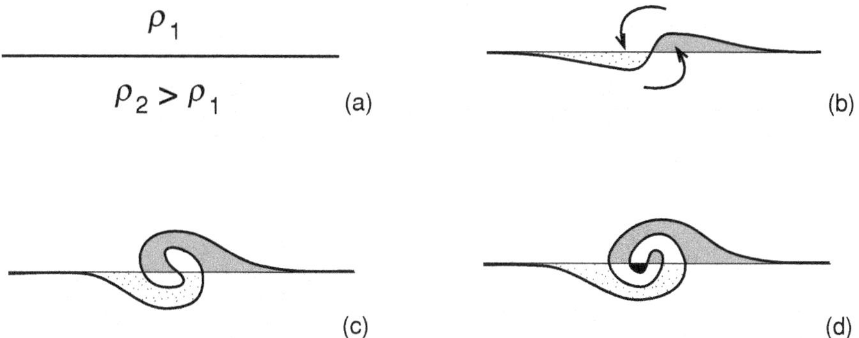

Figure 1.15. An overturning eddy on an interface. In (b)–(d) water from the less dense upper layer that is carried below the initial level of the interface shown in (a) is stippled whilst that of the lower layer carried upwards is shaded. The overall effect of the eddy is to raise the centre of gravity of the fluid and therefore to increase the potential energy. The billows in this sketch may be compared with those in, e.g., Figs. 2.1 and 4.4.

against buoyancy forces to lift or lower fluid, energy that must be supplied by, and lost from, the eddies.

• The energetics of mixing in a stratified ocean consequently differ from those in a homogeneous ocean, and the Reynolds number no longer provides a sufficient criterion for the onset of turbulence. Parameters that involve a measure of the density stratification are now involved in determining the nature of turbulent motion, as described in Chapter 4.

The mixing of fluid across density surfaces (diapycnal mixing) leads to an unbalanced pressure field that results in collapse and spreading of mixed fluid along isopycnals as illustrated in Fig. 1.16, where isopycnal surfaces (of constant density) are marked by full lines. Initially, at (a), a patch of fluid marked by dots lies between two isopycnal surfaces within a region in which the density, sketched to the left of (a), has a uniform gradient. Turbulence illustrated in (b), perhaps caused by a breaking internal wave, mixes the patch vertically by entrainment and engulfment of surrounding water, and spreads the markers, resulting in a patch of relatively small density gradient within the dashed oval in (c).

The density in section A–A through the centre of the patch is now as shown on the left, and to simplify the following calculation it is assumed that mixing makes the density of the patch uniform. The density profile outside the mixed patch remains unchanged, and is as shown in (a). If the pressure at the top of the mixed region at $z = h$ is p_0, a hydrostatic calculation shows that within the mixed region the pressure is $p_1 = p_0 + \int_z^h g\rho_0 \, dz = p_0 + g\rho_0(h - z)$, and, because the initial density is $\rho = \rho_0(1 - N^2 z/g)$, that in the surrounding ambient water is $p_A = p_0 + \int_z^h g\rho \, dz = p_0 + g\rho_0\{(h - z) - [N^2/(2g)](h^2 - z^2)\} = p_1 - g\rho_0[N^2/(2g)](h^2 - z^2)$, which is less than p_1 at levels $-h < z < h$. There is therefore a horizontal pressure gradient in (c) that causes the water in the mixed region to accelerate, and the region and its markers

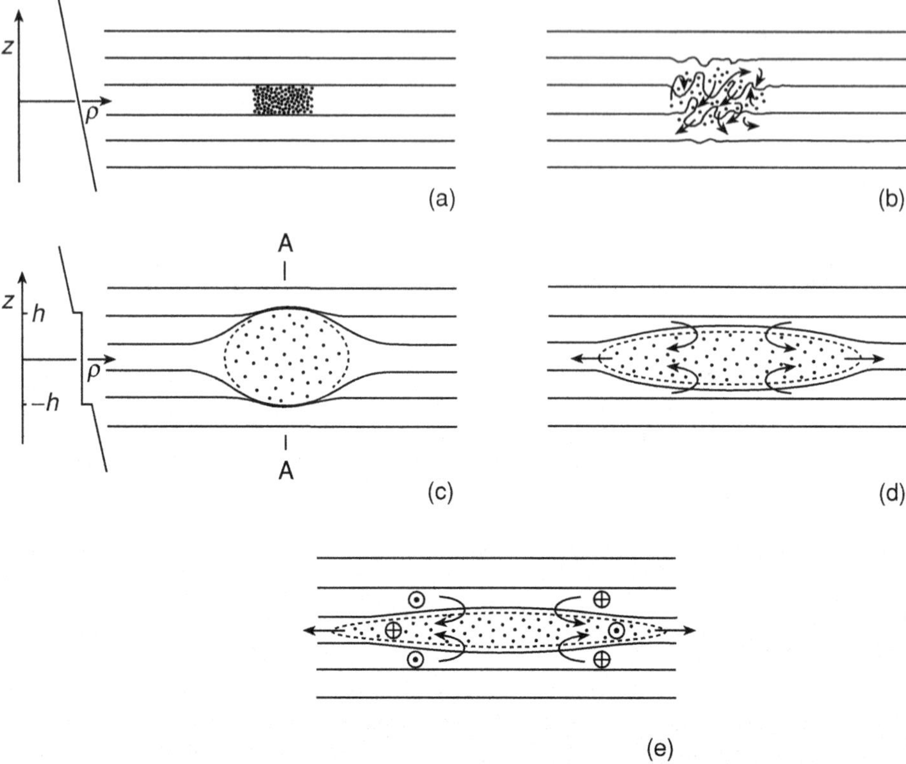

Figure 1.16. Collapse of a turbulent patch in stratified water. (a) On the left is shown the initial density profile with a locally uniform density gradient. Points are marked in a rectangular region where, in (b), turbulent mixing occurs. The effect of the mixing is to produce, in (c), a patch of relatively uniform density. The density profile through it at A–A is shown on the left. (d) The patch collapses under the gravitational pressure forces and spreads horizontally into the surrounding ambient water, forming an intrusion and dispersing the marked water particles (dots) horizontally. (e) After a time of about f^{-1}, the effect of the Earth's rotation (the Coriolis force) begins to produce a deflection of the flow, to the right in the northern hemisphere (as shown), resulting in the formation of counter-rotating vortices.

to spread horizontally,[21] as shown in (d), intruding into the surrounding water along the isopycnal surface that has the same density as the mixed region, and dispersing the markers horizontally. At the same time the mixed region or intrusion collapses vertically, thinning to conserve its volume. After a period of time of order f^{-1} (where f is the Coriolis frequency, equal to $2\Omega \sin \phi$, where Ω is the angular frequency of the

21 If the dots in Fig. 1.16(a) represent small neutrally buoyant particles (solid particles having the same density as the fluid surrounding them) rather than a neutral dissolved tracer, because their density is therefore between either $\rho_0(1 - N^2h/g)$ and ρ_0 (the upper 'step' in the density profile through the mixed patch shown to the left of (c)) or ρ_0 and $\rho_0(1 + N^2h/g)$ (the lower 'step' in the same density profile), they will tend to rise or fall towards, and tend to float on, the boundaries of the dashed oval rather than dispersing uniformly, as would dye, within the mixed region. This effect may influence their horizontal dispersion in the collapsing stage, (e). (But see P1.8.)

Earth's rotation, $7.292 \times 10^{-5}\ \mathrm{s}^{-1}$, and ϕ is the latitude[22]), the effect of the Earth's rotation will cause the lens-like patch to begin to rotate anticyclonically (clockwise in the northern hemisphere) as shown in (e). Inward-moving water above and below the patch will begin to rotate cyclonically. A geostrophic balance in which the pressure forces driving the horizontal spread of the patch are equal to the Coriolis forces will be reached when the ratio of the width of the patch to its thickness is about N/f, where N is the buoyancy frequency in the surrounding water. Its swirl velocity is then of order NH, where H is half the thickness of the patch. Further spreading of a patch may occur as a result of molecular heat and momentum transfer and through further breaking events or interactions with neighbouring collapsing mixed regions. [**P1.9**]

• Diapycnal mixing and 'diffusion', followed by the 'dispersion' of neutrally buoyant or fluid particles, their spreading or 'intrusion' on isopycnal surfaces, and a subsequent rotation or 'spin-up' described by Fig. 1.16, are features common in oceanic mixing and dispersion. In this simplified example, the relatively small-scale turbulence involved in the mixing processes leads to collapse and intrusive motions at relatively larger scales. In practice, these organized motions may, for example by producing shear and strain, affect the decay of the turbulent motions on a smaller scale that produced the diapycnal diffusion. Some of the energy of turbulent eddies may also be lost through the generation and radiation of internal waves from the patch of turbulent motion.

The internal Rossby radius, defined as c_n/f, is an important measure of rotation-affected motions. Here c_n is the speed on long internal waves of the nth mode. Each mode, therefore, has a corresponding Rossby radius. The internal Rossby radius is commonly defined as $L_{\mathrm{Ro}} = c/f$, where c is the speed of long internal waves of the first mode – the mode number of the fastest internal waves. In the ocean, c is of order $1\ \mathrm{m\ s}^{-1}$, so L_{Ro} is typically about 10 km, but varies with the stratification (or N) and generally, since $f = 2\Omega \sin \phi$, decreases with increasing latitude. The effects of rotation are insignificant in waves of horizontal wavelength much less than the Rossby radius, their properties being determined by buoyancy forces, whilst rotation dominates those of horizontal wavelength much greater than the Rossby radius. The Rossby radius is also known as the 'Rossby radius of deformation' or simply as the 'deformation radius'. It is a length scale that characterizes the scale to which a patch, collapsing under gravity, will extend before the pressure gradient becomes equal to the Coriolis force, resulting in a geostrophic balance. (In the case considered in Fig. 1.16, the appropriate phase speed, c_n, is about NH, where H is the patch height, and the horizontal scale at which adjustment to geostrophic balance occurs and collapse or deformation is complete is about $l = c_n/f \approx NH/f$, so $l/H \approx N/f$, as stated above.) The interactions, dynamics and dispersive properties of a number of such rotating

22 At a latitude of 30°, $f = 2\Omega \sin 30°$, so $f^{-1} = 3.81$ h. The related period, $2\pi f^{-1}$, is known as the inertial period and (being half that of a Foucault pendulum) is equal to half a 'pendulum day'. The inertial period is approximately 12 h at the Poles and increases as the Equator is approached.

patches (or vortical modes – see Section 5.4.3) in stratified water depend on the ratio of their radii to the internal Rossby radius. Rotating patches may become baroclinically unstable, with typically two vortices forming around their circumference, if the ratio of their height to their width is less than about $0.2f/N$.

Huge horizontal eddies called 'mesoscale eddies' with horizontal dimensions of typically 100 km are found in the ocean and affect dispersion, as described in Section 5.3.1. The most energetic mesoscale eddies have a modal structure that generally corresponds to that of the first internal wave mode and dimensions that appear to scale with L_{Ro}.

Suggested further reading

Fundamental laboratory experiments

Joule's (1850) and Reynolds' (1883) extraordinary and now famous studies are briefly described above, but their multifaceted papers are well worth reading, if only to gain an understanding of the care required in making laboratory studies and a perspective on the historical development of the study of heat and energy and turbulent motion.

Diffusion and dispersion

Both Eckart's (1948) and Welander's (1955) papers provide further background to understanding the evolution of ideas about the nature of mixing in natural fluid bodies.

Basics of physical oceanography

Gill (1982) gives a comprehensive and informative account of the processes common to the ocean and the atmosphere and is a valuable text that includes reference to ideas, as well as processes, underlying some of the understanding of ocean turbulence.

Further study

Further experimental studies of the Reynolds transition from laminar to turbulent flow have been made by Durst and Ünsal (2006).

Gill (1981) gives a mathematical treatment of the *spread of a mixed patch* in stratified surroundings, and Griffiths and Linden (1981) demonstrate through laboratory experiments how patches become baroclinically unstable. Analysis of the spread and adjustment to the effects of the Earth's rotation in a related problem, that of a spreading body of water in the form of a density or gravity current, is given by Hunt *et al.* (2005).

There is a growing literature on breaking waves, a subject to which we return in Chapter 3. The efficiency of a plunging breaking surface wave (the ratio of energy lost to the original wave energy) is discussed by Rapp and Melville (1990) and acoustic studies of wave breaking and energy loss are described by Lamarre and Melville (1994).

Further information about the processes of ocean mixing is given in Chapter 1 of TTO. The relation between internal waves and turbulence is complex and an important aspect of oceanic turbulence and mixing. An introduction to *internal waves* is given in Chapter 2 of TTO; their breaking is described in Chapter 5 and related fine-structure in Chapter 7.

Problems for Chapter 1

(E = easy, M = mild, D = difficult, F = fiendish)

P1.1 (M) Flux of energy. The speed of the laminar flow in the circular tube of radius, a, in Reynolds' experiment is $u(r) = 2U(1 - r^2/a^2)$ at radius $r \leq a\,(= d/2)$, where U is the radial-average flow. (Check, by integration, that U is equal to the average!) The flow following the transition to turbulence can be regarded as being composed of a mean flow, U, which is approximately uniform (i.e., independent of radius, neglecting very thin viscous boundary layers on the tube walls where the flow is reduced to zero), and a fluctuating turbulent part with zero mean. How large is the change from the flux of the kinetic energy in the steady flow upstream of the transition to the flux of kinetic energy in the mean flow downstream of the transition? How might you account for the change ignoring any work done ($\int pu\,\mathrm{d}z$) by the pressure, p? Estimate the efficiency of the transition from laminar to turbulent flow in producing turbulence.

Hints: the average flow is given by $(\int_0^a 2\pi r u\,\mathrm{d}r)/(\pi a^2)$ and the kinetic energy flux is $(\rho/2)\int_0^a 2\pi r u^3\,\mathrm{d}r$. The efficiency might be defined as the flux of turbulent energy divided by the energy flux in the laminar flow.

(The purpose of this problem is to stimulate ideas about mean and fluctuating flows and about the flux of energy and its conservation, which will be discussed further in Chapter 6.)

P1.2 (E) The criterion for turbulence. Reynolds' experiment shows that turbulence with eddies of size comparable to the tube radius develops when the Reynolds number, *Re*, exceeds a critical value. The mean depth of the Irish Sea is about 60 m and the tidal currents are typically 0.1–1 m s^{-1}. With only this information and assuming that the critical Reynolds number for oceanic flows is of order 10^4, should the tidal flow in the Irish Sea be laminar or should it be turbulent, probably with some eddies of size comparable to the water depth?

• The estimate of *Re* was made by G. I. Taylor, 1919, and used to dismiss earlier calculations of the dissipation of tidal energy in the Irish Sea carried out assuming only molecular viscosity in a laminar flow over the seabed and in the water column.

P1.3 (D) Wave energy. The flux of energy of a wave on the sea surface per unit crest length is equal to its local energy or 'energy density', E (the sum of the mean kinetic and potential energies in the wave field per unit surface area), multiplied by the speed at which this energy is transported, a speed called the wave group velocity,

c_g. Approximately, $E = a^2\rho g/2$, where a is the wave amplitude (half the vertical distance between the wave crest and the trough), ρ is the density of seawater (about 1028 kg m^{-3}) and g ($\approx$9.81 m s^{-2}) is the acceleration due to gravity, and, in deep water, $c_g = c/2$, where c is the wave phase speed (the speed at which wave crests advance), given by the dispersion relation c (m s^{-1}) $\approx 1.56T$ (s), where T is the wave period. (The greater the wave period, the faster do wave crests advance. The wave phase speed is equal to the time taken for a wave crest to advance through a distance of one wavelength.)

(i) If the waves have a period $T = 5$ s, what is their wavelength?

(ii) Waves with a period of 5 s that are 1 m high approach shore from a direction normal to the beach, and dissipate all their energy by turbulence within a surf zone that extends, on average, 40 m from shore with a mean depth of 1 m. At what average rate per unit mass must energy be dissipated within the surf zone?

(iii) What is the mean rate of increase in the temperature of the zone resulting from wave breaking? You should suppose that no energy is lost in heating the sediment, exchanged in the form of heat with the atmosphere, or radiated as sound in the atmosphere and ocean or in the form of microseisms (waves with a frequency of typically 1 Hz that travel through the solid Earth). In practice, these are all ways that some energy is transferred from the surf zone, and the heat budget of the surf zone is generally very difficult to determine precisely.

P1.4 (E) Turbulence in the swath zone. The swath zone is the region at the shoreward edge of the surf zone where waves carry water up and down the beach, covering and then exposing the underlying sediment. If a layer of water that is 2 cm deep runs down a smooth and gently sloping beach in the swath zone, what speed is required to ensure that the layer is turbulent? (Assume a critical Reynolds number of 10^4. In practice, there may be residual turbulence left and carried upslope after earlier wave breaking within the surf zone, and flow over and around stones or shells on the beach may cause eddies that contribute to turbulent motion. From your observations whilst walking along a sandy beach, is this speed required for turbulent motion rarely or commonly exceeded?)

P1.5 (D) A spreading patch. Show that the marked patch in the region of convergent field of motion represented in Fig. 1.10(b) by $u = qx$ and $v = -qy$, where q is a constant, has an extent in the x direction that increases as $\exp(qt)$.

Suppose that, instead of being two-dimensional as envisaged in Section 1.6, the marked patch is a vertical cylinder of radius r and height $2z$, and that convergence, with vertical velocity component $w = -qz$, results in a radial spread of a cylindrical volume in the horizontal x–y plane. How does the radius of a patch then vary with time, assuming that the patch spreads as a cylinder still with vertical sides?

P1.6 (D) Energy in a spherical eddy at an interface. Suppose that a spherical eddy of radius r in solid-body rotation is formed at the interface between two layers of densities ρ_1 (above) and ρ_2 (below) with $\rho_1 < \rho_2$ and $(\rho_2 - \rho_2)/(\rho_1 + \rho_2) \ll 1$, rotates the fluid through 180°, and then stops (Figs. 1.17(a) and (b)). Calculate the increase in potential

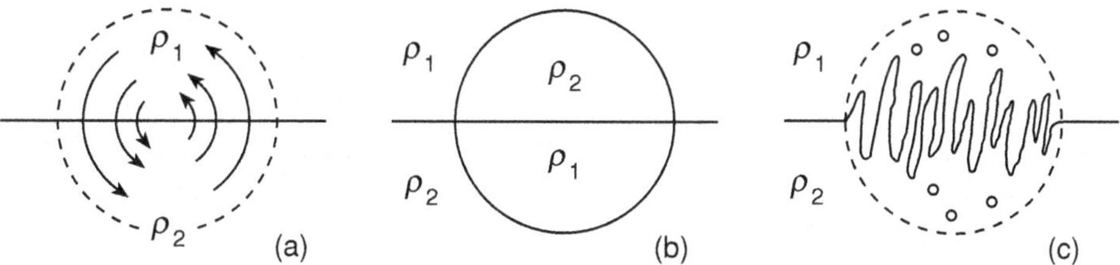

Figure 1.17. A spherical eddy of radius r overturning as a solid body at the interface between two uniform layers of densities ρ_1 and ρ_2. (a) The initial state, but indicating the subsequent rotation about an axis normal to the plane of the paper. (b) After the eddy has rotated by 180°. (c) Convection mixing the eddy to uniform density $(\rho_1 + \rho_2)/2$.

energy. If the water in the eddy is now mixed by turbulent convection (Fig. 1.17(c)) and finally reaches a uniform mixed state, with no kinetic energy remaining, before there is a collapse and lateral spread of the mixed region, how much turbulent energy will have been dissipated, supposing that no energy is lost by viscosity? If this turbulent energy is all dissipated by viscosity, how much will the temperature of the region increase, supposing that the heat is dissipated uniformly within it? In comparison with the difference in densities, $\rho_2 - \rho_1$, how great a density decrease results from the rise in temperature? Provide an answer supposing that $r = 10$ m and that the coefficient of thermal expansion $\alpha = 2 \times 10^{-4}\ \mathrm{K}^{-1}$.

P1.7 (D) Mixing in a cylindrical eddy and breaking internal waves. Suppose that turbulence completely mixes a horizontal cylinder of stratified water of length l and radius r that has an initially uniform stratification and buoyancy frequency N, leaving it in a state of rest and with a final density that is uniform. Ignoring other factors such as the viscous loss of energy and the tendency of the mixed fluid to spread horizontally (as in Fig. 1.16(d)), find the minimum initial kinetic energy required in the eddy to complete the mixing. You should suppose that 20% of the kinetic energy goes into potential energy, the remainder being dissipated in turbulence.

If, instead of causing mixing, this energy is used to lift the unmixed cylindrical body of water without change from its initial level, how far will its centre be raised, neglecting work done against pressure in the translation?

In a linear approximation, the energy density of an internal wave (i.e., the mean energy per unit volume of seawater in the field of motion created by the wave) travelling in a fluid of uniform buoyancy frequency, N, is $a^2N^2\rho_0/2$, where a is the wave amplitude and ρ_0 is the mean density. Suppose that the wave breaks, completely mixing a cylinder of water of radius qa as described above, where q is a constant. For what value of q is the mean wave energy within the cylinder of water equal to that required to bring about complete mixing of the cylinder of water?

Help: $\int_{-1}^{1} x^2(1 - x^2)^{1/2}\,\mathrm{d}x = \pi/8$.

P1.8 (F) Vertical dispersion of floating particles. Suppose that a layer of uniformly stratified water with constant buoyancy frequency, $N = 10^{-2}\ \mathrm{s}^{-1}$, and containing a uniform, but diffuse (and therefore non-interacting), distribution of spherical, neutrally buoyant particles of radius 1 mm is mixed by turbulence to a uniform density, ρ_0, and that immediately after mixing the particle distribution is uniform within the mixed layer. Estimate how long it will be before half of the particles (having been neutrally buoyant in the linear stratification but being no longer neutrally buoyant within the mixed layer) sink or rise to its lower or upper boundary. You should assume that the turbulent motion mixing the layer is rapidly dissipated so that the rise or sinking speeds of the particles are as given in footnote 16 for motion in fluid at rest, and with no clustering of particles.

(The time is found to be large relative to a typical value of f^{-1}, so the accumulation of such neutrally buoyant particles at the boundaries of the layer will not be substantial before the lateral spread of the mixed region leads to the onset of significant anticyclonic rotation.)

P1.9 (M) The energy needed to mix a stratified region. What is the minimum energy required to reduce an initially uniform density gradient with buoyancy frequency N_0 to a final state with frequency $N < N_0$, over a depth of $2h$?

(An application of this calculation is found in the description by Sundermeyer *et al.*, 2005, of the horizontal diffusion of dye in the ocean.)

Chapter 2

Measurement of ocean turbulence

Measurement is at the heart of science. The measurement of turbulence in the ocean has proved difficult, and not all the technical and operational problems have been overcome. In this chapter we review the characteristics that are used to describe turbulent motion and its effects, and describe some of the methods of measuring and quantifying turbulence.

2.1 Characteristics of turbulence

Some of the characteristics of turbulence are described in general terms in Chapter 1. These can provide ways of quantifying turbulent motion as explained in this, and later, sections.

2.1.1 Structure

Figure 2.1 is a shadowgraph image of the development of a turbulent shear flow in a laboratory experiment. It shows large billows formed downstream of a 'splitter plate' dividing two streams of gases with different speeds and densities. As in the photograph of the surf zone (Fig. 1.4), Fig. 2.1 shows that the flow contains patterns or structures – the billows – that recur. Each billow extends over a finite region: the motions within are spatially coherent. Although the billows are transient, they also persist for times long enough to allow them to be identified: they are coherent for short periods of time. The structure within billows varies in detail from one to another, and consists of small-scale turbulent motions that lead to the fine 'texture' visible in the image.

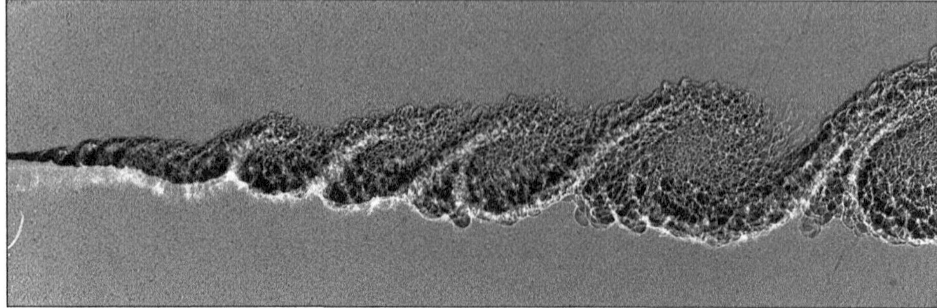

Figure 2.1. Very-small-scale irregular and larger-scale coherent structures coexisting in a turbulent mixing layer between two layers of different gases, the upper less dense gas moving to the right more rapidly than the lower. Eddies or billows are growing downstream of a splitter plate, off the figure to the left, and are made visible using a shadowgraph. Parallel light is shone horizontally through the gas layers onto a transparent screen, the gradients in refractive index distorting the light to produce the dark and light tones of the images of the flow in the photograph of the screen. Relatively large-scale coherent structures are often found in turbulent flows – provided, of course, that the means to detect them exist. Billows on fluid interfaces are also shown in Figs. 4.1, 4.2 and 4.4 later. (From Brown and Roshko, 1974.)

Such patterns of relatively large-scale coherent eddies containing small-scale motions are commonly found in turbulent flows. The larger eddies often provide clues to the source of turbulence and, in particular, to its energy supply. Structures similar to the billows illustrated in Fig. 2.1 are visible at the edges of jets and are found, for example, in the wind-mixed layer at the ocean surface, suggesting a similar cause, the presence of shear. The 'boils' shown in Fig. 1.6 provide another example of large turbulent eddies, and in this case the source can be traced to the shear flow over the seabed.

• A field of motion, although turbulent, commonly contains coherent structure. Whilst the velocity measured at any point in the ocean may appear to be highly variable or random, in some average sense motions at points separated in space may be similar, provided that the distance between them is not too great (e.g., scales less than that of billows).

Turbulent structure can be quantified, for example by determining from measurements how the cross-correlation (a measure of coherence) of fluid velocity or temperature fluctuations at two points varies as a function of their distance apart or by measurements from which energy spectra can be derived as described in Section 2.3.6. Such quantities can provide useful measures of turbulence and its dynamical effects.

Figure 2.1 suggests a further idea, that the small eddies are a consequence of the motions associated with the larger billows or that there is a cascade of energy from larger to smaller scales. This important concept is discussed further in Section 2.3.5. The mean size of the larger eddies, if it can be quantified, may provide information from which other quantities can be derived. One example is that the size of eddies overturning in stratified waters is related to the rate of loss of turbulent kinetic energy, as explained in Section 4.4.1. Another example is involved in the use of acoustic Doppler current profilers to estimate the same dissipation rate (Section 2.5.4).

A clear distinction needs to be recognized (particularly in interpreting 'images of turbulence') between the coherent structures such as eddies that, at any moment, are found in a field of turbulent motion, and the patterns, such as filaments (Section 5.2.2), that result from the past action of turbulence in spreading patches of dye, particles or tracers embedded within a turbulent fluid of motion. Both have structure, but the latter represent the relative displacements or dispersion caused by the velocity field of the former.

2.1.2 Stress and flux

Measures of the effect of turbulence can be determined from the rates of transfer of momentum and heat (or other fluid properties) by turbulent motion. Momentum transfer is equivalent to the application of a stress. The transfer of heat (or another property) involves a flux. These are discussed in Section 2.2. Stress is particularly important as a measure of the effect of turbulent water motion on a sedimentary (e.g., sandy) seabed, and its magnitude determines whether and how sediment is moved by the flow of water.

• Turbulence can lead to stress and heat flux, or to the transfer and spread of other water properties such as salinity and other constituent dissolved chemicals at rates that generally greatly exceed those of molecular transport.

2.1.3 Dissipation

It was explained in Section 1.3 that turbulence is a form of motion containing kinetic energy that is dissipated and transferred into heat much more rapidly than in a laminar flow with the same mean velocity. This is of vital importance because it represents an irreversible loss of kinetic energy from the ocean.

• The measure of this dissipation is usually expressed as the rate of dissipation of turbulent kinetic energy per unit mass, commonly denoted by ϵ with units of W kg^{-1} or equally m^2 s^{-3}. (The latter usefully reveals that the dimensions of ϵ are L^2T^{-3}, where L is length and T is time, but W kg^{-1} are the more usual units.)

The energetics of turbulent motion are discussed in Section 2.3.

2.2 Transport by eddies

2.2.1 Reynolds stress

In a paper published in 1895, Reynolds[1] recognized that a fundamental property of turbulence is that its irregular motions transfer momentum, and equally (but not

1 This is again Osborne Reynolds (see Section 1.2). In addition to his elegant and basic studies of the onset and the nature of turbulent motion in homogeneous (and stratified) fluids, Reynolds contributed to the understanding of ocean turbulence, for example by proposing a model of vortex

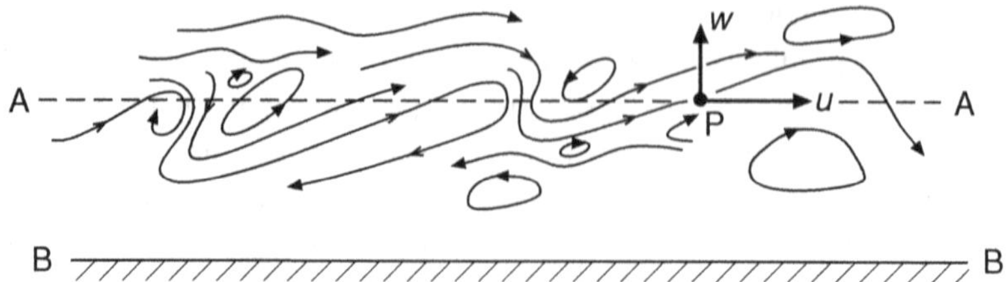

Figure 2.2. A sketch of turbulent motion, illustrating the Reynolds stress. A particle of fluid, P, with density ρ and horizontal velocity u, has a horizontal momentum per unit volume of ρu. On moving upwards across the surface A–A at speed w, horizontal momentum is transported upwards into the overlying fluid at a rate $\rho u w$. Some particles on the surface move (and carry their horizontal momentum) downwards, and some have negative horizontal speeds, u, but the average of $\rho u w$ over the surface A–A (or in time if u and w are the horizontal and vertical components of velocity measured at a fixed point) is equal to the mean vertical transport of momentum per unit area from below the level A–A to the region above, or, equal $\tau = -\langle \rho u w \rangle$, the 'Reynolds stress' of the fluid above A–A on the region below. In flows commonly encountered in the ocean, ρ varies very little from the mean density, ρ_0, in comparison with the relative variations in velocity about the mean, so $\tau \approx -\rho_0 \langle uw \rangle$. The two components of flow may be written as $u = \langle u \rangle + u'$ and $w = \langle w \rangle + w'$, representing a steady mean flow $(\langle u \rangle, \langle w \rangle)$ and the fluctuations (u', w') induced by the turbulence (and by waves, if they are present). If the mean flow is horizontal (e.g., in a fluid bounded by the rigid horizontal plane B–B), there will be no mean vertical velocity, so $\langle w \rangle = 0$, and, by virtue of the definition of the mean, $\langle u' \rangle = 0$ and $\langle w' \rangle = 0$. The product, $\langle uw \rangle$, or $\langle (\langle u \rangle + u')(\langle w \rangle + w') \rangle$, is then equal to $\langle u'w' \rangle$, so the Reynolds stress $\tau \approx -\rho_0 \langle u', w' \rangle$. (There are some misleading aspects of the sketch. For example, it does not represent the three-dimensional field of turbulent motion. The lines and arrows suggest steady streamlines of motion, perhaps continued over a period of time, rather than instantaneous particle motion at the time at which the average over the surface A–A is determined.)

necessarily at directly corresponding rates) heat and salinity, across surfaces, usually conceived of as planes fixed in space at positions where properties are to be quantified, but having no physical substance – they are not solid surfaces. The rates of transfer are given by the temporal averages of the product of the transferred properties and the component of the turbulence velocity normal to the surface.

Taking as an example the horizontal momentum of fluid particles as the transferred property (as illustrated in Fig. 2.2), the mean downward vertical rate of transfer of horizontal momentum in the x direction, ρu, by the upward vertical component of their velocity, w, across a horizontal plane surface at the level marked A–A in Fig. 2.2 is τ, given by

$$\tau = -\langle \rho u w \rangle. \tag{2.1}$$

The symbols with angular brackets, such as $\langle Q \rangle$, imply that an average is taken of the quantity Q, usually over time at a fixed point rather than in, say, one or more coordinate

formation by raindrops impacting on the sea surface and creating motions that might act to reduce the intensity of wave breaking – the 'knocking down of the sea', a long-known, but little studied, calming effect of heavy rainfall (Reynolds, 1900).

directions at a fixed time, for observational convenience. The average will normally be a function of the position of the point, e.g., its height above the seabed. Often, and particularly in flows in oceanic boundary layers described in Chapter 3, the stress is assumed to be independent of horizontal position, (x, y), and to depend only on the vertical coordinate, z, of the surface of measurement.[2] The symbol τ represents a stress or force per unit area with dimensions $\mathrm{ML^{-1}T^{-2}}$. It is now called the 'Reynolds stress', the turbulence stress imposed by the fluid above the horizontal plane on that below, hence the negative sign in (2.1). Since the density varies far less than do the velocity components, the stress is approximately equal to $-\rho_0\langle uw\rangle$, where ρ_0 is a constant average or reference density. A zero correlation between u and w leads to a zero Reynolds stress (e.g., when the two are 90° out of phase as they are in some wave motions). [**P2.1**]

• The Reynolds stress is equal to the mean rate of transfer of momentum across a surface by turbulent motion.

The stress should strictly be written as a matrix quantifying transfers of three $(\rho u,\ \rho v,\ \rho w)$ components of momentum across the three orthogonal (x, y, z) plane surfaces, i.e., a matrix of averaged products of the three $(u,\ v,\ w)$ components of velocity. Often, however, because of observational difficulties, only a single component is estimated from measurements. It is sometimes important, particularly in ocean models representing the effects of the large mesoscale eddies referred to later (in Section 5.3.1), to account for the horizontal component of Reynolds stress, $-\rho_0\langle uv\rangle$.

The transfer of momentum is usually represented through a parameterization that relates the x–z Reynolds stress to the mean velocity gradient $\mathrm{d}U/\mathrm{d}z$ (or, using the $\langle\ldots\rangle$ notation, $\mathrm{d}\langle u\rangle/\mathrm{d}z$) through a coefficient known as the 'eddy viscosity', K_ν, with dimensions $\mathrm{L^2T^{-1}}$:

$$\tau/\rho_0 \approx -\langle uw\rangle = K_\nu\,\mathrm{d}U/\mathrm{d}z. \tag{2.2}$$

Strictly, like Reynolds stress, the eddy viscosity should be represented by a matrix, derived from the mean products of pairs of velocity components in all three directions and, as defined, K_ν is the vertical component of the eddy viscosity. The value of K_ν can be determined by measurements of the mean shear, $\mathrm{d}U/\mathrm{d}z$, and both the u and the w velocity components in the benthic boundary layer as explained later.

• 'Eddy coefficients' are commonly used in parameterizing turbulent fluxes, with an eddy coefficient multiplied by the gradient of the transferred quantity being set equal to the turbulent flux of the quantity. The coefficient relating to momentum transfer is known as the eddy viscosity because, when turbulence is parameterized by such a term, it plays a role in the equations of motion similar to that of kinematic viscosity in laminar flow.

2 The Reynolds stress may be uniform, unchanging with z, in steady turbulent boundary layers as, for example, is explained in Section 3.3.

2.2.2 Heat and buoyancy flux

Just as momentum is transferred from one level in a fluid to another by eddies, so can heat (or salinity and other properties) be transferred. The vertical flux of heat depends on the ability of eddies to carry water of a higher temperature upwards or water of a lower temperature downwards, which is often inhibited by buoyancy forces derived from stratification.

Suppose that T' represents the difference between the temperature of water measured at a fixed point at time t and a reference temperature, commonly taken as the mean temperature at the level of the point. If the convention that upward fluxes are positive is adopted, the vertical upward heat flux per unit horizontal area at the point can be expressed as the average of the product ρc_p multiplied by the temperature variations T' (the product, $\rho c_p T'$, is the heat fluctuation per unit volume of fluid corresponding to the temperature change, T') and also by the speed, w, at which the temperature fluctuations are carried upwards:

$$F = \langle \rho c_p w T' \rangle. \tag{2.3}$$

The vertical flux of heat carried by turbulent motion is often expressed as

$$\langle \rho c_p w T' \rangle = -\rho c_p K_T \, \mathrm{d}T/\mathrm{d}z, \tag{2.4}$$

using the analogy with the rate at which heat is carried by molecular conduction, or approximately as

$$\langle w T' \rangle = -K_T \, \mathrm{d}T/\mathrm{d}z, \tag{2.5}$$

where $\mathrm{d}T/\mathrm{d}z$ is the mean vertical temperature gradient.

• Equation (2.5) defines the parameter K_T, the vertical eddy diffusion coefficient of heat or the eddy diffusivity of heat, with dimensions $\mathrm{L}^2\mathrm{T}^{-1}$.

Eddy diffusivities of other properties of seawater, such as density (or mass), K_ρ, and salinity, K_S, are defined in a similar fashion, e.g., $\langle w\rho' \rangle = -K_\rho \, \mathrm{d}\rho/\mathrm{d}z$, where ρ' is a density fluctuation from the mean and $\mathrm{d}\rho/\mathrm{d}z$ is the mean density gradient. The eddy diffusion coefficients of heat, K_T, and of density, K_ρ, are equal when density variations are dominated by those of temperature, as in a freshwater lake, and may also be equal if there is a monotonic relationship between density and temperature.[3] Estimates of K_T can be obtained from measurements of ocean microstructure as described in Section 4.4.2. [**P2.2**]

In a fluid that has no salinity variations, (1.4) gives $\rho = \rho_0(1 - \alpha T')$ and so the density variation, ρ', from the ambient, $\rho - \rho_0$, is equal to $-\rho_0 \alpha T'$. From its definition, (2.3), the heat flux can therefore be written as $F = -c_p \langle w\rho' \rangle / \alpha$. But the turbulent flux of the buoyancy, $b = g(\rho_0 - \rho)/\rho_0 = -g\rho'/\rho_0$ (defined in Section 1.7.2), is

3 Care is needed! In conditions under which molecular transports are important (e.g., where double diffusive convection is possible as described in Section 4.8), the coefficients K_T, K_S and K_ρ differ from one another.

defined as

$$B = \langle wb \rangle, \tag{2.6}$$

so

$$B = -g\langle w\rho' \rangle / \rho_0. \tag{2.7}$$

This implies that B has dimensions L^2T^{-3}.

• When salinity variations can be neglected, the turbulent buoyancy flux, B, is therefore related to the turbulent heat flux, F, by

$$B = g\alpha F/(\rho_0 c_p). \tag{2.8}$$

The importance of this relation will be apparent when we come to discussing the consequences of heating in the boundary layers of the ocean in Chapter 3.

2.3 Energetics

2.3.1 Turbulent dissipation, ε, and isotropy

The surf zone is briefly described in Section 1.4. Much of the energy lost by the breaking waves and hydraulic jumps is transferred into turbulence. But where does the turbulent energy go? Some of the energy may be dissipated in eroding sediment and in repeatedly lifting sediment from the seabed, so raising its potential energy. Some may be advected in rip-currents and carried out of the surf zone into deep water. However, as in Joule's experiment, much of the dissipation occurs locally in the surf zone through viscosity, and for this to be effective in the relatively short times between re-supply of energy in successive waves, high shears are required, as explained below.

The rate of loss of the kinetic energy of the turbulent motion per unit mass through viscosity to heat is usually denoted by ε. This can be expressed in general as

$$\varepsilon = (\nu/2)\langle s_{ij}s_{ij} \rangle, \tag{2.9}$$

where the velocity is written as (u_1, u_2, u_3) in three orthogonal directions $x = x_1$, $y = x_2$ and $z = x_3$, and ν is the kinematic viscosity. The tensor s_{ij} is given by

$$s_{ij} = (\partial u_i/\partial x_j + \partial u_j/\partial x_i), \tag{2.10}$$

and products are taken in (2.9) over repeated suffices i, $j = 1$ to 3.[4] The complexity of the expression for ϵ implies that its precise measurement is generally extremely difficult! The presence of spatial gradients of velocity components in (2.10) means that the rate of loss of energy in turbulent motion results from shear and is enhanced

4 The term $s_{ij}s_{ij}$ in (2.9) is equal to the sum of all the products, $s_{11}s_{11} + s_{12}s_{12} + s_{13}s_{13} + s_{21}s_{21} + s_{22}s_{22} + \cdots + s_{33}s_{33}$, where $s_{11} = \partial u_1/\partial x_1 + \partial u_1/\partial x_1 = 2\,\partial u/\partial x$, $s_{12} = \partial u_1/\partial x_2 + \partial u_2/\partial x_2 = \partial u/\partial y + \partial v/\partial x$, etc.

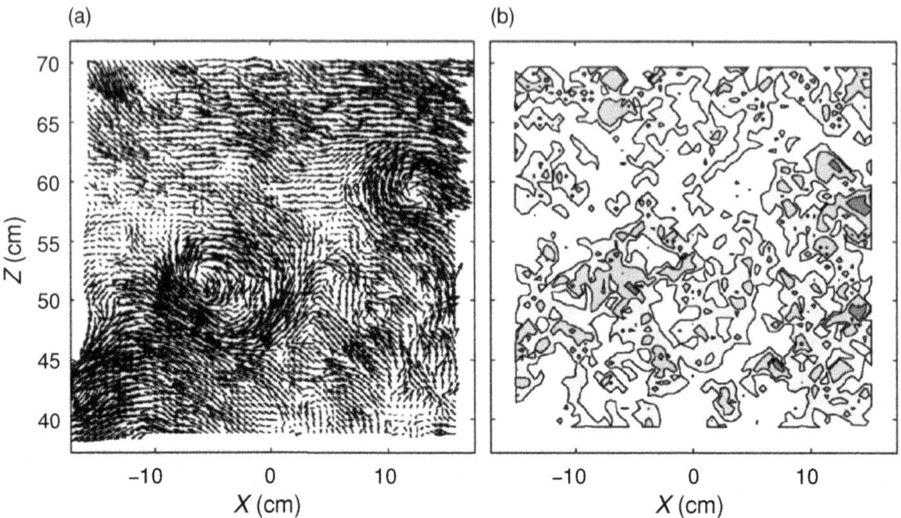

Figure 2.3. Particle image velocimetry (PIV) measurements of currents made in the sea. Here two dimensions, those in the plane of the mean flow, are shown to illustrate the complexity of motions and dissipation rates found in a small area of the flow field at a height of 0.4–0.7 m above the seabed. A mean horizontal current of 11.3 cm s^{-1} to the right and an upwards vertical motion of −0.2 cm s^{-1} have been subtracted from the speeds shown in (a). Thin shear layers and eddies of diameter 5–10 cm are visible. Part (b) shows the corresponding dissipation rates estimated assuming isotropy. The contour levels shown are (0.1, 0.3162, 1.0 and 3.162) × 10^{-5} W kg^{-1}. Three-dimensional images at frequencies of 25 Hz or more are now attainable and, in principle, will allow not only direct estimates of ε to be made using (2.9), but also study of isotropy and turbulent structure. See P3.6 for further discussion. (From Nimmo Smith *et al.*, 2005.)

by high rates of shear. These are generated by the interacting field of small-scale eddying motion characteristic of the turbulence observed, for example, in Reynolds' experiment and illustrated in Fig. 2.1. [**P2.3**]

Making the measurements of all the velocity gradients required to determine ϵ from (2.9) is, technically, highly demanding and rarely achieved. Methods of measuring velocity based on particle image velocimetry (PIV) have recently become possible in the ocean and have provided valuable information about eddy structure as well as dissipation; an example is given in Fig. 2.3. Considerable simplification to (2.9) is possible, however, for conditions under which the properties of the turbulent motion, including its velocity gradients, are the same in all directions, when turbulence is said to be 'isotropic'. (The ideas underlying the assumption of isotropy are discussed further in Section 2.3.5.)

- In isotropic motion the mean square gradients of quantities such as $\partial u/\partial y$, $\partial v/\partial z$, $\partial w/\partial x$, and their mean products, are equal and (2.9) reduces to the much simpler equation

$$\varepsilon = (15/2)\nu\langle(\partial u/\partial z)^2\rangle, \tag{2.11}$$

where the average value of any spatial derivative in a direction normal to its direction may be taken. (The derivative, $\partial u/\partial z$, of the horizontal component of current, u, in the vertical direction is chosen in (2.11) because of its relevance to measurement by sensors or 'probes' described later that are mounted on free-fall instrument packages.) The assumption of isotropy, which is commonly made (although often without justification!), reduces the measurement requirement substantially, since only one component of the shear needs be found in order to estimate ϵ.

2.3.2 The range and observed variation of ε

The oceanic values of ε have a vast range, extending over nine orders of magnitude, from about 10^{-10} W kg^{-1} in the abyssal ocean to 10^{-1} W kg^{-1} in the most actively turbulent regions, such as the surf zone and in fast tidal currents through straits. Rather than expressing values of ε in units with SI prefixes (see page xviii; e.g., its range as 10^{-1} nW kg^{-1}, or 100 pW kg^{-1}, to 1 dW kg^{-1}), it is customary to express ε with powers of 10 given explicitly, and we adopt this convention here. [**P2.4**]

Figure 2.4 shows examples of ε profiles, together with the potential density, σ_θ, and east and north components of current, u and v, respectively, from three different ocean regions. Figure 2.4(a) is from the Straits of Florida, running almost due north between the coast of Florida and the Bahama Banks, where data were obtained using a freely falling (or free-fall) instrument called the Multi-Scale Profiler (MSP). The σ_θ profile shows the near-surface mixed layer, the thermocline from about 50 m to 200 m, and a homogeneous near-bed benthic boundary layer below 530 m. The major flow in the Florida Current is northwards, its magnitude reaching about 2.3 m s^{-1} near the surface where dissipation is relatively high, with values exceeding 10^{-6} W kg^{-1}. Dissipation rates of similar magnitude are found near the seabed. Generally, however, values of ε in mid-water are patchy and smaller, of order 10^{-9}–10^{-8} W kg^{-1}: the Florida Current is not a location of extremely intense turbulence.

Figure 2.4(b) shows data from the Camarinal Sill in the Strait of Gibraltar obtained using the Advanced Microstructure Profiler (AMP; see Fig. 2.13(a) later). The orientation of the Strait is roughly east–west. The σ_θ profile shows a number of regions of small thickness in which the density decreases with depth, indicating static instability. (The low value at about 0.58 MPa, a depth of about 58 m, appears to be erroneous.) The upper layer has a velocity, u, of about 0.6 m s^{-1} and is composed of Atlantic water moving eastwards through the Strait into the Mediterranean Sea. The lower, westward-moving layer is composed of Mediterranean water. Although Mediterranean water is warmer than the overlying Atlantic water, it is much saltier (a consequence of the dominance of evaporation over precipitation in the Mediterranean and adjoining seas) and, as a result, denser. At the time these profiles were obtained, the rate of turbulent dissipation, ε, reached about 10^{-6} W kg^{-1} in the sheared interface between the two water masses. Dissipation rates as large as 10^{-4} W kg^{-1} are found in mid-water at other phases of the tidal flow through the Strait, and thicker regions of static instability are observed when the shear leads to the mixing illustrated in Fig. 4.13. (The Mediterranean water subsequently flows into the Gulf of Cadiz, mixing with the water of the Atlantic as it does so, and spreads westwards as a tongue with relatively high salinity at depths between 600 m and 1500 m as shown in Fig. 5.1 later.)

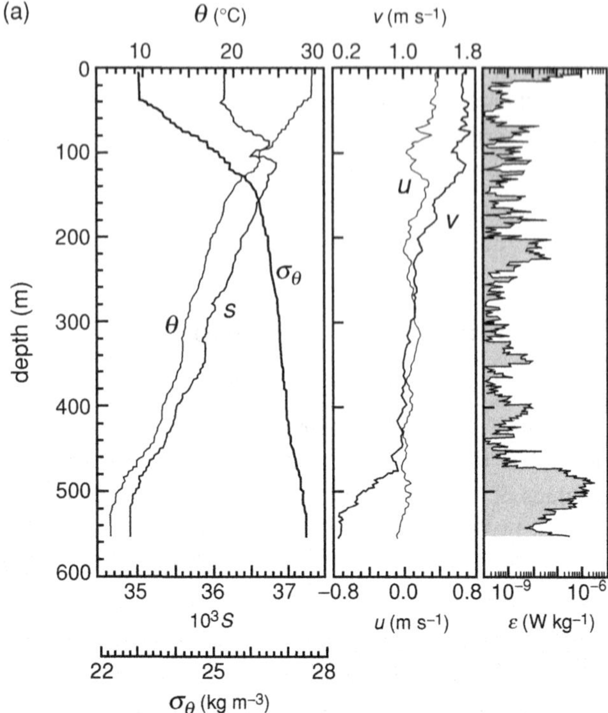

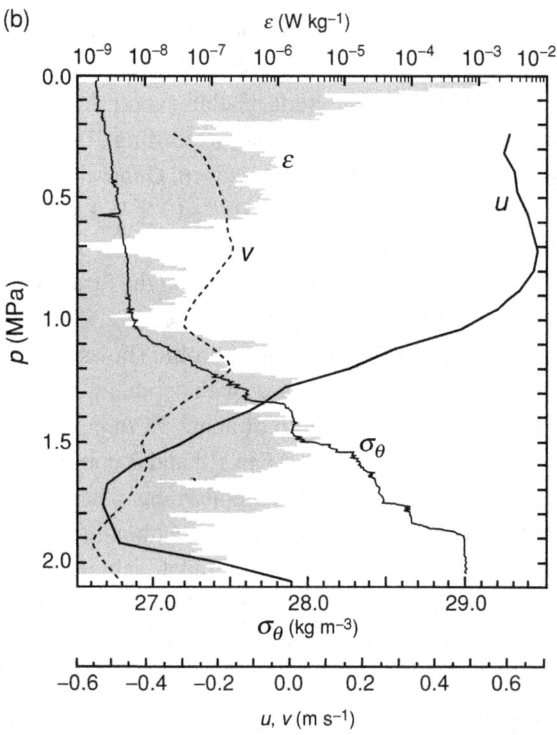

Figure 2.4. Profiles of current, density and dissipation rate. Measured profiles of eastward velocity, u, northward velocity, v, the potential density, σ_θ, and the rate of dissipation of turbulent kinetic energy, ε, (a) in the Florida Strait; θ is the potential temperature and S is the salinity in psu (from Winkel *et al.*, 2002); (b) in the Strait of Gibraltar (from Wesson and Gregg, 1994); and (c) at the Equator; W is the fall speed of the probe (from Moum *et al.*, 1995).

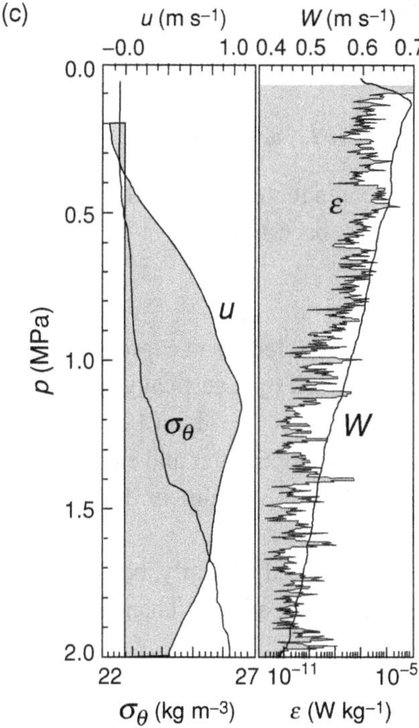

Figure 2.4. *(cont.)*

The third region is on the Equator in the Pacific Ocean at 140° W. Figure 2.4(c) shows a profile of ε, again obtained with the AMP. The northward v component of the current is not shown, but the eastward u component shows the presence of the Equatorial Undercurrent, with maximum eastward speed at 1.2 MPa (depth about 120 m), contrary to the westward flow near the surface. The fall speed of the AMP, W, is displayed; it decreases from a peak of about 0.7 m s^{-1} near the surface to 0.45 m s^{-1} at 2 MPa (depth about 200 m). Values of ε range from about 10^{-5} W kg^{-1}, near the surface, to 10^{-10} W kg^{-1}, which is found below the depth of the maximum speed in the eastward-going Equatorial Undercurrent.

2.3.3 The rate of loss of temperature variance, χ_T

Velocity or velocity shear is often much more difficult to measure in the ocean than is temperature. Temperature is measured not only because its variability is of importance in its own right through its effect on density and relation to heat energy, but also because its spatial variation can provide important information about the presence and action of turbulent motion. Measurements are commonly made from microstructure profilers recording at frequencies of tens of hertz, or over sub-centimetre distances.

• A measurable parameter describing the effect of turbulence on the fluid temperature field, the smoothing out of the temperature variations by the molecular conduction of heat, is the rate of loss of temperature variance,

$$\chi_T = 2\kappa_T \langle (\partial T'/\partial x)^2 + (\partial T'/\partial y)^2 + (\partial T'/\partial z)^2 \rangle, \quad (2.12)$$

which, in isotropic turbulence when the mean square gradients of temperature (and other properties) are the same in all directions, becomes

$$\chi_T = 6\kappa_T \langle (\partial T'/\partial z)^2 \rangle. \quad (2.13)$$

As in (2.11) – and for the same reason – we have selected an expression in terms of z: vertical gradients in temperature may be measured by a vertically falling instrument. Values of χ_T in the ocean range from 7×10^{-10} K^2 s^{-1} to about 10^{-4} K^2 s^{-1}, with the higher values generally in regions of strongly stratified water and energetic turbulence. (The temperature variance is related to K_T via the Cox number defined by (4.7): see P4.5.)

The rate of loss of salinity variance, χ_S, is defined similarly, but is harder to determine accurately because of problems in measuring salinity changes over small distances.[5]

2.3.4 The Kolmogorov length scale, l_K

On dimensional grounds, the length scale of the turbulent motions at which viscous dissipation becomes important must depend on factors that provide measures of the turbulent motion and of its viscous dissipation, that is on ε, characterizing the turbulent motion, and ν, characterizing viscous effects; generally, no other dimensional and relevant quantities are available that might affect the scales at which energy is lost or that might characterize the length scale. It must, therefore, to have the correct dimensions – that of a length – be proportional to $(\nu^3/\varepsilon)^{1/4}$ (ε has dimensions L^{2}T^{-3} and ν has dimensions L^{2}T^{-1}, where L is length and T is time). The coefficient of proportionality is about unity.

• The scale is known as the Kolmogorov[6] length scale, $l_K = (\nu^3/\varepsilon)^{1/4}$.

The oceanic values of ε mentioned in Section 2.3.2 lead to a range of l_K from about 6×10^{-5} m in very turbulent regions to 0.01 m in the abyssal ocean. Measurement of velocity gradients at the smaller of these scales is generally impossible, and consequently it appears that estimates of ε cannot be obtained directly using (2.11), at least

5 As mentioned in Section 1.7.1, salinity is generally inferred from conductivity with a temperature correction, and consequently from two sensors, one measuring conductivity and the other temperature, that are slightly separated in space, with the consequent introduction of uncertainty in the estimated values.

6 A. N. Kolmogorov (1903–1987) was an outstanding Russian mathematician who contributed immensely to the theoretical understanding of turbulent flow. His work in fluid mechanics, and particularly his important study of the form of the spectrum of turbulence published in the early 1940s (Section 2.3.6), is described by one of his students, A. M. Yaglom (1994). Several other Russians made significant advances in the knowledge of turbulence during the period 1940–1980, among them A. S. Monin, A. M. Obukov and R. V. Ozmidov.

not without some further information about the size of the length scales of the mean square shears leading to the dissipation or some necessary correction to the shears that can be resolved by measurement. The coherent structures identified in Section 2.1.1 have sizes and durations that are, however, usually much larger than l_K and velocity gradients can usually be measured over some, but not all, of the range of length scales over which motion is turbulent and relatively unimpeded by viscosity.

2.3.5 The turbulence cascade and the structure of turbulence

The idea of a cascade in energy from larger to smaller scales goes back to the second decade of the twentieth century.

It is supposed that energy is supplied, introduced or produced in the fluid at a relatively large scale (e.g., that of the mean dimension of eddying motions, sometimes – as in Fig. 2.1 – recognizable as coherent eddies), and is successively passed by interactions between eddies or their instability through a spectrum of smaller and smaller eddies within which inertial forces,[7] rather than viscosity forces, are dominant, finally being conveyed to eddies of size comparable to l_K, where viscosity is effective in transferring their kinetic energy into heat. The scale at which turbulent energy is introduced is often equal to or a little larger than that which contains most of the kinetic energy, and eddies of this size are consequently referred to as the 'energy-containing eddies'.

The concept was given vivid imagery by Richardson,[8] who, in 1922, perhaps following Jonathan Swift's poem referring to fleas rather than to eddies or 'whirls', coined the ditty

Big whirls have little whirls that feed on their velocity,
And little whirls have lesser whirls and so on to viscosity
– in the molecular sense.

A closely related concept is that, although motion may be very anisotropic at the scale at which energy is provided, any directional asymmetry is diminished as eddies interact with one another, passing energy to successively smaller scales, until eventually the motion field becomes independent of direction and therefore isotropic at some scale sufficiently small in comparison with that at which energy is put into the turbulent motion. (It is of note that, if the Reynolds stress is non-zero in one direction but zero in the other two, an overall directionality is implied and so, strictly, the fluid motion cannot be isotropic, at least at the length scales of motion that contribute most to

7 The inertial forces are those represented by the terms $\rho\boldsymbol{u}\cdot\nabla\boldsymbol{u}$ in the Navier–Stokes equation of Eulerian motion.

8 L. F. Richardson (1881–1953) is another scientist who contributed some profound ideas about turbulent motion. He and H. Stommel were the first to obtain measurements of turbulent dispersion in the sea (Section 5.3.1) and his name is given to parameters that describe turbulence and its onset in stably stratified fluids (Sections 4.2, 4.4.2 and 4.5). He is also known for his ideas about the numerical forecasting of weather, long before computers made this feasible. A description of his life and achievements is given by Ashford (1985). Ashford suggests that Richardson's ditty may be based on lines written by a nineteenth-century mathematician, Augustus De Morgan.

the transport of momentum, which are found to be large in comparison with l_K. This applies, in particular, to flows in turbulent boundary layers.)

In practice, it is found that, in the stratified regions of the ocean, isotropic motions do exist over some range of scales bounded at their lower end by l_K, provided that the non-dimensional isotropy parameter,

$$\bullet \quad I = \varepsilon/\nu N^2, \tag{2.14}$$

exceeds about 200, where ε and N are the local mean values of the dissipation rate and buoyancy frequency, respectively.[9] It is commonly the case that $I > 200$ in much of the weakly stratified or small-N, near-surface, mixed layer, where turbulence is therefore isotropic over some range of scales $>l_K$. The fraction of water in which $I > 200$ is often <50% in the pycnocline, and smaller still at greater depths. There the fraction of water within which turbulent motion is isotropic is very small.

The concept of an energy cascade carrying energy towards dissipation at small scales, however, requires particular qualification in stratified and rotating systems, of which the ocean is one. A description is given in Section 1.8.2 of the development of a relatively large eddy from the collapse of a region mixed by small-scale turbulent motions as illustrated in Fig. 1.16, a transfer from small to larger scales of motion counter to the energy transfer of the cascade described above. A further example of the transfer of energy from small to larger scales, the pairing of neighbouring eddies (or billows) with vorticity of the same sign, is given in Fig. 1.9(c) and in Section 4.2. Unlike the three-dimensional eddies in isotropic turbulence, anisotropic two-dimensional eddies (and the mesoscale eddies described in Sections 5.3.1 and 6.5) can interact to carry energy to larger scales.

There is also some doubt about whether the structure of turbulence at small scales is always, or for some purposes, best described as eddy-like or more nearly resembles interacting vortex filaments, localized jets, or sheets of high vorticity; and about whether – when unsteady and not homogeneous – it has a universal form or may differ according to the circumstances of its generation. Such matters are of importance in the ocean, where turbulence is generated in many different ways, is often patchy and transient and, beyond the mixed layer, is often anisotropic even at small scales. The structure of the variable flow can affect the rates of sinking and collision of small particles (which are often of a size smaller than l_K) and their amalgamation into flocs. In severe turbulence, the survival of small planktonic organisms can be at risk. High shear in turbulent motions inflicts physical damage to flagella, and turbulence enhances the rates of encounter of plankton with its predators. The relative locations of groups of organisms (moved around by turbulence) and regions of dynamic hazard depend on the structure of the turbulent motion. (See also footnote 6 of Chapter 5.)

9 It is shown in Section 4.4.1 that, if I is large, the size of the largest overturning eddies is significantly greater than l_K.

2.3.6 The Taylor hypothesis and the spectrum of turbulent energy

At a fixed point of measurement past which turbulent eddies are being carried (or advected) at a mean speed U, the small eddies, passing in relatively short times, will cause fluctuations in the measured characteristic of turbulence (e.g., a velocity component or temperature fluctuation) of higher frequency than those caused by large eddies. The time taken for eddies of size l to pass a fixed point is $T = l/U$, and the corresponding measured frequency, $\sigma = 2\pi/T$, is related to the eddy wavenumber, $k = 2\pi/l$, by $\sigma = kU$. It is possible to translate temporal measurements described in terms of frequency into spatial measurements of wavenumbers, provided that, as they are carried past the measurement point, eddies do not change or evolve significantly, i.e., when the turbulent structure is 'frozen' and passive as it is advected past the measurement sensor. (In contrast, waves propagate through the water. The relation between their frequency, σ, and wavenumber, k, is the wave dispersion relation as explained in Section 1.8.1.)

• The hypothesis that this is so is known as the *Taylor hypothesis*.[10] When it is valid, the measured frequency (σ) spectra derived from time (t) series measurements can be converted into wavenumber (k) spectra with distance, x, being set equal to Ut and wavenumbers, k, given by σ/U. A necessary condition for the validity of the hypothesis is that the gradient, $\mathrm{d}U/\mathrm{d}x$, of the mean relative speed is such that the smallest wavenumbers are much greater than $(2\pi\ \mathrm{d}U/\mathrm{d}x)/U$.

If the fluctuations measured at a point are caused by the advection of turbulent eddies (rather than, for example, waves), the measured characteristics of the motion will be composed of a combination of fluctuations derived from the range of wavenumbers that correspond to the sizes of the turbulent eddies. An energy frequency spectrum derived from analysis of the amplitude of the time variation of velocity at the fixed point can be converted to a wavenumber spectrum using Taylor's hypothesis. This spectum shows how the contribution of the energy of the turbulent eddies to the overall kinetic energy of the variable flow is distributed in wavenumber (or in eddy size). The wavenumber spectrum is in fact proportional to Fourier transforms of spatial cross-correlation coefficients[11] of fluctuations, a relationship established by Taylor in 1938.

10 The Cambridge scientist G. I. Taylor (1886–1975) made numerous contributions to science, many in fluid dynamics, including several very important discoveries about turbulent motion, particularly about turbulent dispersion – see Chapter 5. His approach was one based on reducing problems to their essentials and of testing his theoretical results by often very elegant laboratory experiments. Taylor's life, his achievements and the legacy of scientific method he left are described by G. K. Batchelor (1996), who was himself an expert and important contributor to the theory of turbulence. The apparent frequency, $\sigma = kU$, of a frozen field should be compared with the frequency shift of propagating waves, the Doppler shift, referred to in Section 1.8.1.

11 Cross-correlation coefficients. Suppose, for example, that the fluctuation of the x-directed velocity components about the mean, U, is u, so that the x-velocity component is $U + u$, with $\langle u \rangle = 0$. The cross-correlation function, $R_u(x)$, of speeds u, measured as time series at two points at locations x_1 and x_2 separated by a distance $x = x_1 - x_2$, is given by $R_u(x) = \langle u(x_1, t)u(x_2, t)\rangle / \langle (u(x_1, t))^2 \rangle$, where the average values, $\langle \ldots \rangle$, are obtained from the time series over times, $t = T$, sufficiently long to ensure that $R_u(x)$ is independent of T. In the equation for $R_u(x)$, $u(x_1, t)$ is simply the x-velocity component at the position x_1 and at time t, and $R_u(x)$ is a function just of the separation

It was, however, Kolmogorov who first predicted how the energy in homogeneous and isotropic turbulent motion is distributed as a function of the wavenumber of eddies, and therefore obtained a quantitative description of the effects of the conjectured cascade of energy described in the previous section.

For observational convenience, or simply because simultaneous measurements in three dimensions are often impracticable, the variations of velocity (or of other turbulence quantities in oceanic flows) are usually measured in a single direction, either as a sensor attached to a recording instrument is towed horizontally, lowered or falls through the water, or as turbulence is carried by a mean flow past a stationary probe at a fixed point on a mooring or bottom-mounted rig. If the characteristic Reynolds number, *Re*, is very large (of order 10^7 or more) and turbulence is homogeneous and isotropic, Kolmogorov showed – and it is generally observed – that the kinetic energy per unit mass per unit wavenumber bandwidth, the 'spectral kinetic energy density' associated with a single component of the velocity fluctuations, is given by

$$\Phi(k) = q\varepsilon^{2/3}k^{-5/3}, \tag{2.15}$$

within a 'subrange' of wavenumbers, $k = 2\pi/(\text{eddy diameter})$, a range where inertial forces dominate over viscous. This range lies between the wavenumbers of the energy-containing eddies in the turbulent field and the Kolmogorov dissipation scale, $2\pi/l_K$, although, as we show below, (2.15) is not valid across the whole of this wavenumber range. The inertial subrange of wavenumbers within which (2.15) applies increases with *Re*. The factor q in (2.15) is a non-dimensional constant.

• This is the famous Kolmogorov minus five-thirds power law of the inertial subrange of the energy spectrum of homogeneous turbulence. If it is supposed that viscosity plays an insignificant part in processes at scales much greater than l_K, then, except for the size of q, (2.15) can be deduced on dimensional grounds. [**P2.5**] It is found empirically that the constant, q, is approximately equal to 0.5.

Figure 2.5 is an example of spectra of the velocity component transverse to the vertical direction of fall of a free-fall instrument (the Advanced Microstructure Profiler or AMP) and expressed as a function of the vertical wavenumber, k_3. The spectra derived from measurements are fitted to spectra compiled by Nasmyth, the Nasmyth universal spectra that are shown by the thin lines, to obtain estimates of ε (effectively using relations like (2.15) fitted to the observed spectra to derive ε).

The two measured spectra in Fig. 2.5 correspond to rates of dissipation ε equal to 1.0×10^{-8} W kg^{-1} for the lower spectrum and 5.4×10^{-5} W kg^{-1} for the upper. The corresponding Kolmogorov wavenumbers, $(\varepsilon/\nu^3)^{1/4}$, are about 316 cyc m^{-1} for the lower spectrum and 2710 cyc m^{-1} for the upper. (Units of cyc m^{-1} are used in

distance, x. (It is supposed that the mean conditions are steady and that $\langle(u(x_1,t))^2\rangle = \langle(u(x_2,t))^2\rangle$.) $R_u(0) = 1$, and the value of x at which $R_u(x)$ first equals zero as x increases provides a measure of the distance at which the x-fluctuations of velocity become decorrelated or incoherent. In a steady mean flow, U, the spatial cross-correlation coefficient, $R_u(x)$, in the direction of the mean flow may be calculated from a time series at a single point (a time-lagged autocorrelation function), by putting $u(x_2,t) = u(x_1, t - x/U)$ using the Taylor hypothesis.

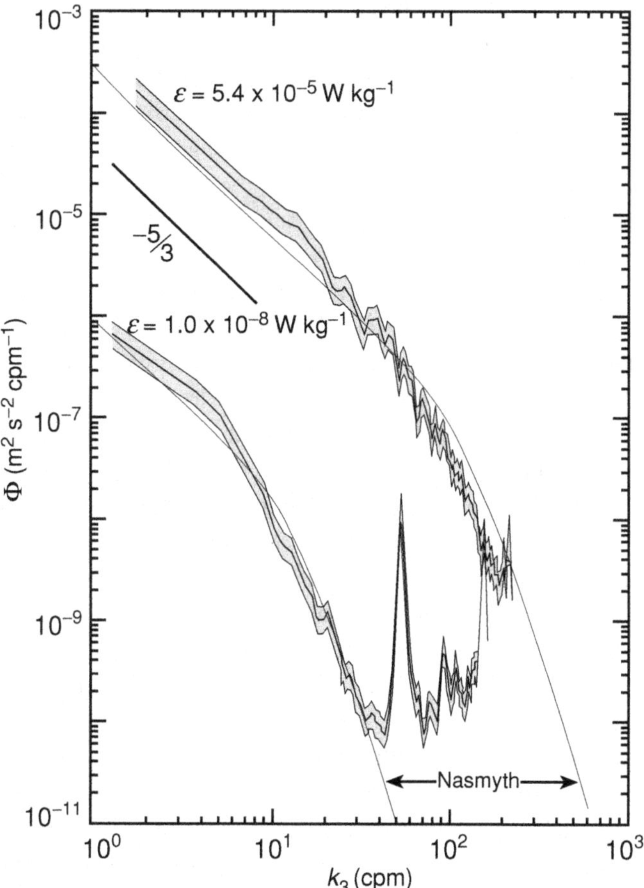

Figure 2.5. Two energy spectra showing the Kolmogorov $-5/3$ law relationship in their lower wavenumber range. The spectra are derived from data collected using the free-fall AMP in the Strait of Gibraltar. The figure shows the variation of energy of the transverse component of velocity with vertical wavenumber, k_3. The dissipation rate, ε, is equal to 1.0×10^{-8} W kg^{-1} for the lower spectrum and 5.4×10^{-5} W kg^{-1} for the upper. The spectra have been averaged over successive 0.5-m intervals in layers 20 m thick (i.e., the averages are of 40 spectra, each determined from measurements made over 0.5 m), within which there is relatively uniform dissipation. The shading indicates the size of the variations of the estimated values (95% confidence limits). Thin lines marked 'Nasmyth' are fitted curves from which ε is estimated as explained in Section 2.5.2. (From Wesson and Gregg, 1994.)

Fig. 2.5, rather than rad m^{-1}.) The slopes of the spectra at low wavenumbers, namely wavenumbers less than about 6 cyc m^{-1} for the lower spectrum and 50 cyc m^{-1} for the upper, are close to Kolmogorov's $-5/3$, but the slopes become more negative at larger wavenumbers, exceeding those of the $-5/3$ range. Deviation from the $-5/3$ law generally occurs at wavenumbers that are, as here, less than the Kolmogorov wavenumber by a factor of about 16π. The lower spectrum has a sharp peak at about 55 cyc m^{-1} that is caused by vibration of the instrument. (It corresponds to the first bending mode of the AMP free-fall instrument.) The effect of this vibration in the upper spectrum is masked within the more intense turbulent motion.

It is important to recognize that, as shown in Fig. 2.4, turbulence is highly variable in the ocean and that, as in Fig. 2.5, spectra have to be derived from data obtained over relatively large distances or periods of time, averaging being done by putting data together in collections or 'ensembles' of data, usually repeated samples, each taken over distances that are large relative to that of the energy-containing eddies. Careful statistical averaging is implicit and necessary in this process, particularly because single sets of data, each collected over a relatively short period of time or over a short distance, may differ considerably. It is, for example, found that the histogram or probability distribution function (pdf) of ε in the near-surface mixed layer is commonly log-normal (see Fig. 2.15 later). This means that the pdf of $\log \varepsilon$ is normal or 'Gaussian', with zero skewness (the normalized third moment of the distribution about the mean) and a kurtosis (the normalized fourth moment) of 3, but that the pdf of ε is positively skewed; high values of ε are relatively rare, and will not be adequately sampled if data sets are too short. The existence of a broad band of energy within a spectrum obtained by such ensemble averaging does not therefore imply that over some relatively short period of time or in some small volume of water the motion field will be broad band in frequency or wavenumber, or will *simultaneously* contain all the range of temporal or spatial motions of an ensemble spectral description derived over longer times or greater spatial extent.

In practice in the stratified ocean, both turbulence and internal waves may affect the energy spectrum. Although, as hinted in Section 1.8.1, it is sometimes possible to use frequency to distinguish between internal waves with frequencies $<N$ and turbulence at higher frequencies (as in the example, Fig. 2.6), both the waves and the turbulence are transient and may change their nature in time. The internal wave field, for example, may at times be dominated by the passage of groups of waves generated near the sea surface that pass through the sampling region, breaking and periodically generating or intensifying turbulence. [**P2.6**]

2.4 The terms in the energy balance equation

Turbulence in the ocean derives its energy from a variety of sources, some of which are described in later chapters. The turbulent energy equation expresses the rate of increase of the kinetic energy of turbulent motion as a result of the energy generated by these sources, the rate of transfer of energy from kinetic to potential energy and the rate at which kinetic energy is lost through molecular viscosity. Three terms usually dominate the various contributions to the rate of change of the mean kinetic energy of the turbulent flow per unit volume,[12] and their balance or inequality leads to turbulence being sustained or to its growth or decay.

12 Other terms in the energy equation are the rate of working of pressure fluctuations on the turbulent motion and the diffusion of turbulent kinetic energy by viscosity but, except for example very near rigid boundaries, where viscous effects may be large, these are usually smaller than at least two of the three described in (2.16). Not specifically included in (2.16), however, is the contribution of

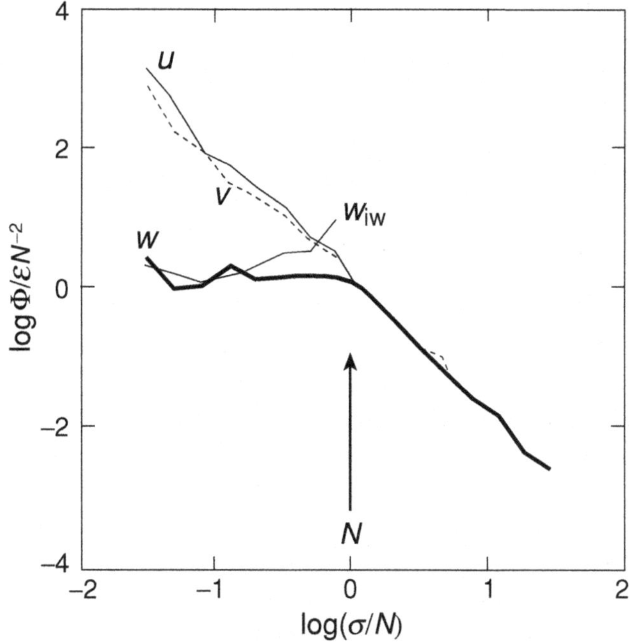

Figure 2.6. Lagrangian frequency spectra of waves and turbulence. The data from which the spectra are derived are obtained by tracking the motion of neutrally buoyant floats designed by D'Asaro (2001), and hence deriving the water motion in a Lagrangian frame of reference (i.e., measurements of speed following the path of fluid particles, in contrast to the more conventional Eulerian measurements of velocity as fluid passes the fixed locations of current meters on moorings). The spectra of the two horizontal components (u and v) and of the vertical velocity, w, shown here derive from measurements in a region of energetic mixing in stratified water over the sill of Knight Inlet, British Columbia. The spectra have units of velocity squared, $m^2\ s^{-2}$, per frequency increment in s^{-1}, or $m^2\ s^{-1}$, and are non-dimensionalized by dividing by εN^{-2}, where ε is the dissipation rate ($m^2\ s^{-3}$) and N is the buoyancy frequency (s^{-1}). The frequency, σ, is non-dimensionalized by dividing by N. (The scales are logarithmic, so that the slope of straight lines would reveal power-law relationships between the quantities plotted.) At frequencies less than N, the spectra of u (full line) and v (dotted) are approximately equal (to within the 95% confidence limits of their estimates), and exceed that of w. The latter is consistent with the spectra W_{iw} (shown by the thin line) estimated by assuming that the horizontal spectra are entirely due to internal gravity waves. At frequencies above N where the propagation of internal waves is precluded, the spectra of u, v, and w are equal, indicating isotropy within a frequency range of turbulent motion. (After D'Asaro and Lien, 2000.)

• The terms that are usually dominant in the turbulent energy equation can be expressed as

$$\mathrm{D}E/\mathrm{D}t = \text{rate of production by the mean flow} + \text{buoyancy flux} - \text{rate of dissipation} \tag{2.16}$$

turbulent energy that may come from breaking waves at the sea surface and, as mentioned in Section 3.3, care is needed in applying the conclusions of Section 2.4 very close to the sea surface.

where DE/Dt represents the mean rate of change of the kinetic energy of turbulence per unit volume as it is carried by the mean flow.

The terms in (2.16) represent averages over some large volume or period of time, not local instantaneous values. The terms on the right-hand side (rhs) of (2.16) are discussed in the following sections. In a steady state the turbulent kinetic energy does not change, so $DE/Dt = 0$ and the terms on the rhs sum to zero.

2.4.1 The rate of production of turbulent kinetic energy by the mean flow

The first term on the rhs of (2.16) is the rate at which turbulent kinetic energy is produced by the mean flow, by which is meant the flow, perhaps varying relatively slowly in time, which contains the turbulent motion. A mean flow can provide a source of turbulence, as do transient flows over the seabed in the surf zone, as described in Section 1.4.[13]

• The term is expressed as the product of the Reynolds stress and the mean shear. In oceanic boundary layers, for example the near-surface mixed layer and the benthic boundary layer, this can be written as $\tau\, dU/dz$ or $-\rho_0\langle uw\rangle dU/dz$, the rate of production of turbulent kinetic energy by the mean shear flow. It can be interpreted as the rate of working of the stress, τ, defined by (2.1), that is exerted by the turbulent motion on a mean shear flow, dU/dz.

The mean flow (in contrast to the turbulent motion) will lose energy at this rate, and will consequently be unsteady and will decelerate if not sustained and continually supplied with energy in some way, for example through a pressure gradient.

2.4.2 The turbulent potential energy

The second term in (2.16) is equal to (minus) the rate at which the turbulent motions increase the potential energy of the fluid or (plus) the rate at which turbulent motions are provided with energy as a result of a loss of potential energy. (The idea that turbulent mixing increases the potential energy in the stratified ocean, and therefore provides a supply of potential energy, is introduced in P1.6.)

Potential energy is a quantity measured relative to some initial or final state. It is usually the potential energy which is 'available' to do work and change the kinetic energy that is of relevance to the dynamics; it is, for example, the potential energy lost as the weights in Joule's experiment descend from their highest to their lowest position that drives turbulent mixing and its eventual dissipation to produce heat within the insulated cylinder in Fig. 1.3. The potential energy of the turbulent motion can be represented as a sum over fluid particles (or small volumes of fluid) of terms $g\rho'\eta$,

13 A distinction is being drawn here between laminar flows that lead to a transition to turbulence and those mean flows that help sustain or diminish turbulence once it has been generated. Here, and often in the ocean, we are concerned with the latter. In Reynolds' experiment the laminar mean flow is unstable, leading to a transition to turbulence. Downstream of the transition, the velocity profile of the mean flow is changed (see P1.1) and the flow past the inner boundary of the tube may now act to help sustain turbulence.

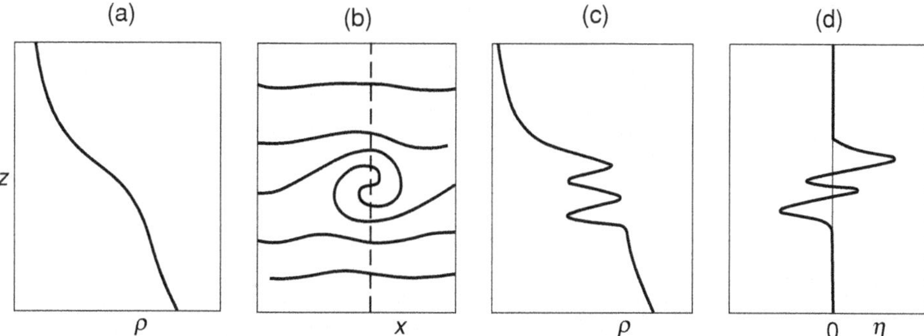

Figure 2.7. Displacements caused by an overturning eddy in stratified water. (a) The initial profile of density with density increasing with depth. (b) The distortion of isopycnals by the eddy and the path (dashed) of a free-fall instrument through the eddy. (c) The density profile measured from a sensor on the free-fall profiler. (d) The vertical displacements of density, η, resulting from the eddy. In practice, the initial density profile is unknown. Statically stable density profiles may be reconstructed from several observed profiles by reordering those measured into ones in which density increases with depth, supposing that no change in density has been brought about by molecular diffusion during the time over which the eddies develop, and then by finding the average depth of a set of selected values of the density in the several reconstructed profiles. (The latter procedure removes the smearing of wave-disturbed density interfaces that would be obtained by averaging densities at a set of fixed depths. Consider, for example, how to find the mean density profile across the wave-disturbed sea surface: averaging densities measured at fixed levels will smear the profile across a thickness equal to the wave height and lose the sharp change that, in reality, occurs at the sea surface.) The rms vertical displacements required in such a reordering provide measures of the vertical scale of eddies and of their potential energy.

where η is a vertical displacement of a fluid particle with measured density fluctuation ρ' from some prescribed level: $\rho' = \rho - \rho_0$, as in Section 2.2.2. For example, η may be the height to which a fluid particle with a measured density difference from its surroundings of ρ' is carried by turbulent motion from its equilibrium position in a stable stratification in which density increases downwards.

An estimate of the stable stratification from which fluid is disturbed by turbulent eddies may in practice be found from a measured vertical density profile by reordering the sequence of values of density measured at known depths into a new sequence in which the density increases everywhere with depth (Fig. 2.7). The rms vertical distance through which the original density values are moved in this rearrangement, or reordering, is then a measure of the turbulent displacement of fluid particles, η. Unless it is breaking, an internal wave will simply raise fluid up and down, perhaps distorting the density gradient but, like the surface waves shown in Fig. 1.5, not overturning or producing any statically unstable regions and displacements in a reordered sequence.[14]

14 That is, no reordering is needed and $\eta = 0$ in a non-turbulent, non-breaking internal wave field. As defined, the displacements are different from those characterized by $\zeta = \rho'/(\mathrm{d}\rho/\mathrm{d}z)$ representing a fluctuation in density, ρ', at a point as a vertical density gradient, $\mathrm{d}\rho/\mathrm{d}z$, is moved vertically past it, e.g., by a non-breaking internal wave.

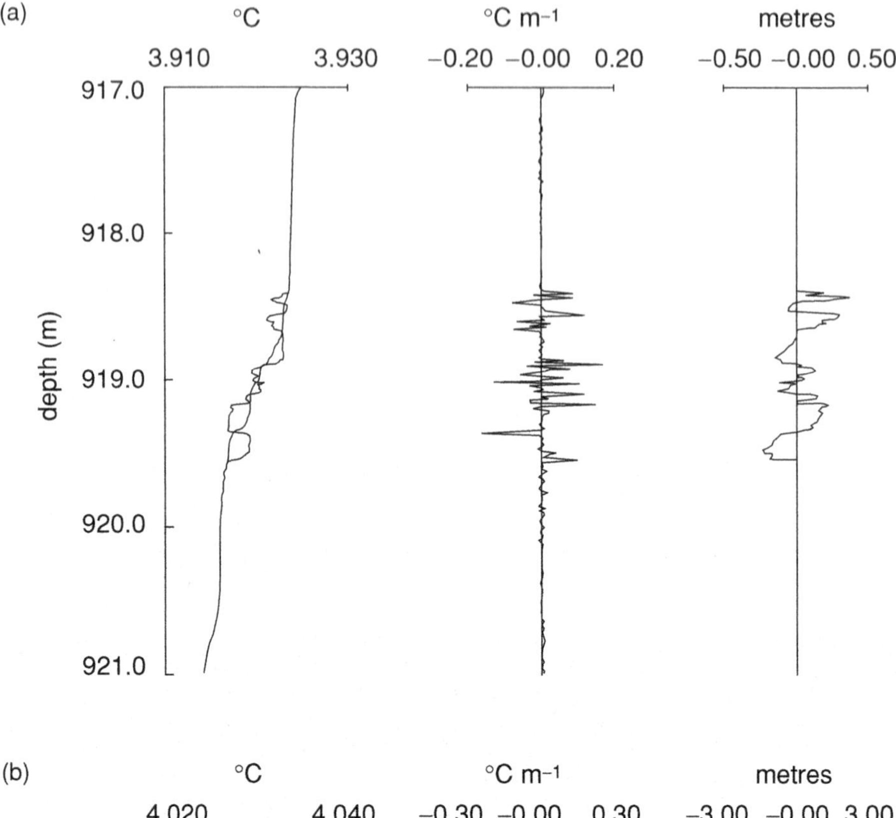

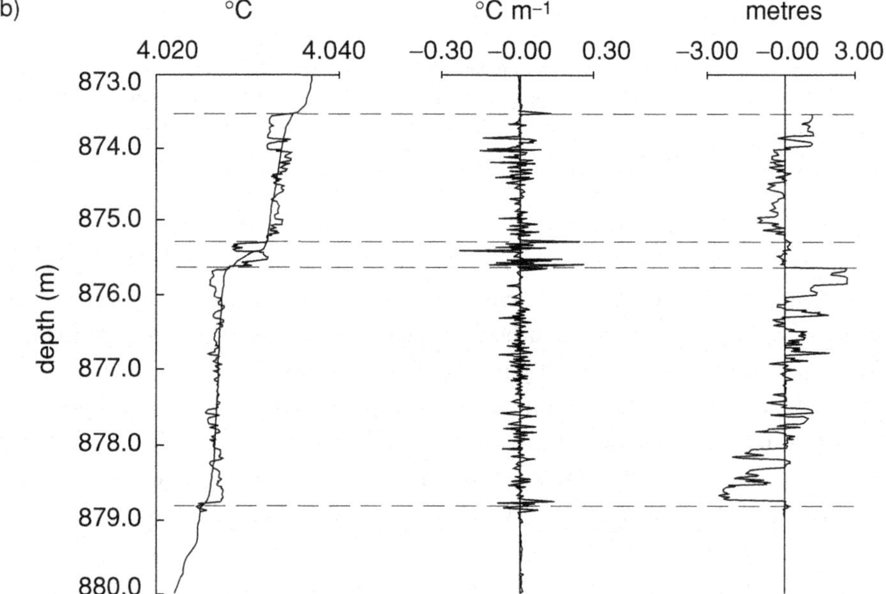

Figure 2.8. Profiles of temperature and displacements. Two examples of (left) temperature profiles and reordered profiles with temperature increasing with depth, (centre) the vertical gradient of temperature, and (right) the calculated displacements. Temperatures were obtained using a glass rod thermistor of diameter 0.5 mm carried by a free-fall microstructure profiler in the central Pacific Ocean. (a) shows a single overturn at an interface at a depth of about 919 m, whilst (b) shows three regions of mixing separated by dotted lines and defined by positive displacements at the top and negative displacements at the bottom. The lower of the three is about 3 m thick. (From Gregg, 1980.)

An example of displacements and a reordered profile, using temperature as an indicator of fluid density, is shown in Fig. 2.8.

• The rms displacement determined from the measured η values, $L_T = [\langle\eta^2\rangle]^{1/2}$, with the average taken over some vertical distance and usually over several density profiles, provides a measure of turbulent mixing that is used in Section 4.4.1 to derive an estimate of ε.

The mean rate of change of the potential energy of a fluid particle of density fluctuation ρ' at a measurement level z is determined by calculating the mean of the product of $g\rho'$ with the rate at which the vertical position of the fluid particle is changing, $\partial\eta/\partial t$, i.e. its vertical velocity, w. The rate of change of the potential energy of the density field resulting from turbulence is therefore $g\langle w\rho'\rangle$ or, from (2.7), $-\rho_0 B$. The rate at which turbulent kinetic energy is *lost* through transfer to potential energy is therefore $-\rho_0 B$, and the term appearing on the rhs of (2.16), contributing to a rate of *increase* of turbulent kinetic energy, is therefore $+\rho_0 B$.

• The buoyancy flux, *B*, is related to the rate of change of the potential energy of the turbulent flow: the mean rate of change of the turbulent potential energy per unit mass is equal to $-B$. It is evident from (2.8) that a heat flux, *F*, is also directly related to a rate of change of potential energy.

2.4.3 The rate of dissipation

The final term, the rate of loss of turbulent kinetic energy per unit volume, is approximately $\rho_0\varepsilon$, the rate of loss of turbulent kinetic energy per unit volume, through the effect of viscosity. (Energy per unit volume is approximately ρ_0 times energy per unit mass.)

The buoyancy flux may sometimes be negligible in the almost unstratified conditions that are found in the mixed boundary layers at the sea surface and seabed. In steady conditions ($\mathrm{D}E/\mathrm{D}t = 0$), ε may then be determined from the balance of the first and third terms on the rhs of (2.16), provided that the turbulence stress and the mean shear, and hence the rate of production of the turbulent kinetic energy, can be measured. This is discussed further in Sections 2.5.4 and 3.3. Similarly, in steady conditions under which the Reynolds stress is very small, the balance of the second and third terms on the rhs of (2.16) may provide another means of estimating ε, as described in Section 3.2. [**P2.7**]

2.5 Measurement techniques and instruments

The preceding sections have suggested some measures that may characterize and quantify turbulent motion, including the rate at which turbulence dissipates kinetic energy, ε, the Reynolds stress or eddy viscosity and the spectrum of velocity fluctuations. Figures 2.4, 2.5 and 2.8 show examples of measured values. How are these measurements of turbulence made?

2.5.1 The first measurements of turbulence: spectra

Making simultaneous measurements at points separated in space to find the spatial gradients required to estimate ε using (2.11) or to calculate wavenumber spectra has generally in the past been impractical (but is now becoming possible – see Fig. 2.3). Instead estimates of spatial gradients in turbulent flows are usually made indirectly, by measuring temporal variations, either at a fixed point past which turbulent fluid is advected at a measured mean speed U or by using a sensor that is translated rapidly at a known mean speed U measured relative to the water. Spatial gradients are then found by dividing the time derivatives of the measured variables by U, making the Taylor hypothesis.

Measurement of small-scale turbulent motion remote from the sea floor began with the studies made by Grant, Stewart and Moilliet in the early 1960s in experiments designed to test Kolmogorov's minus five-thirds power law, (2.15), and to estimate the unknown constant, q. Observations were made in the unstratified and tidal Discovery Passage off the west coast of Canada, where the Reynolds number is of order 10^8, a value far higher than those then obtainable in the laboratory. Hot-film anemometers were mounted on the bow of a research vessel and on a streamlined towed body (Fig. 2.9) to measure the fluctuations in flow speed.[15] The turbulence was so intense that the ship was moved by the larger turbulent eddies! Wavenumber spectra at scales down to wavelengths of about 5 mm were derived from the time series data of the anemometers. The $k^{-5/3}$ dependence of the one-dimensional energy spectrum was confirmed over a range of several decades with a value of the constant q in (2.15) equal to 0.47 ± 0.02.

2.5.2 The air-foil probe: the measurement of ε

• Not until the mid 1970s did it become possible to make measurements of the low levels of dissipation in the weak stratification at depths below the seasonal thermocline. The sensor that was designed, constructed and tested by Osborn, the 'air-foil' probe shown Fig. 2.10, is still in common use and provides the principal means of measuring turbulence at small scales in the stratified ocean.

The probe incorporates a piezoelectric crystal, similar to those once used in gramophone pick-ups, protected from direct contact with seawater by a moulded rubber sheath with a diameter of about 6 mm. The crystal provides electrical signals proportional to the changes in one component of the lateral force. The probes are mounted on an instrument package (e.g., as in Figs. 2.11–2.13), and fall or are carried through the water at a known speed, typically 0.5–2 m s^{-1}. The force to which the probe

15 The conducting film on each anemometer is a few millimetres in width. It is slightly heated by an electrical current passed through it but cooled on exposure to flowing water at a rate proportional to its speed. The cooling causes the electrical resistance of the film to diminish and the current to vary if it is not controlled. The variations in electrical current (or control) provide a measure of flow speed. This method of measuring turbulent velocity fluctuations provides reliable measures of turbulence only in very energetic flows with large gradients in turbulent velocity components and is not suited to measuring turbulence in stratified conditions.

Figure 2.9. The towed paravane carrying the hot- and cold-film thermistors, visible at the top on the left, used to obtain the first records of small-scale turbulence and to determine its spectrum. (From Grant *et al.*, 1962.)

responds is that caused by the relative lateral water speed produced by the turbulent eddies in the water through which the probe is traversing. The probes are calibrated to convert the rate of change of the force into the rate of change of the component of the relative lateral speed of the water in the direction of the measured force component, or the shear. Measurements are made at frequencies greater than 200 Hz, corresponding to horizontal distances of about 2.5–10 mm. A probe moving horizontally in the x direction at speed U, perhaps mounted on a submarine (Fig. 2.11) or autonomous underwater vehicle (AUV: the mounting of probes on one such AUV, 'Autosub', is illustrated in Fig. 2.12), may measure fluctuations in $\mathrm{d}w/\mathrm{d}t$ from which spatial derivatives $\mathrm{d}w/\mathrm{d}x = (1/U)\mathrm{d}w/\mathrm{d}t$ are found.[16] In conditions of isotropic turbulence, it is then possible, *in principle*, to use (2.11) to determine ε with a single air-foil probe.

16 The mean horizontal speed through the water, U, is measured by more conventional instruments, e.g., the ADCP described in Section 2.5.4. For vertically falling profilers, the speed through the water is generally given to a good approximation by the rate of change of depth, which is obtained from the pressure measured by sensors on the profilers.

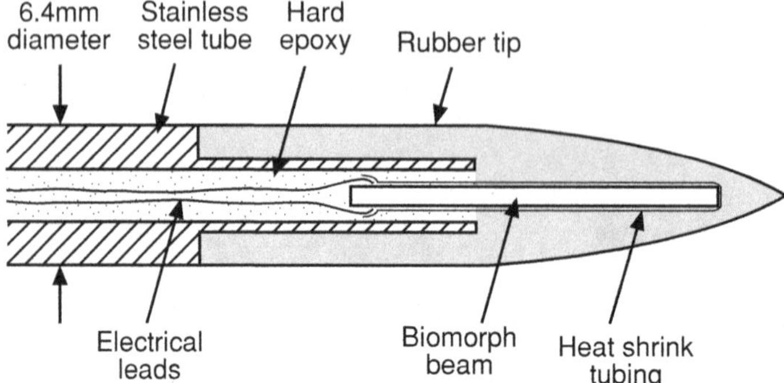

Figure 2.10. The piezoelectric air-foil shear probe designed by Osborn (1974). It has a diameter of 6.4 mm. Lateral forces caused by turbulence motion as the probe moves steadily through the water are converted into electrical signals that are calibrated to determine the relative turbulent velocity normal to the probe as a function of time. Together with measurements of movement through the water (e.g., depth for free-fall instruments or relative mean flow speed for probes mounted on moorings or carried horizontally by an AUV), these are used to determine the spatial derivative of the relative lateral velocity (e.g., $\partial v/\partial x$ and $\partial w/\partial x$ in Fig. 2.14). (From Gregg, 1999.)

Figure 2.11. The air-foil probe mounted on a tripod on a submarine. The turbulence package supporting the probe is visible on its mount at the top of the tripod needed to provide a stable non-vibrating mounting and to carry the probe above the water disturbed by the presence and forward motion of the submarine's hull. (With kind permission of Dr T. R. Osborn.)

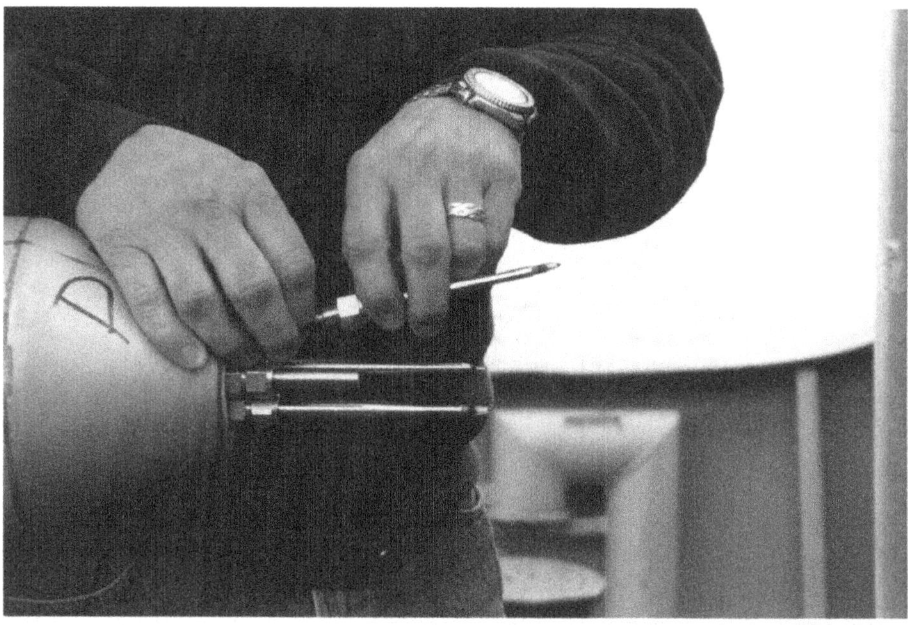

(a)

(b)

Figure 2.12. Air-foil and temperature probes, (a) being mounted on an AUV, 'Autosub'. During deployment, (b), the probes are protected by a cowl that retracts once the AUV is subsurface. The self-propelled and internally recording AUV can be preset to run legs between selected waypoints at about 1.25 m s^{-1} and at constant depth below the surface. It operates unattended for periods of about 100 h. (Photographs kindly provided by Mr Alan Hall.)

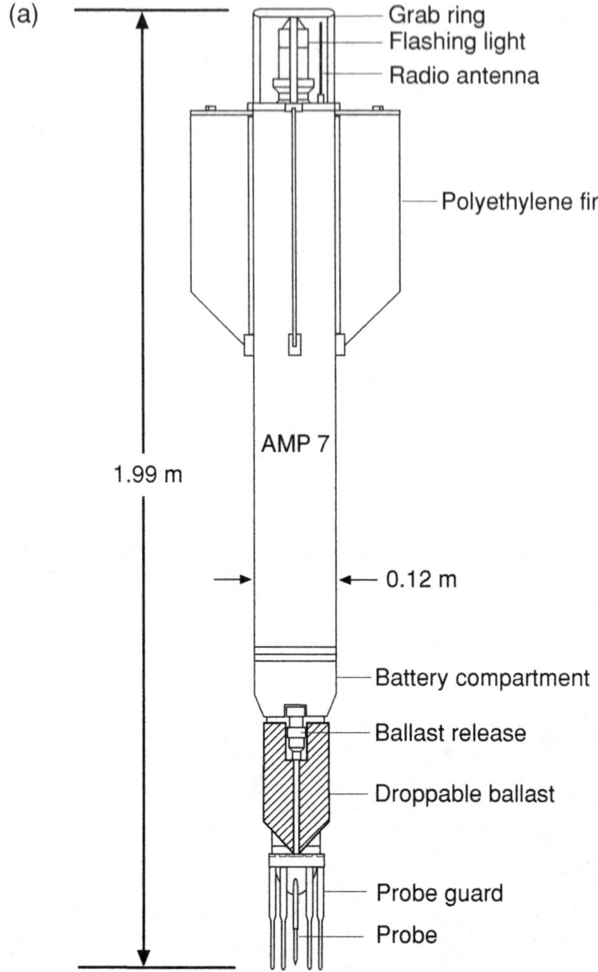

Figure 2.13. Free-fall instruments. (a) A schematic diagram of the Advanced Microstructure Profiler (AMP). It is generally tethered to the research vessel by a 2.7-mm-diameter line with strong Kevlar fibres to aid recovery and with an optical fibre for data transmission. The design is unique, but illustrates the typical arrangement of such semi-free-fall instruments. Sensors are mounted at the base of the AMP to sense motions and temperatures in water undisturbed by the falling instrument, and fall speeds are generally about 1 m s^{-1}. (From Moum *et al.*, 1995.) (b) Another profiler, the High Resolution Profiler (HRP), being assembled on deck prior to launch. The delicate sensors are visible on the bottom of the instrument. The projecting vanes at the other end (the top) of the instrument are to reduce its speed of fall through the water. (Photograph kindly provided by Dr Ray Schmitt of the Woods Hole Oceanographic Institution, USA.)

Streamlined free-fall microstructure instruments, such as the University of Washington's Advanced Microstructure Profiler (AMP) shown in Fig. 2.13(a) or the Woods Hole Oceanographic Institution's High Resolution Profiler (HRP), Fig. 2.13(b), have been designed to fall smoothly and freely thorough the water, being recovered by a loose tether to a ship or on returning to the sea surface after having released a weight.

(b)

Figure 2.13. *(cont.)*

In these cases the air-foil probes mounted on struts below the falling instrument, as shown in Fig. 2.13(b), provide a measure of the shear, $\partial u/\partial z$.

• Temperature and conductivity (from which salinity and hence density may be found) are also commonly measured alongside shear at the scales of 'ocean microstructure' of 1 cm or less.

The size of the air-foil probe is generally too large to resolve fluctuations down to the Kolmogorov scale, l_K, and it is often difficult to resolve spatial variation to scales much smaller than those of about $10l_K$ at which the spectrum of ε is known to reach its maximum value. (Much of the energy in the wavenumber spectrum of the dissipation lies in the range $(0.08-0.8)(\varepsilon/\nu^3)^{1/4}$, or at eddy scales of $2\pi l_K(0.08-0.8)^{-1} \approx (7.8-78)l_K$.) Equation (2.11) cannot therefore be used directly to determine ε because the small-scale velocity gradients are not fully resolved. Spectral estimates of spatial gradients are therefore fitted to a known universal spectrum, the 'Nasmyth universal spectrum', as in Fig. 2.5, and the fitted and interpolated spectrum is then used to find ε, as explained in Section 2.3.6, or integrated to obtain an estimate of $\langle(\partial w/\partial x)^2\rangle$ or $\langle(\partial u/\partial z)^2\rangle$.[17]

17 The accuracy of the estimates of ε depends on whether the turbulence is really isotropic, the success in fitting the universal spectrum to data and the ability of the supporting package to move through the water without itself causing disturbance or vibrating. Processes are available to remove the effects of vibrations from the signal, provided that their frequency band is relatively narrow. Accelerometers are also commonly incorporated into measuring instruments to differentiate real

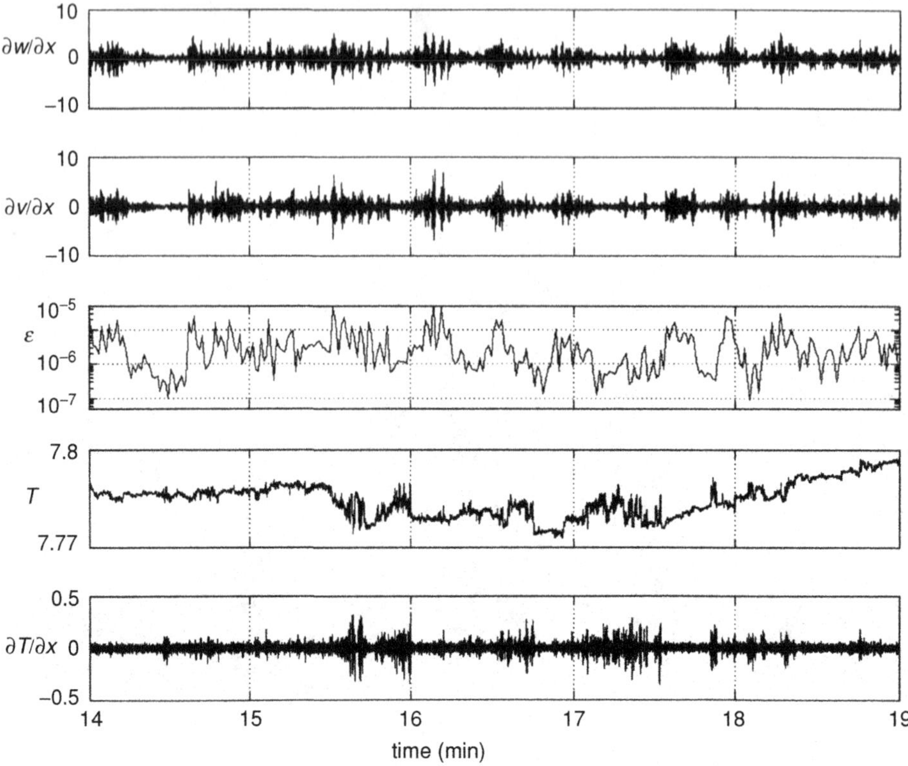

Figure 2.14. Records obtained using air-foil probes and temperature sensors. A 5-min record of shear, dissipation, temperature and temperature gradient obtained using the AUV, 'Autosub', moving along a horizontal track directed (in the x direction) into a wind of 12.4 m s^{-1} at a speed of 1.25 m s^{-1} and a mean depth below the sea surface of 2.23 m. From top to bottom: the vertical shear, $\partial w/\partial x$ (s^{-1}); the horizontal shear, $\partial v/\partial x$ (s^{-1}); 1-s average values of ε (W kg^{-1}); temperature, T (°C); and temperature gradient, $\partial T/\partial x$ (°C m^{-1}). The two shears are almost indistinguishable, suggesting isotropy. High dissipation, ε, corresponds to high shear. Although there are times at which shear, dissipation and temperature gradient are all high (e.g., 15.5–15.8 min), there are others when the temperature gradient is high but shear and dissipation are low (e.g., 17.1–17.4 min) and when shear and dissipation are high but the temperature gradient is relatively small (e.g., 14.0–14.2 min), suggesting the occurrence of different processes of turbulence generation or that shear and temperature variations decay at different rates. (The data and processing methods are described in more detail in Thorpe *et al.*, 2003. Figure kindly provided by Professor T. R. Osborn.)

Figure 2.14 is an example of a short time series of shear, ε, temperature and temperature gradient obtained from an AUV moving along a straight track about 2.23 m below the surface in deep water. The occasional, but variable and irregular, correlation of shear and temperature gradient suggests that several different processes may contribute to

from body-induced fluctuations, and to make appropriate corrections. The resulting estimates of ε, usually averaged over intervals of about 1 s or 1 m, commonly have an uncertainty of about 50%. This may appear large but, since ε generally ranges over several decades in any set of measurements, the uncertainty is often less important than it at first appears, an exception being when an objective is to assess the relative importance of measured sizes of terms in (2.16).

the generation of turbulence. Figure 2.15 shows the probability distribution function (pdf) of log ε calculated from such data close to the surface where breaking waves contribute to the generation of turbulence. The pdf is approximately log-normal both for the whole record and for those parts where bubble clouds, marking the probable presence of turbulence caused by breaking waves, are observed.[18]

2.5.3 First measurements of Reynolds stress, and the related dissipation per unit area

The first measurements of Reynolds stress in the sea were made in the 1950s by Bowden and Fairbairn in Red Wharf Bay in the Irish Sea using electromagnetic current meters mounted on a frame lowered onto the sandy seabed in depths of 12–22 m (Fig. 2.16(a)). Electromagnetic current meters[19] are able to provide simultaneous measurements of the fluctuations in two components of velocity, usually a horizontal component, u, and the vertical component, w, on small, $O(0.1\text{ m})$, length scales, from which the Reynolds stress, $\tau = -\rho_0\langle uw\rangle$, can be estimated.

Figure 2.17 shows an example of the two components of currents and the derived product uw, where u is the deviation of the current from a mean current of 0.49 m s^{-1} measured at a height of 1.5 m above the seabed. The stress is dominated by bursts of higher or lower current lasting only a few seconds, marked S or E for 'sweeps' and 'ejections' as explained in the figure caption.

• The scale of the eddies responsible for transporting momentum is relatively large, typically, in a boundary layer, of the order of the distance of the measurement point from the boundary and much larger than the Kolmogorov length scale. It is for this reason that measurements of stress can be made by sensors that are large in comparison with, say, the air-foil probes used to measure turbulent dissipation.

18 The bubbles, typically of radius 10–100 μm, of which the bubble clouds are composed may be regarded as a tracer of water that has been involved in the process of wave breaking (albeit a non-conservative tracer, because the bubbles dissolve and rise to the surface). The depth of observation in Fig. 2.15 is about twice the significant wave height, H_s, the mean crest-to-trough distance of the highest one-third of the waves. (The height, H_s, increases with wind speed.) At this depth below the surface the higher-than-average dissipation in the bubble clouds is related to turbulence generated by the breaking waves that produced the bubbles, rather than to turbulence generated by the rising bubbles themselves.

19 The electromagnetic current meters used by Bowden and Fairbairn are 10 cm in diameter in the form of oblate spheroids or discuses with minor axes of 3.8 cm. A circular solenoid is contained within the discus, lying in its plane, and this produces a magnetic field. The flow of the conducting seawater parallel to the faces of the discus passing through this magnetic field generates a potential difference (pd) as a result of the Faraday effect, and this is sensed by two orthogonal pairs of electrodes exposed to the water on the face of the discus. The component of flow normal to the line joining a pair of electrodes is proportional to the measured pd. When the discus lies in a vertical plane, the measured pds derived from the two electrode pairs allow the horizontal and vertical flow components to be measured simultaneously. The volume of water sampled is of extent comparable to the solenoid diameter. In later instruments, the solenoid is contained in a circular insulated moulding, and the instrument looks like a flattened ring doughnut, its open centre reducing the flow obstruction caused by the earlier discus form. They are typically 15 cm in diameter. The absence of moving parts avoids frictional problems found in mechanical current meters in low flows, but it has not been found possible to reduce the size of electromagnetic current meters so that they can resolve currents at the centimetre scales typical of turbulent motion. Measurements of the stress on the seabed have been made both in shallow tidal seas and in the benthic boundary layer at abyssal depths.

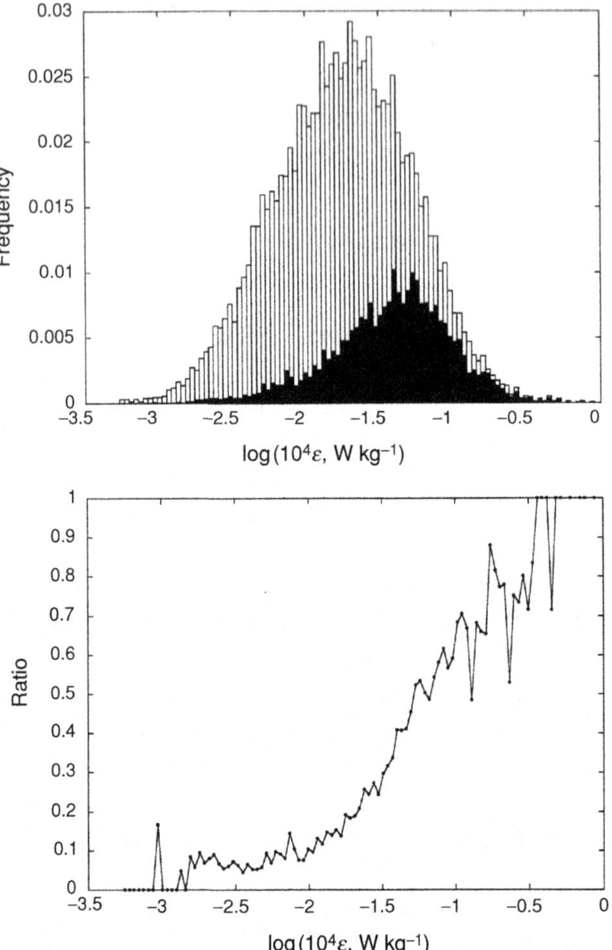

Figure 2.15. Turbulent dissipation rates near the sea surface. Top: the probability distribution function (pdf) of $\log(10^4\varepsilon)$ (ε measured in W kg^{-1}) at a mean depth of 2.22 m in a wind speed of about 11 m s^{-1} and a fetch of about 21 km measured using the AUV shown in Fig. 2.12(b). In black are shown the measurements made in the presence of bubble clouds produced by breaking wind waves. Both pdfs are close to being normal or Gaussian, i.e., with forms $\{1/[\sigma(2\pi)^{1/2}]\}\{\exp\{-[\log(\varepsilon) - \mu]^2\}/(2\sigma^2)\}$, where μ is the mean value of $\log\varepsilon$ and σ is the standard deviation of $\log\varepsilon$. Below: the ratio of the values of the pdf of $\log(10^4\varepsilon)$ in bubble clouds to those in the total sample. The mean level of dissipation is higher than average within the bubble clouds, and most of the higher values of ε are found within bubble clouds. The breaking waves that produce the bubbles make a substantial contribution to turbulence near the sea surface. (From Thorpe *et al.*, 2003.)

Electromagnetic current meters have now been replaced by smaller and less intrusive acoustic instruments, e.g., those on the rig in Fig. 2.16(b).

• One purpose of making stress measurements is that they allow estimation of the drag coefficient, C_D, a parameter relating the stress of the turbulent water on the seabed and the mean current, U, at a specified height, commonly 1 m, above the boundary using an equation of the form

(a)

Figure 2.16. Instruments used to measure Reynolds stress. (a) The rig carrying two electromagnetic current meters used by Bowden and Fairbairn to measure the turbulent stress on the seabed, about to be lowered to the bed. The rig has an attached 'sail' to orientate it into the mean flow, typically about 0.25–0.5 m s^{-1}. The views are (i) from the side showing the 0.1-m-diameter current meters; and (ii) from the front, looking in the direction of the flow. From this direction the current meters can be seen to be supported on horizontal struts. Observations were made at heights from 0.75 to 1.75 m off the bed. Recording was made using photographic recorders and the records were digitized by hand at 1-s intervals. (From Bowden and Fairbairn, 1956.) (b) A more modern rig, STABLE (Sediment Transport And Boundary Layer Equipment). Stress measurements are made using three two-component acoustic current meters, visible on the left and having short 'horizontal Y-shaped' struts. Data are recorded at 25 Hz within the relatively massive recording and power-supply package above. STABLE also carries an acoustic backscattering system operating at 1, 2 and 4 MHz to detect sediment particles. (Photograph kindly provided by Dr A. Souza of the UK Proudman Laboratory.)

$$\tau = \rho_0 C_D U^2, \tag{2.17}$$

from which the stress may be derived from knowledge only of the current, U. The value of C_D is found to be about $(2.5 \pm 0.5) \times 10^{-3}$.

The Reynolds stress on the seabed is particularly important because sediment begins to be eroded and brought into suspension when the stress exceeds a value that depends on sediment type but is typically about 0.16 N m^{-2}. With knowledge of C_D, maps of the bottom stress can be derived using (2.17) from charts of tidal streams and therefore, since sediment erosion depends on the magnitude of the stress, the regions of stable (or immobile) and unstable sediment can be predicted.

The stress in the water column is also important because of its relation to the production of turbulent kinetic energy as explained in Section 2.4.1.

(b)

Figure 2.16. (*cont.*)

The Reynolds stress is a factor that determines the nature of the dynamics of the oceanic boundary layers. Dissipation is proportional to U^3, being the product of the bottom stress (proportional to U^2) with the current, U. The term $\rho_0 C_D U^2|U|$ (the stress multiplied by flow speed, averaged over time and integrated over unit area of the seabed) is equal to the rate at which tidal energy is dissipated by turbulent motion per unit area of the seabed or, equivalently, the rate at which turbulent kinetic energy is produced per unit area. This is one of the two methods used by G. I. Taylor in 1919 to estimate tidal dissipation in the Irish Sea and provided the first calculation of ε in the ocean. [**P2.8**] (It was more for the convenience of a local site than because of its possible relevance to Taylor's calculation that Bowden and Fairbairn – from the nearby UK Liverpool University – made their observations in the same body of water.)

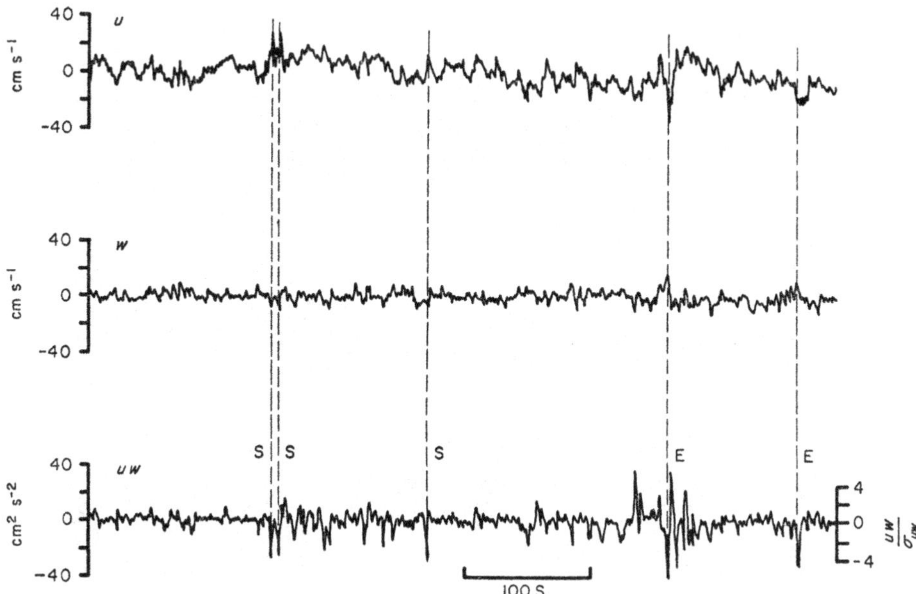

Figure 2.17. Turbulence near the seabed. A typical record of the horizontal, u, and vertical, w, components of current fluctuations and of their product, uw, from which the Reynolds stress is determined in the near-bottom boundary layer. This is at a height of 1.5 m from the seabed and the mean current of 0.49 m s^{-1} has been subtracted from the horizontal component. Higher than average values of u associated with negative w (downward moving sweeps of faster moving water, 'S') and lower than average u and positive w (ejections of slower moving water, 'E', from near the seabed) are indicated. These are suggestive of the active and coherent stress-transferring processes described in Section 3.4.4. (From Heathershaw, 1979.)

2.5.4 Estimates of Reynolds stress and ε using an ADCP

Figure 2.18 shows a conventional acoustic Doppler current profiler (ADCP) with four transducers. They emit short (typically of duration 0.5 ms or, with a typical speed of sound in seawater of 1500 m s^{-1}, about 0.75 m long) pulses of high-frequency (usually 10 kHz or more) sound directed to travel along four narrow beams, each inclined at the same angle to the vertical. The sound is partially reflected from particles being carried through the beams by the flowing water. The reflections of a sound pulse in a particular beam are received back by the transducer that emitted the sound pulse. The reflected sounds of each beam are recorded and analysed separately. Because the reflecting particles are moving, the sound reflected in a beam has a frequency different from that emitted, and the difference in frequency, the Doppler shift, provides a measure of the component of the speed of particles in the beam direction. Knowing the speed of sound in the seawater, the range along a beam from which sound is reflected can be determined from the time between the transmission and return of sound pulses.

Two of the four ADCP beams are shown in Fig. 2.19, beam 1 being aligned in the direction of the mean flow and beam 2 in the opposite direction. The speeds measured by the ADCP at a height z (or range $r = z/\sin\phi$, where ϕ is the inclination of the beam

Figure 2.18. An acoustic Doppler current profiler (ADCP). This instrument, photographed on the seabed, has four circular transducers. The aluminium plate on which the ADCP is mounted is 0.3 m across and the diameter of the transducers is about 0.07 m. Each transducer is tilted at 20° from the vertical. This particular model is the RDI workhorse Zedhead operating at 1200 kHz. (Photograph taken by and reproduced with kind permission of Jens Larsen, NERI, Denmark.)

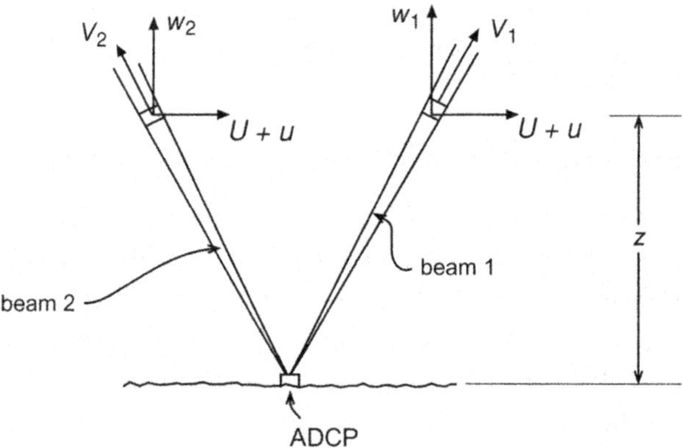

Figure 2.19. ADCP beams. Only two of the four beams are shown, beam 1 aligned in the mean direction of the flow, U, and beam 2 in the opposite direction. Analysis of the Doppler returns from the individual beams allows the terms $\langle uw \rangle$ and $\mathrm{d}U/\mathrm{d}z$ to be estimated, their product providing a measure of the rate of production of turbulent kinetic energy.

to the horizontal) are the two components of speed, V_i, along the beam directions. If the vertical component of velocity is w and the horizontal is $U + u$, where U is the mean flow at height z, then $V_1 = (U + u)\cos\phi + w\sin\phi$, whilst $V_2 = w\sin\phi - (U + u)\cos\phi$. If the mean fluctuation in horizontal speed $\langle u \rangle = 0$ (so that $\langle U + u \rangle = U$) and the mean vertical speed $\langle w \rangle = 0$ (provided that the bottom is horizontal, the water surface is not rising and the particles move with the water), it is then found that $\langle V_1^2 - V_2^2 \rangle = 2\langle uw \rangle \sin(2\phi)$, and (apart from the density term, which is reasonably well known) the Reynolds stress can be estimated from the measured values of V_1 and V_2. This assumes, however, that the turbulent eddies that contribute most of the stress are large relative to the separation between the beams at height z so that the vertical and horizontal components of the velocity fluctuations produced by these eddies in the two beams are equal. The mean current, U, at height z can be found from $\langle V_1 - V_2 \rangle = 2U \cos\phi$, and its gradient, $\mathrm{d}U/\mathrm{d}z$, can be found from differencing measurements taken at different ranges.

• The rate of production of turbulent kinetic energy, the product $\tau\, \mathrm{d}U/\mathrm{d}z$ (Section 2.4.1), can therefore be estimated from measurements using an ADCP.

If the measurements are made in a layer of seawater that is well mixed and of uniform density, as will often be found in the coastal zone, where tidal or other flows are strong, and if conditions are steady, then the first and final terms in the turbulent energy equation will be equal, so that $\varepsilon = (\tau/\rho_0)\mathrm{d}U/\mathrm{d}z$. This provides a means to estimate the dissipation rate, ε, from the ADCP-determined estimates of turbulent energy production rate.

A second acoustic method of estimating ε that has been used in surface mixed layers derives from the reasoning by G. I. Taylor that the presence of the large energy-containing eddies with characteristic velocity u and dimension l will lead to a dissipation rate

$$\varepsilon \propto u^3/l. \tag{2.18}$$

Measurements of the eddy scale l with estimates of the rms vertical velocity w obtained from a single, vertically pointing, ADCP beam have been compared with estimates of ε from air-foil probes and, to a good approximation (and in accord with Taylor's reasoning),

$$\varepsilon = cw^3/l, \tag{2.19}$$

where c is a constant roughly equal to unity. The value of l may be obtained by estimating the mean distance between the locations at which w crosses zero in profiles of vertical velocity obtained using the single vertical ADCP beam.

Suggested further reading

First measures of turbulence

The paper in which G. I. Taylor (1919) calculated the rate of dissipation of tidal energy in the Irish Sea is a classic. Two different methods are used in the calculation, one as described in Section 2.5.3 and the second based on estimates of the flux of tidal energy into and from the Irish Sea. Remarkably (in view of the approximations involved) the two methods lead to similar values. (See also Section 6.3.1.)

Bowden and Fairbairn (1956) describe the first measurements of Reynolds stress near the seabed.

Grant *et al.* (1962) succeeded in making microscale measurements of turbulence in mid-water from a towed body. Subsequent measurements were later made from a more stable body, a submarine (Grant *et al.*, 1968).

Spectra and the nature of turbulent flow

There are several textbooks that provide very useful explanations of the basics of turbulence, including Hinze (1959) and Tennekes and Lumley (1982). The latter is frequently used in courses on turbulence.

Free-fall instruments, turbulence sensors and data processing

Oakey (2001) gives a general introduction to microstructure sensors and techniques used to derive ε.

Further study

This chapter provides a very incomplete introduction to the present understanding of the nature of turbulent flow and to its measurement. There is much more to be known about data processing and the production and interpretation of spectra, some of which may be found in Oakey (1982) and Moum *et al.* (1995). Several types of microstructure profilers are now available, although reliable operation requires 'back-up expertise' and few are ready to use 'off-the-shelf'. Further methods of determining measures of turbulence are described in the following chapters. Methods involving ADCPs, now commonly being used in coastal waters to determine mean currents and turbulence stress, and to estimate turbulent dissipation, are described by Gargett (1999) and Lu and Lueck (1999).

Detection of coherent structures

Both shadowgraph (see caption to Fig. 2.1) and Schlieren (caption to Fig. 1.14(a)) optical methods have been developed to observe the structure of the density variations resulting from turbulence on scales of a few centimetres or less in the ocean, for example by Williams (1975), Kunze *et al.* (1987) and Karpen *et al.* (2004), but neither is in common use.

Free-fall instruments, turbulence sensors and data processing

Moum *et al.* (1995) describe the free-fall AMP and the versatile towed instrument CAMELION, so named because its sensing package can easily be changed and is therefore, like the lizard (chameleon in its English spelling), adaptable to its operational environment. Winkel *et al.* (1996) provide a description of the MSP.

Technical problems involved in deriving ε and χ_T are described by Gregg (1999).

Isotropy

Gargett *et al.* (1984) made measurements from a manned submersible, 'Pisces', in a stratified and turbulent flow within Knight Inlet, British Columbia, and give a detailed and illuminating account of the analysis of data to examine the conditions under which turbulence is isotropic.

Smyth and Moum (2000) describe an interesting numerical model study of the isotropy of flow resulting from the instability of a stratified shear flow (see Fig. 4.5 later).

The *measurement of ocean turbulence* is the subject of Chapter 6 in TTO.

Problems for Chapter 2

(E = easy, M = mild, D = difficult, F = fiendish)

P2.1 (E) Reynolds stress in surface waves. The horizontal and vertical velocity components in the water beneath an irrotational surface wave of small amplitude, a, wavenumber, k, and frequency, σ, are $u = a\sigma\ \exp(kz)\sin(kx - \sigma t)$ and $w = -a\sigma\ \exp(kz)\cos(kx - \sigma t)$, respectively, where z is measured upwards from the mean level of the water surface. Show that the Reynolds stress, $-\rho_0\langle uw\rangle$, is zero. (The wave velocity components are 90° out of phase.)

P2.2 (E) Reynolds stress in internal inertial gravity waves. The velocity components of a plane internal wave of amplitude a, wavenumber vector (k, 0, m) and frequency σ, travelling in the x–z plane through a fluid at rest in a rotating system of Coriolis frequency f, in a region of constant buoyancy frequency N, are $u = (am\sigma/k)\cos(kx + mz - \sigma t)$, $v = (afm/k)\sin(kx + mz - \sigma t)$ and $w = -a\sigma\cos(kx + mz - \sigma t)$. Show that the y-component of the Reynolds stress, $-\rho_0\langle vw\rangle$, averaged over a horizontal plane, is zero, but that the z-component, $-\rho_0\langle uw\rangle$, is non-zero, but independent of z.

(Since the momentum transferred to the mean flow in a horizontal layer depends on differences in stress on its two sides – i.e., the divergence, $\partial/\partial z$, of the stress – internal waves do not generally drive an Eulerian mean flow unless they break, although, like surface waves, they may have an associated Stokes drift.)

Internal waves induce density variations at fixed points. If the density depends solely on temperature rather than salinity, then the temperature variation corresponding to the wave-induced velocity fluctuations is $T' = -[N^2a/(g\alpha)]\sin(kx + mz - \sigma t)$, where α is the thermal expansion coefficient. Show that the mean vertical heat flux, and consequently the eddy diffusion coefficient of heat, resulting from the motion produced by internal waves is zero. (N is the buoyancy frequency given by $N^2 = g\alpha\,\mathrm{d}T/\mathrm{d}z$, where $T(z)$ is the mean temperature at level z.)

• Internal waves do not produce a vertical transfer of heat, unless they break and drive a turbulent heat flux.

P2.3 (M) Energy dissipation in shear and convergent flow. In the shear flow illustrated in Fig. 1.10(a), $\mathrm{d}u/\mathrm{d}z \neq 0$ and u does not vary in x or y, but the other two velocity components, v and w, are both zero. In the convergent flow of Fig. 1.10(b), $u = qx$ (independently of y and z) and $v = -qy$ (independently of x and z), where q is a constant rate, whilst $w = 0$. In each case find the terms s_{ij} and use (2.9) to derive ε if, in the shear flow, $du/dz = 4\ \mathrm{s}^{-1}$, and, in the convergent flow, $q = 3\ \mathrm{s}^{-1}$. You should suppose that $\nu \approx 1 \times 10^{-6}\ \mathrm{m}^2\ \mathrm{s}^{-1}$.

The assumed rates, du/dz and q, are quite high, involving, say, speed differences of 2 and 1.5 cm s^{-1}, respectively, over 5 mm. The rates are also steady, and you should find that they are independent of position, whereas the equation for ε in (2.9) is expressed as an average ($\langle\ldots\rangle$) in time or space of the product of spatial derivatives of the variable and turbulent flow. The divergence of the vector flow speed $\boldsymbol{u}$ is equal to $\partial u/\partial x + \partial v/\partial y + \partial w/\partial z$, and is zero in both the shear and convergent flows, as must be the case in an incompressible fluid.

P2.4 (M) Root mean square shear and heating by turbulent dissipation. (a) Given the observed range of ε and supposing that turbulence is isotropic, what range of values of the root mean square shear, $[\langle(du/dz)^2\rangle]^{1/2}$, is found in the ocean? (b) What is the corresponding range of the rate of change of temperature resulting from the dissipation of turbulent kinetic energy? (c) Show that, even if $\varepsilon = 10^{-4}\ \mathrm{W\ kg}^{-1}$, an extremely high value for the stratified ocean, the rate of increase in temperature resulting from the viscous dissipation of turbulent energy is less than 0.1 mK h^{-1}.

• Since such high rates of dissipation are both rare and short-lived, their effect in heating the ocean or in changing its density is negligible.

P2.5 (M) The minus five-thirds spectrum. Derive (2.15) using a dimensional argument.

P2.6 (D) The Lagrangian spectrum. If the frequency spectrum of the vertical component of velocity, $\Phi_w(\sigma)$, measured in a Lagrangian frame of reference in stratified turbulent flow depends, in an inertial range, only on the frequency, σ, and the rate of dissipation of turbulent kinetic energy, but – in the range dominated by inertial rather than buoyancy forces – is independent of buoyancy frequency, N, find how Φ_w varies with σ. Use Fig. 2.6 to determine any unknown constant.

P2.7 (M) Turbulent energy balance. Show that the three terms on the rhs of (2.16) and defined in Sections 2.4.1–2.4.3 all have the same dimensions, equal to those of the term on the lhs of (2.16). Estimate the rate of dissipation of turbulent kinetic energy per unit mass in an unstratified turbulent shear flow through a channel where the (constant) shear is $dU/dz = 0.2\ \mathrm{s}^{-1}$ and the Reynolds stress is 12 N m^{-2}, supposing that the flow and the total turbulent kinetic energy are steady (i.e., unchanging in time) and the density of seawater is 1028 kg m^{-3}.

P2.8 (M) Dissipation by bottom friction in a shallow sea. Estimate the mean rate of dissipation of tidal energy per unit mass in an unstratified well-mixed shallow tidal sea where the tidal currents measured at a point are sinusoidal, given by $U = U_0 \sin(\sigma t)$, where $U_0 = 0.8\ \mathrm{m\ s}^{-1}$ and σ is the frequency of the M_2 tide, about $2\pi/(12.43\ \mathrm{h})$, and in which the mean depth is 75 m.

Chapter 3

Turbulence in oceanic boundary layers

3.1 Introduction: processes, and types of boundary layers

This chapter is about turbulence in the two, very extensive, boundary layers of the ocean, the upper ocean boundary layer or region near the sea surface that is directly affected by the presence of the overlying atmosphere, and the benthic or bottom boundary layer (bbl) that lies above the underlying solid, but possibly rough and (in strong flows) mobile, seabed. Two fluxes imposed at the bounding surfaces have direct effects, those of buoyancy and momentum. The former is often dominated by a flux of heat, related by (2.8) to a flux of buoyancy. This is sometimes supplemented at the sea surface by the entry of buoyant freshwater in the form of rain or snow.[1] The flux of momentum can be equated to the stress, as explained in Section 2.2.1. This stress, or horizontal force per unit area, may be exerted by the wind on the sea surface or, for example, by the frictional forces of an immobile sedimentary layer composed of sand or gravel on a current passing over the ocean floor.

It is useful here, as in preceding chapters, to refer to processes. By a 'process' is meant a physical mechanism, one that can be described in terms of its effects and its associated spatial and temporal structure, that generally involves the transfer of energy from one scale to another or from one part of the ocean to another. Although energy transfers from small to relatively large scales do occur, the transfer of energy of most interest here is usually from some larger-scale motion to one of smaller scale that may

1 The loss of water through the sea surface in the form of spray and aerosol generation may also contribute to buoyancy flux, although care is required to account properly for the effect of salinity. Heat is transferred at the sea surface by radiation, by evaporation, requiring energy in the form of latent heat, and through conduction in the form of molecular transfer known as 'sensible heat flux'.

Figure 3.1. Eddies forming on the top of cumulus clouds on a warm summer day.

be described (generically) as 'turbulence'. A process may involve motions that are themselves best described as turbulence.[2]

The structure associated with a process will commonly have a stochastic nature. The structure may in practice be that of the large eddies supplying energy at the low-wavenumber end of a turbulence spectrum and fuelling a cascade of energy towards dissipation scales. A vivid example is that of convection in the atmospheric boundary layer on a warm summer day, with thermals forming over extensive and relatively uniform countryside, leading to and made visible by the appearance of cumulus clouds. The process is one of convection. The associated structure is that of the growing thermals drawing energy from the unstable stratification in the atmospheric boundary layer over the relatively warm land. As can be seen from the structure of the tops and sides of the cumulus clouds, the process leads to turbulent motions of much smaller scale that entrain surrounding air. Figure 3.1, for example, shows eddies entraining air into the tops of cumulus clouds in wind shear.

It is sometimes helpful, as an aid to identifying the sources of energy leading to mixing, to classify processes that produce turbulence according to whether they are driven from within or outside the oceanic region under consideration, i.e., whether the processes involve an internal or external supply of energy (Fig. 3.2). The distinction between internal and external forcing is not absolute; the concept is useful in stimulating ideas about energy flux and the effect of processes, rather than in providing a definitive analytical tool or classification. The sources of energy that drive the ocean are external to it, consisting of the winds, solar and geothermal heating, and the lunar

2 Langmuir circulation, a process of mixing in the upper ocean boundary layer described in Section 3.4.3, provides an example of motions that involve a degree of variability. It is now regarded as part of the near-surface turbulence even though it has a clearly recognizable structure or 'signature' with which there are strongly associated temporal and spatial variations in ε.

and solar tidal forces. More is said of these in Chapter 6. Although the tides produce body forces driving motions throughout the ocean depth, many of the processes that result in turbulence and mixing are found to act at, or near, the ocean boundaries, receiving energy directly from the external sources. Other processes occur far from the boundaries and deep within the ocean, receiving energy from the forcing at the sea surface only after it has already been transferred through the action of a chain of other processes.

Much of the turbulence induced in the surface and benthic boundary layers is driven by processes resulting from the fluxes of buoyancy and momentum through the nearby boundary, and these can be identified as *external processes*. The wind-driven breaking waves and convection produced by air–sea buoyancy flux may be regarded as external processes, driven by forcing from the overlying atmosphere. The stress exerted by a seabed on a passing flow of water (for example a tidally driven current), and the geothermal heat flux, similarly drive turbulence-generating 'external processes'; they are driven by sources of energy outside the benthic boundary layer itself. The stress and the buoyancy flux also provide measures that determine the processes as well as the nature and intensity of much of the turbulence within the boundary layers.

Internal processes leading to turbulence and mixing, for example shear across a density interface, breaking internal waves and double diffusive convection, are topics addressed in Chapter 4.

• Buoyancy and momentum fluxes lead to three main types of boundary layers adjoining horizontal boundaries, those in which

(a) the stress is negligible (for example at the sea surface when there is no wind) and in which turbulent motion is derived from unstable stratification and convection;
(b) there is no flux of buoyancy and where the turbulent motion is driven by the stress on the boundary (for example in a tidal flow over the seabed, through which there is no geothermal heat flux emanating from the Earth's core or from decomposing matter buried in the sediment); or
(c) there is both a stress and a buoyancy flux.

These three types[3] are considered separately in the following sections. The order of (a) and (b) is not significant, except that it may be easier to visualize convective motions than those in shear, and therefore to consider convection first. The ocean is rarely, if ever, exactly in either of the states represented by types (a) and (b), and for this reason many of the examples of turbulence in the boundary layers are not described until type (c) has been considered in Section 3.4.1.

Before describing the nature of the types of boundary layers, it is important to recognize that, whilst boundaries may lead to processes that produce turbulence, the physical presence of the boundaries also modifies the turbulence in their vicinity. Near the almost static or immobile seabed, viscous forces tend to reduce the components of velocity parallel to the boundary, but this does not apply near the 'free' sea surface.

3 A further type of boundary layer is that which occurs in stratified waters on the sloping sides of the ocean and around seamounts and islands (see Further study).

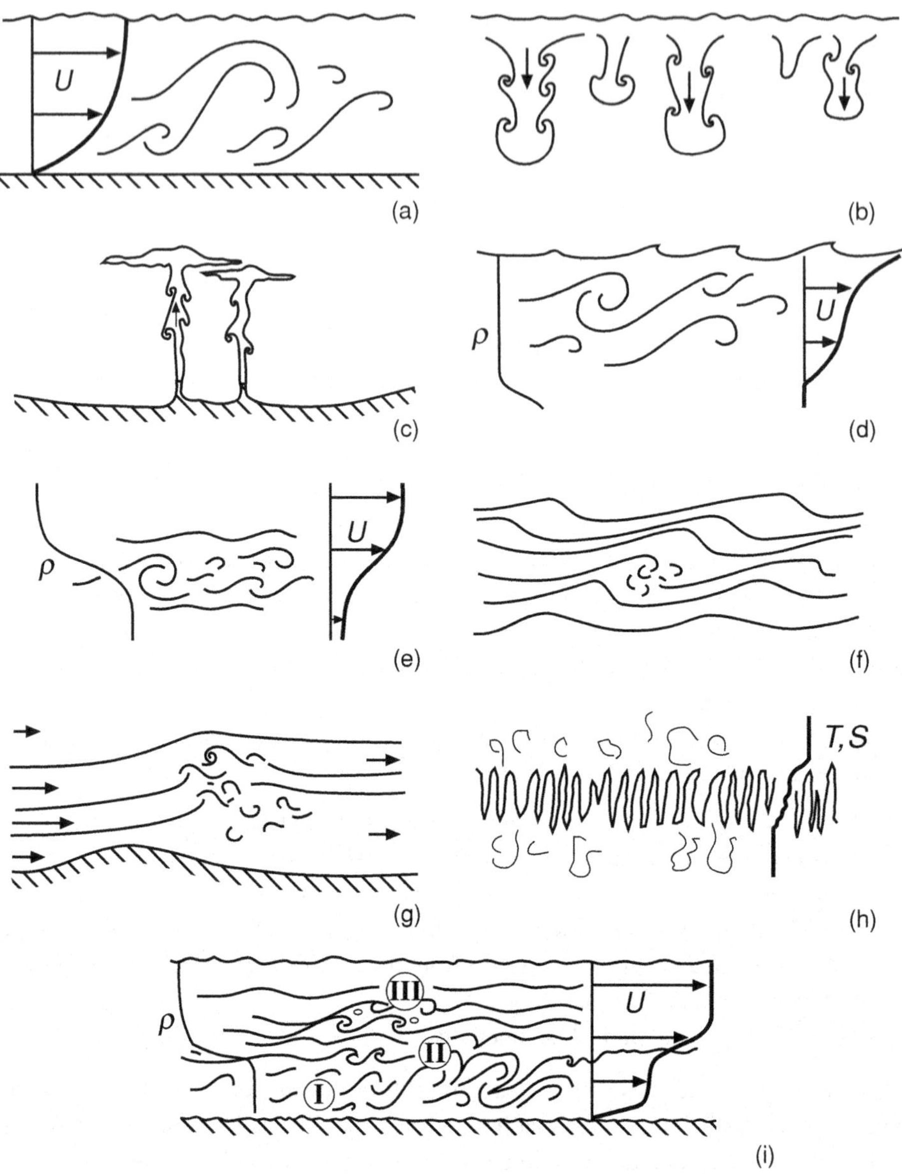

Figure 3.2. Classification of mixing processes according to their immediate source of energy. Parts (a)–(d) are examples of mixing caused by processes with an external supply of energy. (a) Turbulence produced by tidal or wave-induced flow over a solid boundary. (b) Convection resulting from cooling at the sea surface. (The small-scale eddies entraining water into the plumes might, however, be regarded as deriving energy from the kinetic energy of the convective motion – an internal source – rather than *directly* from the surface buoyancy flux.). The same applies in (c), mixing in a hydrothermal plume issuing from a vent, its buoyancy derived from geothermal heating; and (d), turbulence in a wind-generated near-surface shear flow. (Recall, however, that the flow may be a result of a loss in the momentum of waves through their breaking.) Parts (e)–(h) are examples of mixing resulting from an internal energy source. (e) Mixing in a shear flow, perhaps at the foot of the near-surface mixed layer. (f) Mixing resulting from breaking internal waves. (g) Mixing in an internal

There turbulence leads to surface flows that are approximately two-dimensional, lying in the plane of the sea surface: the patterns of foam visible in Figs. 1.1 and 1.4 are largely a consequence of these turbulent two-dimensional flows. The presence of a boundary constrains and modifies the form and scale of turbulence within an adjoining boundary layer by suppressing the normal components of water velocity in its vicinity. At the sea surface this constraint is, however, lost in breaking waves or other processes that cause fluid to be carried through the surface, or from the bounding surface into the underlying water column; processes of 'flow separation' or 'subduction' result in motions that are no longer confined to the two dimensions of the mobile surface. Viscous forces very close to the seabed cause the tangential components of velocity to fall to zero at a rigid boundary and almost to zero when the bed material is mobile. Flow separation may, however, occur there as it does at the sea surface, for example as water moves over the crests of ripples on a mobile sandy bed, sometimes forming lee eddies and carrying mobile sand particles from the bed into suspension.

3.2 Convection in the absence of shear

3.2.1 Convection below a cooled surface or over a heated seabed

In 1900 Bénard showed that regular and steady cellular motions occur at the onset of convective motion when the lower surface of a thin horizontal layer of fluid in a laboratory container is raised to a temperature sufficiently exceeding that of its upper surface. The heat increases the fluid's temperature, so reducing its density and resulting in the buoyancy forces that drive convection.[4] Similar steady convective cells

Figure 3.2. (*cont.*) hydraulic jump caused, perhaps, by flow over a sill in an abyssal channel connecting deep ocean basins (Section 6.8). (h) Mixing caused by double diffusive salt-finger convection with the formation of a step-like structure of temperature (*T*) and salinity (*S*) (Section 4.8). Part (i) illustrates the complexity and linkage of mixing processes that can occur in the ocean, mixing in (I) being caused externally by shear stress on the seabed as in case (a), that in (II) being internal mixing resulting from shear (as in (e)) and the impact of turbulent eddies with a stratified layer overlying the bottom boundary layer; and (III) indicates the internal mixing caused by breaking internal waves (as in (f)) radiating energy upwards from the interface into an overlying stratified layer.

4 The condition from the onset of convection is that the Rayleigh number, $Ra = g\alpha\,\Delta T\,d^3/(\nu\kappa_T)$ (where d is the thickness of the fluid layer of kinematic viscosity ν and molecular conductivity of heat κ_T, $\alpha > 0$ is the thermal expansion coefficient and ΔT is the positive temperature difference between the bottom and the top of the layer), exceeds a value that depends on the nature of the boundary but is typically about 10^3. Static instability, $\alpha\,\Delta T > 0$, is not sufficient to ensure the occurrence of cells or a dynamical instability. As *Ra* is increased beyond this critical value the motion becomes irregular, possibly with vacillations (oscillations that periodically come and go). At a sufficiently high *Ra*, the motion becomes turbulent, although some vestige of the cellular structure at which convection began may be retained. This transition to turbulence as *Ra* increases can be regarded as passing through a set of discrete stages. The cells observed by Bénard in his experiments were probably affected by variations in surface tension as well as buoyant convection.

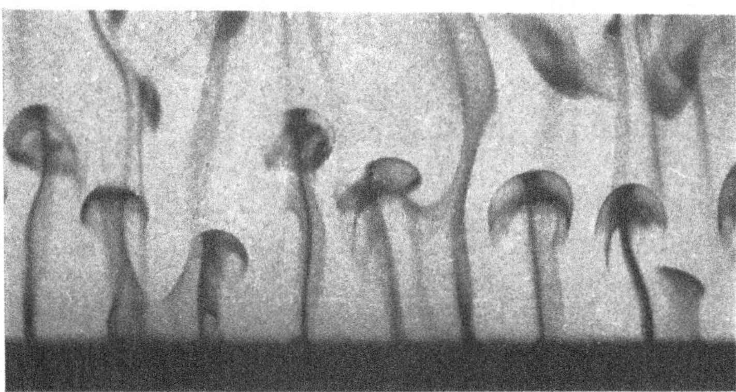

Figure 3.3. Buoyant plumes or thermals. The plumes or thermals, made visible by dye in the two photographs, rise from a heated surface in the laboratory in the absence of shear. The transient near-vertical rising plumes carry dye from the bottom and are topped by circular vortices, like that illustrated in Fig. 1.9(a), of width greater than that of the plumes. There must be descending fluid between the plumes to conserve volume. The spacing of plumes is generally less regular than suggested in these experiments. (From Sparrow *et al.*, 1970.)

are produced if the upper surface of a thin layer of fluid is cooled. Top heating and bottom cooling produce similar, but inverted, convective motions.

The ocean, however, is deep. Cooling of its upper surface can lead to an unsteady or transient convective motion in which plumes of water form and periodically transport water from the boundary, carrying the negatively buoyant fluid into the interior of the water, much as illustrated (although inverted) in the laboratory experiment shown in Fig. 3.3, where, as in Bénard's experiment, the bottom boundary of the fluid has been heated. These plumes are like thermals in the atmospheric boundary layer that are made visible by the formation of cumulus clouds. They form the larger eddies of a range of generally smaller-scale, three-dimensional, turbulent motion.

Periodic cooling of the sea surface leads to substantial changes in density structure and turbulence within the mixed layer. The diurnal day–night cycle of heating and cooling of the mixed layer results in a cycle of turbulent motion that is described in Section 3.4.2. In a few regions, notably the Labrador and East Greenland Sea, and the Gulf of Lions in the northwest Mediterranean, cooling in winter results in convective mixing to depths of 1–3 km, sometimes reaching the seabed. Cold plumes of diameter 300–500 m are observed during winter convection in the Gulf of Lions, sinking at speeds reaching 0.05 m s^{-1}, with smaller, but definite, upward motion between. The processes leading to small-scale turbulence in oceanic convection and the nature of the turbulent motion itself are, however, yet to be thoroughly investigated.

The ocean receives a geothermal flux of heat through the seabed from the Earth's core at an average rate estimated to be about 87.8 mW m^{-2}, but with a smaller mean flux, about 46 mW m^{-2}, through the bed of the abyssal plains. [**P3.1**] Although this is a significant contribution to heat within the abyssal ocean, it generally has little effect on the dynamics of the boundary layer, as shown in Section 3.4.1. There are, however, local exceptions, places where the heat flux is very large, that are referred to in the following section.

In steady conditions with turbulence dominated by convection, there is an approximate balance between the second and third terms in the energy balance equation, (2.16). The rate of loss of potential energy in convective plumes, equal to B, provides the energy to support the dissipation rate, so that approximately

$$\varepsilon = B. \tag{3.1}$$

In steady conditions with no change in temperature, the divergence (or d/dz) of the vertical heat flux must be zero, so that $B = B_0$, where B_0 is the flux of buoyancy through the sea surface, and $\varepsilon \approx B_0$. (This can be deduced by dimensional argument: the basis is that the dimensions of B_0 are, like ε, L^2T^{-3}, where L is length and T is time. In the absence of any natural length scale – and none can be derived from B_0 and ε alone – neglecting molecular effects, those involving ν, κ_T etc., the relationship implies that ε is approximately constant with depth.)

3.2.2 Buoyant plumes and entrainment

One particular, but dramatic, case of convection in the ocean is that which occurs as a result of the continuous release of very hot and buoyant fluid, sometimes in excess of 300 °C, from hydrothermal vents on the ocean floor within ocean ridges, producing rising plumes of water that is much less dense than its surroundings. An example is shown in Fig. 3.4 and sketched in Fig. 3.2(c).

The edges of the plumes entrain water from outside. The local rate at which fluid is entrained into a rising turbulent plume is found to be proportional to the mean speed at which the fluid in the plume is rising through the ambient fluid. This rate is in accord with the 'entrainment assumption' introduced in the 1950s by Morton, Turner and Taylor in connection with their studies of convective plumes. They suggested that entrainment occurs through the process of fluid engulfment and entrainment by turbulent eddies, and that, although the eddies must be involved in the eventual

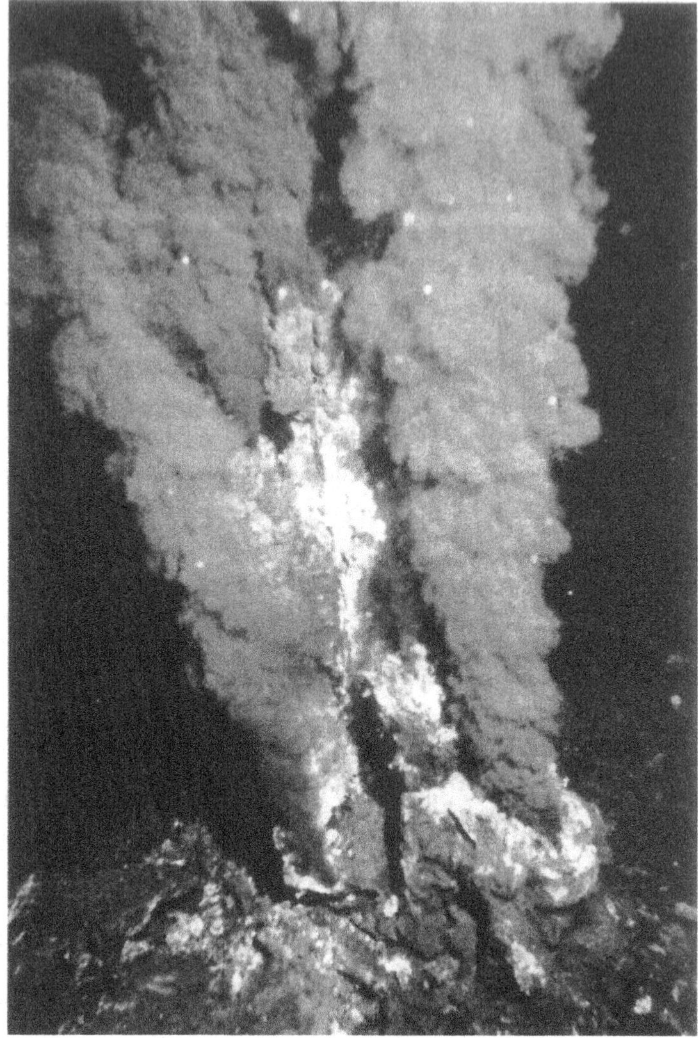

Figure 3.4. Hot fluid rising from black-smoker hydrothermal vents in a mid-ocean ridge. The processes of entrainment are identical to those at the edge of the smoke plume in Fig. 1.8.

homogenization of fluid entrained, viscosity and molecular conductivity play an insignificant role in the process, provided that the turbulent motion in the plume is sufficiently energetic. The only velocity scale on which the speed of the local inflow of fluid into a rising plume can depend is then the difference in speeds, Δu, of the plume and the ambient fluid. (This must also characterize the speed of circulation of entraining eddies.) It follows that the rate of entrainment, an entrainment velocity, w_e, must be proportional to Δu. It is found empirically that $w_e \approx 0.1\ \Delta u$. This assumption has proved successful in many applications, including the predictions of plume rise and spread both in the laboratory and in the atmosphere, from scales of a few centimetres to many kilometres. [**P3.2**, **P3.3**]

In the ocean, the height to which buoyant plumes rise from hydrothermal vents, typically about 200 m, is determined and limited by the density stratification in the water into which the plumes ascend. As a plume rises, the entrainment of the denser surrounding water into the plume causes the plume's density to increase whilst, because the overall stratification is stable, the density of the surrounding water decreases with height above the seabed. At some height the (increasing) density of the water in the plume and the (decreasing) density of its surroundings become equal. At this 'equilibrium level', however, the plume continues to rise for a short distance (or to 'overshoot') because of its vertical momentum, before finally sinking and spreading along isopycnal surfaces at about the equilibrium level (e.g., see Fig. 5.2 later). The part played by plumes in ocean mixing is discussed further in Section 6.9. [**P3.4**]

3.3 Stress and no convection; the law of the wall

When the mean flow and the turbulence are steady, so there is no mean flow acceleration and $\mathrm{D}E/\mathrm{D}t = 0$, and when the buoyancy flux is negligible, the balance of terms in the turbulence energy equation is principally between the rate of production of turbulent kinetic energy by Reynolds stress in the presence of shear (the shear-stress) and the turbulent dissipation, ε, the first and third terms on the rhs of (2.16). Such conditions are commonly found at small distances, z (typically from 0.05 m to 2–5 m), from a plane rigid boundary, but beyond a distance at which viscous effects at the boundary, and its texture or small-scale roughness, have important influences on the flow. As deduced in Section 2.5.4, the balance of terms implies that, approximately,

$$\varepsilon = (\tau/\rho_0)\mathrm{d}U/\mathrm{d}z, \tag{3.2}$$

where $U(z)$ is the mean flow parallel to the boundary. The absence of acceleration in the steady mean flow implies that there are no changes in stress across any horizontal layer, for such changes would alter its momentum and therefore lead to acceleration. In steady conditions the divergence (or $\mathrm{d}/\mathrm{d}z$) of the Reynolds stress is therefore zero, and the stress is generally constant, not varying with distance from the boundary.

For convenience, the Reynolds stress is often expressed in terms of a virtual velocity: $\tau = -\rho_0\langle uw\rangle = \rho_0 u_*^2$, where u_*, with dimensions of velocity, is known as the *friction velocity*. If, as is usual in a turbulent flow dominated by inertial forces, the motion is independent of kinematic viscosity, the mean velocity shear, $\mathrm{d}U/\mathrm{d}z$, can only depend on u_* and on the distance, z, from the boundary and must, on dimensional grounds, be given by

$$\mathrm{d}U/\mathrm{d}z = u_*/(kz), \tag{3.3}$$

where k is a constant, known as von Kármán's constant[5] and found empirically to be approximately 0.41. On integration, (3.3) leads to an equation describing the variation

5 See Further study.

of the mean velocity with distance from the boundary,

• $$U(z) = (u_*/k)[\ln(z) - \ln(z_0)], \tag{3.4}$$

where z_0 is a length scale derived as a constant of integration, and known as the *roughness length* because it depends on the size of the boundary roughness. This remarkable relationship, (3.4), was first announced by von Kármán in 1930. The mean flow, U, varies logarithmically with distance, z, from the boundary with a constant of proportionality u_*/k. The Reynolds stress, $\tau = \rho_0 u_*^2$, can be estimated from measurements of the mean velocity, $U(z)$, at, at least, two distances, z_1 and z_2, above the boundary by using (3.4) to determine u_*, i.e., by eliminating $\ln(z_0)$ to obtain $u_* = k[U(z_1) - U(z_2)]/\ln(z_1/z_2)$.

Equations (3.2) and (3.3) give

• $$\varepsilon = u_*^3/(kz). \tag{3.5}$$

The rate of dissipation of turbulent kinetic energy per unit mass is inversely proportional to distance from the boundary and increases with the Reynolds stress to the power 3/2. (If the characteristic velocity of the energy-containing eddies in a turbulent boundary layer is proportional to u_* – and this is usually the only dimensionally correct scale – then comparison of (3.5) with (2.18) implies that the size of the dominant turbulent eddies, l, increases in proportion to z, the distance from the boundary.) [**P3.5**, **P3.6**]

Equation (3.4) (and sometimes related equations such as (3.5)) is referred to as the '*law of the wall*'.

Equations (3.3) and (3.5) lead towards singularities as the boundary is approached, dU/dz and ε tending to infinity as z tends to zero, and fail to describe flow close to the boundary, where its roughness or the effects of viscosity become dominant. Figure 3.5 shows the variation of measured current speeds with the logarithm of the distance from the seabed on the Oregon Continental Shelf. The bed is composed of fine silt and sand, and, although flat in the immediate locality of the observations, is affected by burrowing organisms producing small burrows and surrounding crater-like mounds. The flow is close to zero at the seabed, and increases to about 6 cm s^{-1} at a height of 1 cm through a viscous sublayer in which the overlying flow adjusts to the boundary. Above this sublayer, the linear variation is consistent with the law of the wall. Above a height of about 0.11 m, however, a second but different linear trend is observed, suggesting that a regime of different friction velocity u_* (and of different Reynolds stress) is reached, possibly reflecting the effects of drag caused by the presence of bed roughness caused by burrowing organisms or by shells and stones beyond the immediate vicinity of the measurement site.

The relation (3.5) with z equal to the depth below the level of the mean sea surface also appears to hold in the surface mixed layer when z is greater than about four times the significant wave height, H_s.[6] Much higher rates of dissipation are found at

6 The significant wave height, H_s, is the statistical measure of surface waves defined in footnote 18 of Chapter 2.

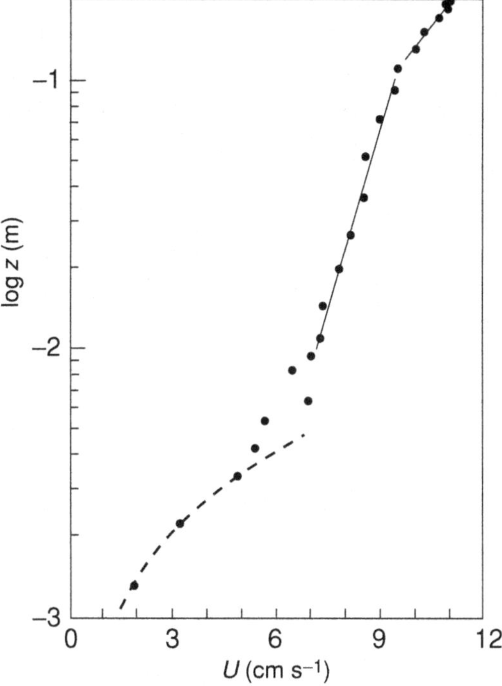

Figure 3.5. The logarithmic bottom layer. The mean horizontal flow speed, U, with z, is plotted on a $\log_{10}$ scale, where z is the distance from the seabed. There is a viscous sublayer close to the boundary ($\log_{10} z$ (m) < -2, or $z < 1$ cm) below the lower of two layers in which $U \propto \log_{10} z$. (From Chriss and Caldwell, 1982.)

shallower depths as a result of turbulence generated by breaking waves (Fig. 2.15); this intermittent and unsteady process of generation introduces further dimensional scales (such as H_s and the speed of breakers, c_b) into the problem of describing the turbulent motion.

3.4 Stress and buoyancy flux

3.4.1 The Monin–Obukov length scale

It is common for both a buoyancy flux and stress to be present in the surface and the benthic boundary layers.

The stress at the seabed is usually represented, or parameterized, using a drag coefficient, C_D, as in (2.17): $\tau = \rho_0 C_D U^2$, where U represents the speed of the flow at some specified distance, commonly about 1 m, from the seabed or sometimes the depth-averaged current in the shallow seas of the continental shelf. The value of C_D is about 2.5×10^{-3}. Since τ is equal to $\rho_0 u_*^2$, the friction velocity in the near-bed logarithmic layer, u_*, is equal to $C_D^{1/2} U$ or approximately $0.05U$.

The wind drag on the ocean surface is generally represented in a similar way but with a different drag coefficient as $\tau = \rho_a C_{Da} W_{10}^2$, where ρ_a is the density of air (about 1.2–1.3 kg m^{-3} at the sea surface, but varying with temperature, pressure and humidity) and W_{10} is the wind speed measured at a standard height of 10 m above the level of the mean sea surface. The drag coefficient at the sea surface, C_{Da}, increases with wind speed above about 5 m s^{-1} and ranges from about 1×10^{-3} to 2.7×10^{-3}.[7] In conditions under which little net momentum is being transferred into the surface waves, the Reynolds stress of the wind on the sea surface is approximately equal to the stress in the water below the surface, so that, equating the two stresses, the friction velocity in the water, $(\tau/\rho_0)^{1/2}$, is given by $[(\rho_a/\rho_0)C_{Da}]^{1/2} W_{10}$.

The dimensions of the available parameters at the boundary, a buoyancy flux, B_0 (dimensions L^2T^{-3}), and surface stress or τ/ρ_0 (dimensions L^2T^{-2}), can be combined to provide only one length scale, in addition to the distance, z, from the boundary, on which turbulence can depend.

• The dimensional scale characterizing turbulence in convective boundary layers with an applied stress is known as the Monin–Obukov length scale:

$$L_{MO} = -(\tau/\rho_0)^{3/2}/(kB_0) \tag{3.6}$$

or

$$L_{MO} = -u_*^3/(kB_0), \tag{3.7}$$

where $u_* = (\tau/\rho_0)^{1/2}$ is the friction velocity in the water and k is von Kármán's constant.

The usual sign convention is that the surface heat flux, F_0, is positive for an upward flux of heat (and, by virtue of (2.8), so is B_0). The Monin–Obukov length scale, L_{MO}, is therefore negative when $F_0 > 0$. These are destabilizing conditions, favouring convection, i.e., when the sea surface is losing heat or when the benthic boundary layer is warmed by a geothermal heat flux. Conversely, L_{MO} is positive for stabilizing conditions where $F_0 < 0$, so the flux contributes towards a statically stable stratification, e.g., when there is a downward flux of heat into the sea surface promoting lower density near the water surface.

At distances from the boundary beyond the effects of breaking waves and bottom roughness or viscosity effects but at distances $z < 0.03|L_{MO}|$, buoyancy has little effect and the law of the wall (3.4) is generally valid.

In destabilizing conditions with $L_{MO} < 0$, mixed conditions in which both shear and buoyancy affect turbulence occur within the range of distances $0.03|L_{MO}| < z < |L_{MO}|$. The buoyancy flux has a small, but notable, effect in this range: when $z/|L_{MO}|$ is less than about 0.3, the mean flow speed is given approximately by

$$U(z) \approx (u_*/k)[\ln(z/z_0) + 5z/L_{MO}]. \tag{3.8}$$

7 There is some evidence, however, that the drag coefficient at the sea surface decreases with wind speed as speeds reach hurricane force. The reason why is unknown.

The shear flow, then given by $dU/dz \approx (u_*/k)[1/z + 5/L_{MO}]$, is less than that in conditions of the same stress but in the absence of a buoyancy flux, B_0, because of the enhanced vertical transfer of momentum in the convectively driven motion illustrated schematically as 'unstable' in Fig. 3.6. In contrast, in stabilizing conditions with $L_{MO} > 0$ ('stable' in Fig. 3.6), for example in relatively weak or only moderate wind and when the ocean surface is strongly heated by solar radiation, the vertical momentum transfer is suppressed. Turbulent energy is lost in doing work against the buoyancy forces. In such conditions, the stress required to maintain the same shear, dU/dz, is reduced or, equivalently, the shear produced by a given stress is increased, resulting in conditions at the sea surface sometimes referred to as a 'slippery sea'.[8] A negative buoyancy flux, B_0, at the sea surface, caused by a downward flux of less dense freshwater in the form of rain, may quite rapidly promote stratification near the sea surface and lead to a decay of turbulence in the underlying mixed layer.[9] [**P3.7**, **P3.8**, **P3.9**].

At distances $z > |L_{MO}|$ from the boundary, buoyancy tends to dominate in destabilizing conditions, driving 'free convection' that is independent of surface stress. Transient plumes begin to form as described in Section 3.2.1. When $z > 2|L_{MO}|$ there is an approximate balance between the second and third terms in the energy balance equation (2.16), expressed as (3.1).

Measurements of ε shown in Fig. 3.7 are from a convective layer formed in the upper ocean during an outburst of cold air off the east coast of the USA, when the surface buoyancy flux resulting largely from evaporation reached a value of 3.2×10^{-7} W kg^{-1}, and these are compared with measurements of ε under convective conditions in the atmosphere. The dissipation rate, ε, is approximately uniform with depth as predicted in Section 3.2.1. The mean value of ε in the convective layer is

$$\langle \varepsilon \rangle \approx 0.72 B_0, \qquad (3.9)$$

where B_0 is the surface buoyancy flux. Although buoyancy plays a major part in mixing the upper ocean, the geothermal heat flux generally has a negligibly small effect on turbulence in the benthic boundary layer of the abyssal ocean. [**P3.10**, **P3.11**]

3.4.2 Diurnal and seasonal heat cycling of the mixed layer

The nature of turbulent motion in the mixed layer at a particular time and location usually depends on how rapidly the buoyancy and wind forcing are changing and on the remaining effects of past changes and events, such as storms. Although conditions within the ocean can rarely be described as 'steady' over times when, for example, the changes in fluxes or transfer rates are negligible in comparison with the mean fluxes or rates, some of the ideas derived in the previous section regarding idealized,

8 Similar 'stable conditions' affecting the relation of stress and shear may occur in the bottom boundary layer when the flux of sediment eroded from the seabed into the layer is relatively large.

9 Some aspects of the complex processes of heat transfer at the sea surface are described in problems P3.8 and P3.9.

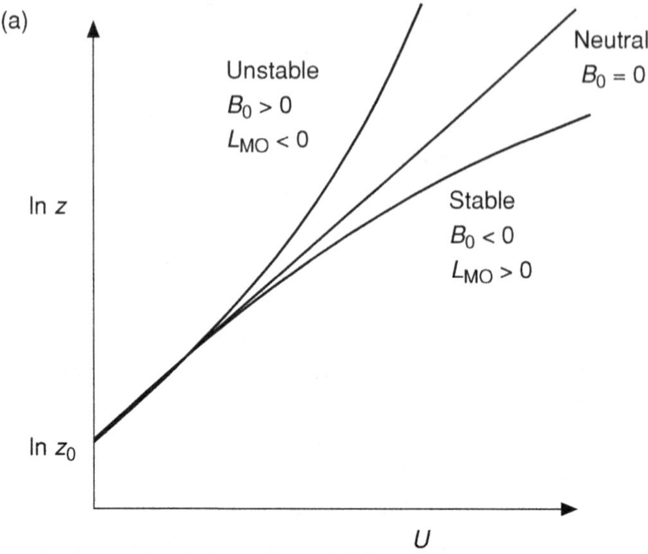

Figure 3.6. A schematic representation of turbulence and flow near a boundary in differing heating regimes. (a) The mean current as a function of ln z, where z is the height above a fixed boundary, in the following three conditions: when there is an upward flux of heat, or buoyancy flux, B_0, through the boundary ('unstable'), when the lower boundary is cooled ('stable') and when there is no heat flux through the boundary ('neutral'). (b) The change in turbulence and shear in the corresponding three conditions at the sea surface, those of sea-surface heating, zero heat flux and surface heat loss, but in similar wind speeds, W_{10}. In conditions of stable heating (top), a stable stratification reduces the vertical transfer of momentum by turbulent eddies, leading to relatively greater shear than in neutral conditions (middle). In conditions of surface cooling (bottom), the tendency for convection to occur enhances the vertical transport of momentum from the surface and reduces the shear.

strictly steady conditions are still found to be instructive (if not useful), as the following example shows.

Observations made in the Bahamas during a pronounced diurnal cycle of strong surface heating and cooling, but fairly uniform wind, are illustrated in Fig. 3.8. The surface buoyancy flux, B_0, is shown below, together with the times (dots) at which the temperature and ε profiles shown in parts 1–6 were obtained. In parts 1–2 nocturnal cooling, with $L_{MO} \approx -17$ m and $B_0 > 0$, causes convective activity that generates turbulence in a deepening layer, eventually reaching to a depth of 100 m. Below this turbulent layer there is a reduction in ε (seen in part 1) of some 1–3 orders of magnitude over a distance of, at most, a few metres. The nocturnal cooling is followed by a period of daytime solar heating, the effects of which are illustrated in part 3. A substantial decrease in ε occurs throughout much of the sampled depth. The heat flux, with $L_{MO} \approx +17$ m, produces stable near-surface stratification, capping what, at night, had been a relatively deep mixed layer, and limiting the depth to which the wind stress at the sea surface causes mixing. Recommencement of cooling in the evening leads again to convection and the erosion of the stratification, resulting in a deepening

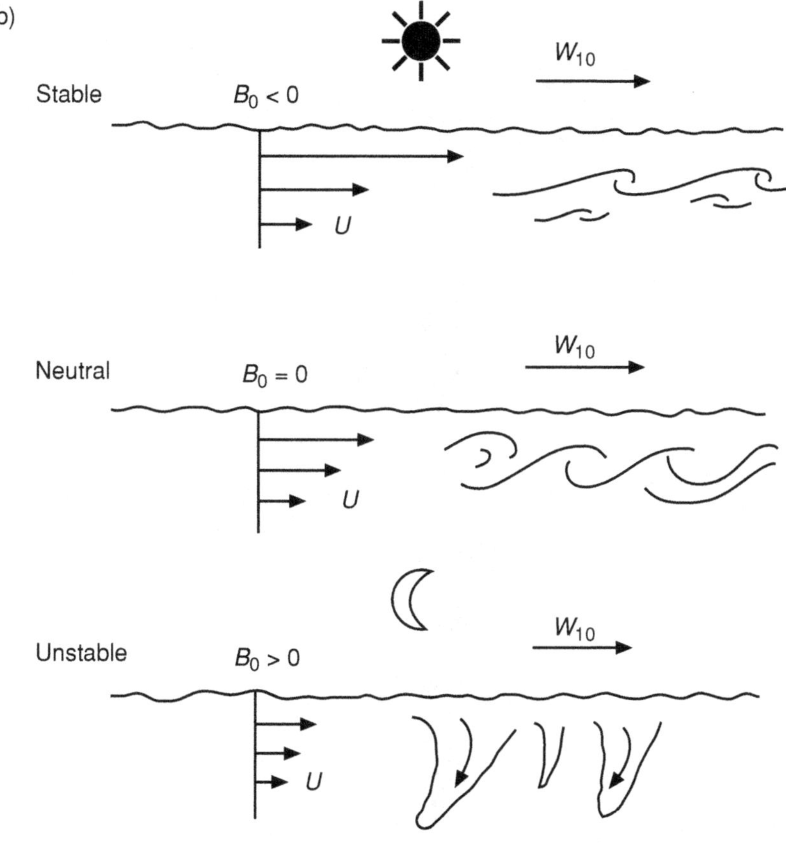

Figure 3.6. *(cont.)*

turbulent layer (parts 4–6). There is sometimes little indication in the vertical profiles of temperature (or density) of how far the surface-generated turbulence extends (e.g., see profiles 1 and 5) and, without measurements of ε (or possibly the rate of loss of temperature variance, χ_T), it is generally impossible to estimate with certainty the thickness of the near-surface layer of active turbulence. During the convective periods the mean dissipation rate in the mixed layer, $\langle \varepsilon \rangle$, is approximately equal to $0.61B_0$, slightly less than is found in the much deeper and more prolonged convective layer shown in Fig. 3.7.

The intricate interactions between the external forcing produced by cooling and wind are illustrated in Fig. 3.9, a 23-day time series during which strong winds (part (a)) produced relatively high surface stress in three periods, with accompanying diurnal cycles in heat flux (part (b)). Particularly notable are the oscillations and gradual deepening of the thermocline during the periods of higher winds, accompanied by cooling of the mixed layer, and the downward progression of regions of (dark shaded in part (d)) high dissipation in the mixed layer, especially during periods of surface heat loss ($F_0 > 0$) between 30 December and 3 January (or days 365–369; dates are indicated at the top of the figure, and year days below), when $N^2 < 0$ in part (e) implies

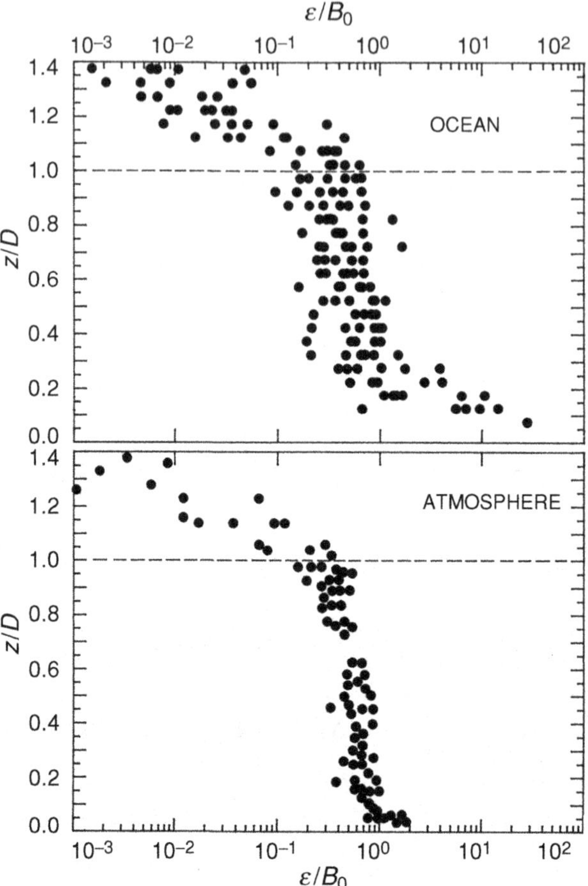

Figure 3.7. Dissipation in the mixed layer in conditions of strong convection. The variation of ε/B_0 with z/D, where D is the thickness of the mixed layer and the coordinate z is the distance from the boundary. Top: in the ocean (with depth, z, upwards). Bottom: for comparison, in the atmosphere (with height, z, upwards). The dashed line is the level where $z = D$. Deeper in the ocean or higher in the atmosphere the fluid is stratified and ε decreases rapidly. Conditions in the ocean were unsteady, with the mixed-layer depth gradually increasing from about 75 to 165 m over a period of about 36 h. The mean (upward) buoyancy flux at the surface during this period was about 3×10^{-7} W kg^{-1} and the mean friction velocity about 1.4×10^{-2} m s^{-1}, leading to a value of $L_{MO} \approx -22$ m and a mean value of $L_{MO}/D \approx 0.2$. Free convection should occur in the range $2|L_{MO}| < z < D$ or $0.4 < z/D < 1$, where ε is observed to be approximately constant. (From Shay and Gregg, 1984b.)

that the stratification is statically unstable. Relatively high dissipation at depths of 40–80 m near the foot of the mixed layer in the period 4–9 January is a consequence of an internal process of mixing resulting from enhanced shear.

Cycles of mixing similar to those of the diurnal cycle are found to occur on seasonal time scales, most notably in the subtropics. They are accompanied by convectively driven, mixed layer deepening in winter, and the reformation of the seasonal

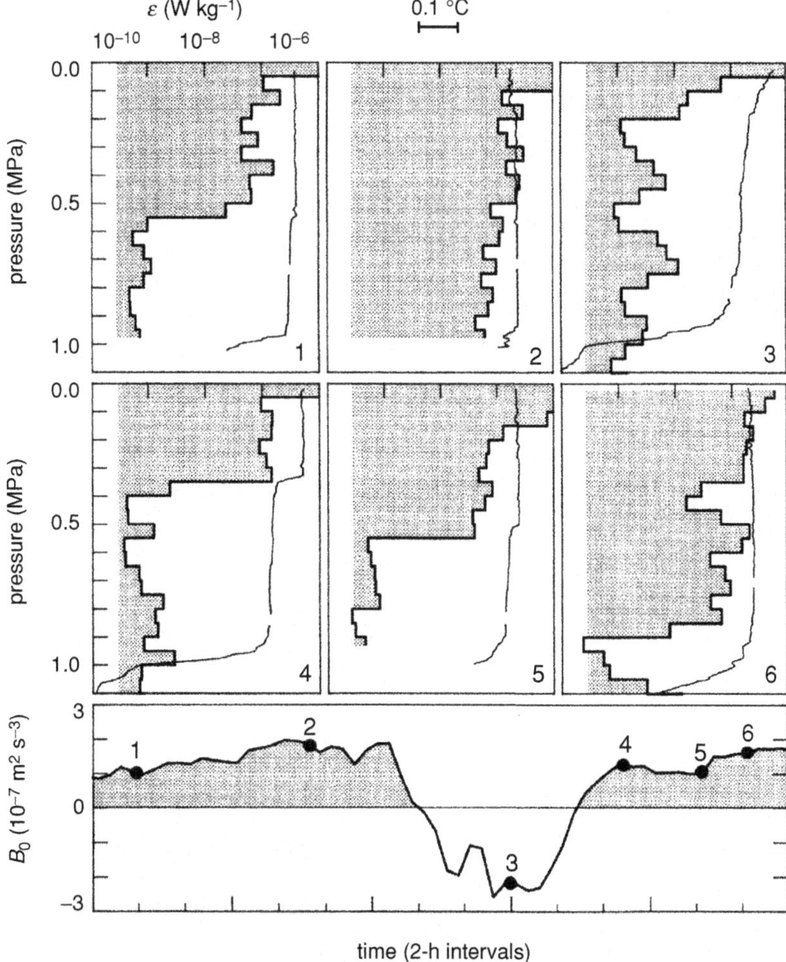

Figure 3.8. A day–night cycle of ε and temperature in the mixed layer: 1–6 show profiles of ε (stippled) and temperature (line). Below is the variation in the surface buoyancy flux, B_0, as a function of time. The times, 1–6, of the profiles are marked by dots, and 3 is at noon. The 30-h period begins with surface cooling and positive buoyancy flux in the early evening of the first day, includes (around time 3) a period of daytime heating when the buoyancy flux becomes negative, and continues into a period of renewed cooling continuing until midnight of the second day. The rate of loss of turbulent kinetic energy, ε, is obtained using a free-fall probe, and the vertical distance over which ε is estimated (about 5 m) accounts for its step-like profile. (From Shay and Gregg, 1986.)

thermocline as solar heating increases in spring. Just as in daytime in the diurnal cycle shown in Fig. 3.8, stratification leads to a reduction in the depth to which surface mixing can penetrate: the water in the deeper part of the winter mixed layer becomes isolated from the surface by the formation of the thermocline in early spring and continues

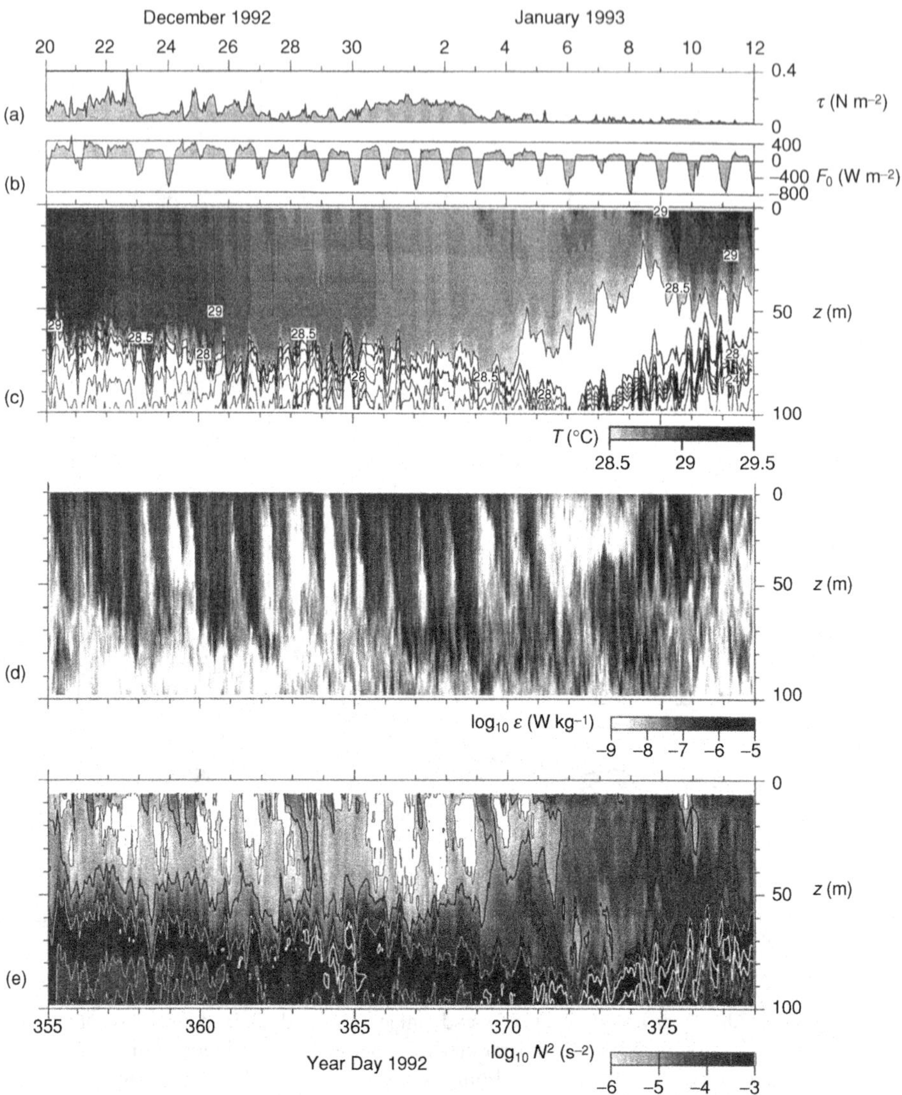

Figure 3.9. The variation of turbulence in the mixed layer during periods of strong wind forcing and diurnal heating and cooling. The data were obtained near the Equator in the Pacific Ocean during a period of westerly wind bursts. (a) Hourly averaged wind stress. (b) The net surface heat flux, F_0, positive values indicating (unstable) periods of surface cooling. Parts (c)–(e) are all from the surface to 100 m with (c) temperature, (d) the rate of dissipation of turbulent kinetic energy, ε, and (e) the square of the buoyancy frequency, $N^2 = -(g/\rho)\mathrm{d}\rho/\mathrm{d}z$. In (e) the density gradient is estimated over 8-m differences of the density averaged vertically over 4 m. White regions are where $N^2 < 0$ and the mean density profile is statically unstable. Black contours are values of $N^2 = 10^{-6}$, 10^{-5} and 10^{-4} s^{-2}, respectively. White contours are at $N^2 = 3.16 \times 10^{-4}$ and 10^{-3} s^{-2}. The microstructure data are obtained from the free-fall package, 'CAMELION', described by Moum *et al.* (1995) (see Further Study, Chapter 2). This probe is tethered to the research vessel by a cable of diameter 0.44 cm and falls at about 0.8 m s^{-1}. It is deployed from the vessel's stern whilst under way at about 0.2 m s^{-1}. (From Smyth *et al.*, 1996.)

to have no direct contact with the atmospheric forcing until it is again reached by convection in the following winter (Fig. 3.10). During summer and early autumn, this 'isolated water' may retain some of the properties, such as its vertical component of vorticity, that were impressed on it during winter, but others may be strongly depleted, e.g., concentrations of oxygen and nutrients through their uptake by marine organisms during summer.

3.4.3 Other mixing processes in the upper ocean

• Processes contributing to the mixing of the upper ocean boundary layer are sketched in Fig. 3.11, and some of these are described below.

The presence of waves makes a strong physical distinction between this boundary layer and that near the seabed. Waves act as a catalyst in the process of transfer of momentum from the wind into ocean currents. The breaking of wind-generated waves creates turbulence that contributes to the mixing of the upper ocean, enhances the rates of dissipation above that predicted by the law of the wall, (3.5), and transfers into the mean flow the momentum previously imparted to the waves from the wind. The energetics of breaking waves and their generation of bubbles and spray are subjects of considerable attention and study particularly because of their bearing on the stirring of the mixed layer, the exchange of gases between the ocean and the atmosphere, and the generation of aerosols.

Foam was mentioned in Section 1.4 as an indicator of subsurface turbulent motions in the surf zone. When the wind exceeds about 4 m s^{-1}, which is sufficient for occasional wave breaking to begin in open water far from shore and to produce bubbles, bands of foam or other flotsam aligned roughly in the wind direction and 5–100 m or more apart can often be seen floating on the sea surface. Bands of foam are also to be seen in moderate to strong winds on the surface of lakes, and an example is shown in Fig. 3.12(a). In the 1920s, Langmuir realized that these bands or 'windrows' are indicative of convergent motions in which water at the surface, although mainly moving downwind, also has spatially periodic cross-wind components with amplitudes of a few centimetres per second that carry floating material into the bands as sketched in Fig. 3.12(b). To preserve continuity, the water moves downwards below the bands at speeds of 1–10 cm s^{-1} (increasing with increasing wind), leaving the buoyant material floating in the windrows on the surface, and subsequently diverges and rises back towards the sea surface. This circulatory motion is now referred to as Langmuir circulation. It appears to be produced by the interaction of waves and shear flow.

Water in and just below the windrows moves downwind more rapidly than the mean flow, resulting in an across-wind variation in downwind flow speed, as indicated in Fig. 3.12(b).[10] The speed of the converging flow at and just below the surface,

10 This cross-wind shear may contribute to the greater downwind than across-wind dispersion observed in dye patches (Section 5.4.1) through the mechanism illustrated in Fig. 1.10(a). The effects of Langmuir circulation on dispersion are described in Section 5.2.7.

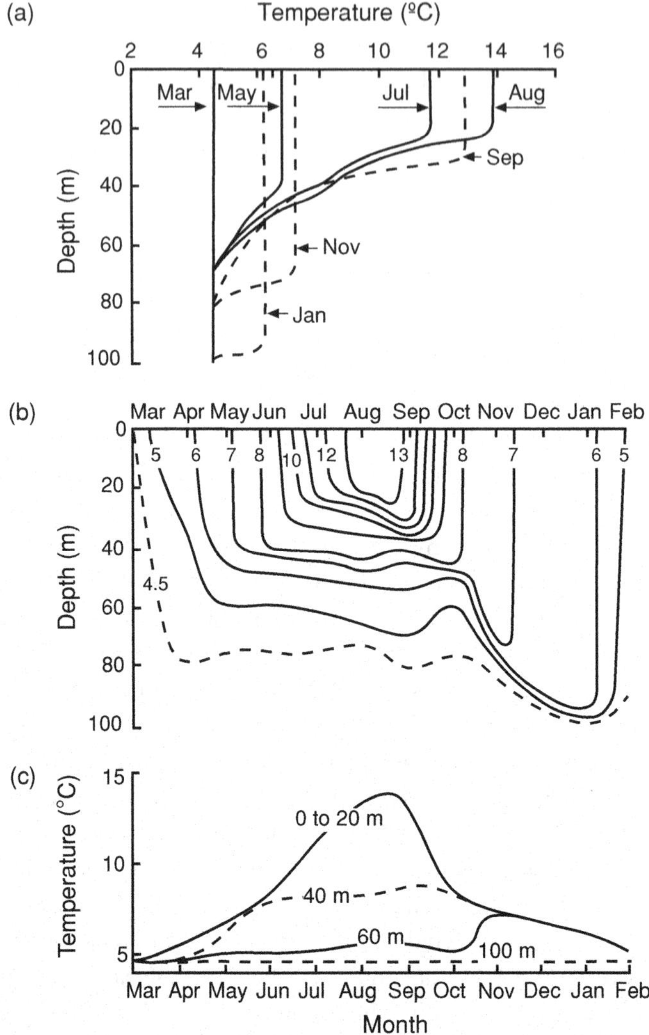

Figure 3.10. The seasonal cycle of heating in the upper ocean. The data are from the northern hemisphere at 50° N, 145° W. (a) Profiles of temperature showing the formation of a mixed layer and seasonal thermocline in May and the subsequent deepening of the mixed layer and winter cooling. (b) Contours of temperature. (c) Temperature variations at four depths. As a result of surface heating during the spring, by May water at depths of 40–60 m becomes isolated from the direct influence of the atmosphere through the formation of the seasonal thermocline. Subsequent surface forcing has little effect on water below 40 m until convective cooling results in a deepening mixed layer in October–November. There is no variation in temperature below 100 m, the maximum depth of winter convection. This maximum depth will vary between winters, depending on the atmospheric forcing. Convection during a second and subsequent winters may fail to penetrate to the deeper water mixed in the first winter and left beneath the thermocline, either because the depth of mixing is reduced because of a decrease in severity of successive winters or perhaps because the water is carried towards the Equator where atmospheric forcing is less and where winter convection does not penetrate so deep. In such cases the deep water retains the 'fossil' remains of some properties impressed on it during its earlier contact with the atmosphere. (From Wijesekera and Boyd, 2001.)

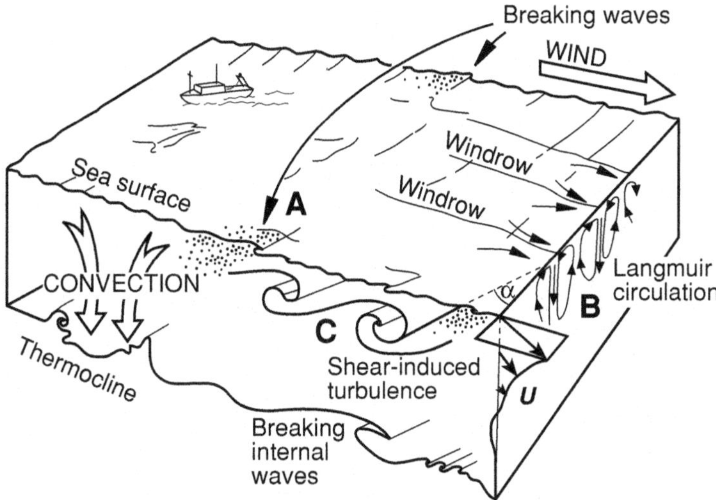

Figure 3.11. A sketch of the processes leading to the mixing of the upper ocean boundary layer. (A): Breaking waves. (B): Langmuir circulation. (C): Temperature ramps. The effect of the Earth's rotation results in a rotation of the mean current direction (to the right of the wind in the northern hemisphere) as sketched. This appears to lead to a rotation of the axes of the large eddies, C, through a small angle, α, as suggested in the sketch, so that the values of skewness in Fig. 3.14 measured in directions at 90° to the wind are non-zero. (From Thorpe, 1985.)

although small, is sufficient to advect clouds of bubbles created by randomly breaking waves to the bands, forming lines of subsurface bubble clouds. The circulation also carries planktonic organisms and contributes to their vertical cycling within the mixed layer, thereby exposing them to varying levels of solar radiation. Water that has been heated or cooled whilst close to the surface and which descends beneath the windrows has a mean temperature typically a few milli-kelvins different from that of the surrounding water in the mixed layer. This has little dynamical effect but provides a coherent thermal 'signature' indicating the presence and vertical extent of the circulation.

Over short periods of time the closed circulation cells may limit the spread of both floating and dissolved material across the wind direction, material released into a circulation cell from a location between bands being able to spread across wind no further than the nearest adjoining band. The cells are not, however, steady, but are transient. Neutrally buoyant floats designed to follow the flow are not observed to cycle repeatedly within the mixed layer. The windrows and associated convergence regions fluctuate in location and, although a cell may persist for some tens of minutes, they are often found to amalgamate with neighbouring cells. One such 'Y junction' is sketched in Fig. 3.12(b).

• The circulation cells, although limited by the presence of the sea surface, have the properties of variability and transience found in smaller-scale turbulent motions, and the circulations are now often described as 'Langmuir turbulence', characterized

(a)

(b)

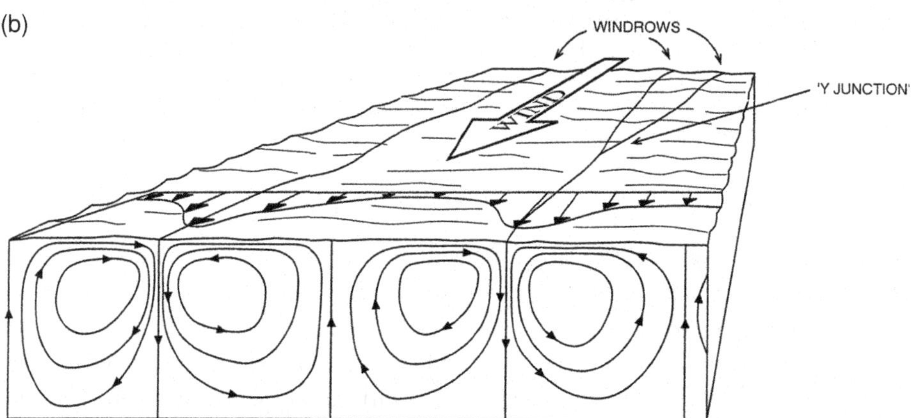

Figure 3.12. Langmuir circulation. (a) Windrows produced by Langmuir circulation on the surface of a lake. The mean distance between windrows is about 8 m. (b) A sketch of the circulation pattern, including a 'Y junction' where neighbouring cells combine. Windrows composed of floating material or foam form as a consequence of convergent motion at the water surface. Their separation is twice the width of the individual Langmuir cells.

by a parameter called the 'turbulent Langmuir number',

$$La_{\text{turb}} = [u_*/(2S_0)]^{1/2}, \tag{3.10}$$

where u_* is the friction velocity and S_0 is the speed of the wave-induced Stokes drift at the water surface. The turbulent Langmuir number has a typical value of about 0.3 in the ocean, but may be respectively larger or smaller in periods of rising or falling winds. The relative effect of Langmuir circulation on the turbulent flow in the mixed layer is enhanced as La_{turb} decreases.

Much of the study of Langmuir turbulence has been carried out using large eddy simulation (LES) models and, whilst comparison with observations appears favourable, more is required. The Langmuir cells are large eddies that supply energy to the turbulence cascade within the mixed layer. Through its advection of water within the mixed layer, the circulation distributes or spreads the turbulent energy produced at smaller scales by other processes, including breaking waves. Langmuir circulation helps sustain the uniform density of the mixed layer. The circulation may also contribute to the process of entraining denser water from the stratified pycnocline and mix previously stratified near-surface water in a period of increasing wind speed. Once a mixed layer has been generated, however, it is less certain that the circulation is very important in the process of entrainment or in the deepening of a mixed layer during periods of persistently strong winds.[11]

The Langmuir cells are not the only coherent structures observed in the mixed layer. Measurements of temperature structure made with arrays of thermistors with 1 mK (or better) resolution reveal patterns of variation with relatively large and coherent gradients in the temperature field that are orientated on surfaces across the wind direction and tilted downwind at typically 45° to the horizontal. They are sketched in Fig. 3.13(a). Figure 3.13(b) shows typical variations of temperature measured by a fixed vertical array of thermistors in a mean flow. The sudden increases in temperature (arrowed in Fig. 3.13(b) and occurring sequentially in time down the array as surfaces of large gradients pass by) correspond to warmer near-surface water reaching the sensors as sketched in Fig. 3.13(a). These 'temperature ramps' are also found in the atmospheric boundary layer, where they are sometimes referred to as 'micro-fronts'. They are advected by the mean flow in the mixed layer, and often extend through vertical distances of about a third of the thickness of the mixed layer.

The ramps are also detected in temperature measurements obtained by towing arrays of thermistors through the water. Their abrupt changes in temperature lead to a skewness of the spatial derivative of the temperature gradient, $S(\mathrm{d}T/\mathrm{d}x)$.[12] As shown in Fig. 3.14, the magnitude of the skewness is of order 1, but it varies approximately sinusoidally with the angle between the tow direction and the wind, as the tilted surfaces are crossed at different angles by the moving temperature sensors. The skewness is of opposite sign in conditions of surface heating and cooling corresponding, respectively, to the stabilizing and destabilizing conditions described in Section 3.4.1. The skewness, $S(\mathrm{d}T/\mathrm{d}x)$, is positive when $B_0 < 0$ and $L_{MO} > 0$, as in Fig. 3.14, and when the tow direction is upwind ($\theta \approx 0°$).

The causes of the temperature ramps in the near-surface mixed layer are not definitely established, but they have been shown to involve the downward displacement of water from near the sea surface, suggesting a surface source, perhaps breaking waves. The existence of ramps in the atmospheric boundary layer over land implies that waves are not the only cause, and it is possible that they are related to the hairpin vortices or

11 The entrainment of water into the near-surface mixed layer is the subject of Section 4.5.

12 The skewness of a variable, X, is defined as $S(X) = \langle (X - \langle X \rangle)^3 \rangle / \sigma_X^{3/2}$, where, as before, $\langle X \rangle$ is the mean value of X, its average in time or space and where $\sigma_X = \langle X^2 \rangle - \langle X \rangle^2$ is the variance of X.

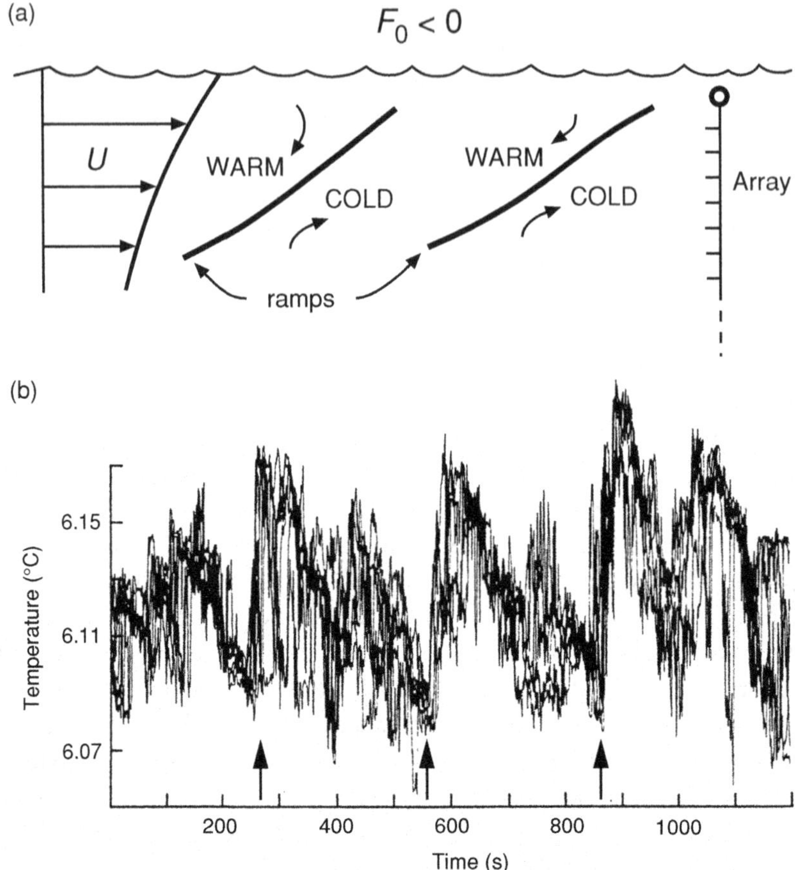

Figure 3.13. Temperature ramps. (a) A sketch showing temperature ramps being advected past the vertical fixed array of thermistors in conditions of surface heating. Abrupt rises in temperature occur as the ramps are advected past the array, at times increasing with the depth of the sensors. (b) The variation of temperature with time measured by a vertical array of five thermistors in the mixed layer where the mean flow is about 8 cm s^{-1}. Ramps are marked by arrows. (From Thorpe and Hall, 1980.)

'bursts' known to cause a vertical transfer of water and momentum within the benthic boundary layer described in the following section.

3.4.4 The benthic (or bottom) boundary layer

• Except near hydrothermal vents (Fig. 3.4), over decaying organic material (e.g., see P. 3.11) or where gases or ground water are released into the overlying water column, the buoyancy flux through the seabed has little effect on the structure of the adjoining boundary layer.

Over very flat surfaces in the laboratory and in the absence of a buoyancy flux, 'streaks' of alternately faster and slower moving fluid, aligned downstream, are found within a viscous sublayer extending a distance of about $z_v = 10\nu/u_*$ from the seabed

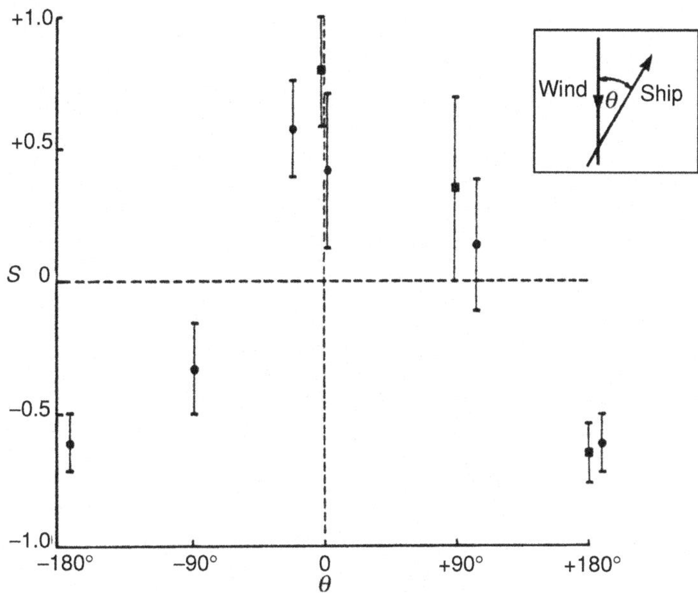

Figure 3.14. The skewness, S, of the time derivative of temperature. Measurements are obtained from towed sensors and are plotted as a function of the direction of the tow, θ, relative to the wind, as defined in the insert, in conditions of stable surface heating, $F_0 < 0$. (From Thorpe, 1985.)

and affecting its mean velocity profile as shown, for example, in Fig. 3.5. The streaks are typically a distance $10z_v$ apart and $100z_v$ in length. [**P3.12, P3.13**] Vortical motions, with their axes aligned in the flow direction and not unlike those in Langmuir circulation, are also observed in the laboratory, but are less regular and of rather smaller size even than the streaks. Similar flow structures will occur in the sea, where the bottom is flat, e.g., where it is composed of very fine sediments or muds. The near-bed flow will generally, however, be strongly affected by roughness when stones, shells or other components of the bed protrude above the mean bed level to more than the thickness of the viscous sublayer (typically a few millimetres), and the effects of the disturbance of such obstacles, the 'roughness elements', or their wakes, will determine the statistical spatial and temporal variability of the flow near the seabed. Relatively large roughness elements such as rocks, or ripples and waves formed in the sediment by the overlying flow, can, as mentioned earlier, result in flow separation and the generation of eddies.

Coherent structures have been identified within the turbulent flow above the layer dominated by viscosity or small-scale roughness elements, the best-described being vortices with shapes resembling that of horseshoes or, in the high-Reynolds-number flows common in the ocean, hairpins, the curved part of the hairpin structure being above and downstream of its trailing legs (Fig. 3.15). These appear to be associated with the 'bursts' of larger than average Reynolds stress in which there are upward motions, marked 'E' (for ejection) in Fig. 2.17, and account for the relatively large scale, mentioned in Section 2.5.3, at which much of the stress is carried. They represent

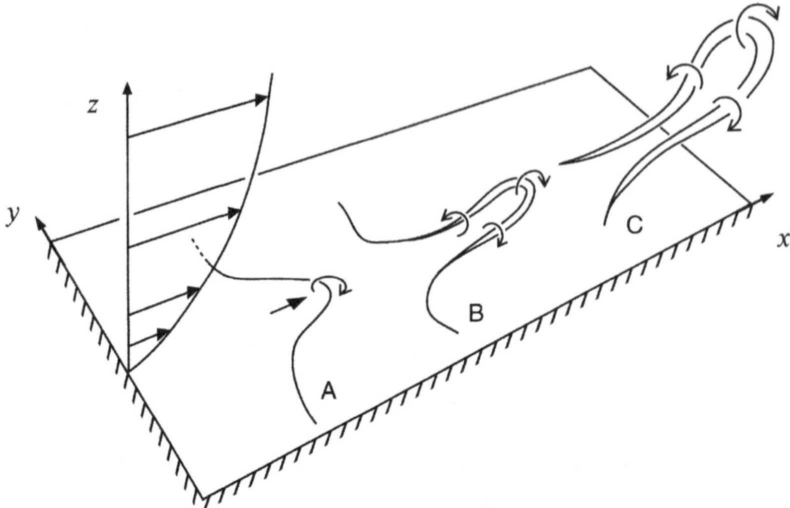

Figure 3.15. A sketch of the development, A to C, and flow pattern of a hairpin vortex in the bottom boundary layer. The legs of the vortex have a form that is similar to the pair of mutually interacting and self-propelling vortices illustrated in Fig. 1.9(b)(i). Groups of three to five hairpin vortices aligned in the direction of the mean flow are sometimes observed.

a process of detrainment of fluid from the near-bed region of the boundary layer. They grow in size as they ascend through the water column, and in sufficiently shallow water they reach the water surface, generating the boils of size comparable to the water depth shown in Fig. 1.6.

Benthic boundary layers 5–60 m thick, with vertically uniform temperature and salinity and maintained in a mixed state by turbulence generated by the shear-stress energy production term, $\tau\, \mathrm{d}U/\mathrm{d}z$, are commonly found overlying the seabed in deep water. 'Benthic nepheloid layers' are those boundary layers that contain elevated levels of sediment concentration, as in the example in Fig. 3.16. These turbid layers of lower light transmission are generated by short-lived and rare periods of relatively high currents that lead to erosion of fine material from the seabed. The sediment sinking very slowing back towards the bed may remain in suspension for several months and be carried by the mean currents far from its source. Other layers of low light transmission are found in mid-water (intermediate nepheloid layers) and near the sea surface (surface nepheloid layers). The former are sometimes produced by the erosion and turbulent suspension of sediment from the sloping boundaries of the ocean and its subsequent spread (much as in Fig. 1.16(d)), whilst the latter may be a result of, for example, plankton blooms.

3.4.5 Tidal mixing and straining in shallow seas

Stratification is often seasonally and spatially variable in the tidally affected shelf seas. Whilst, for example, the southern North Sea is unstratified throughout the year, the

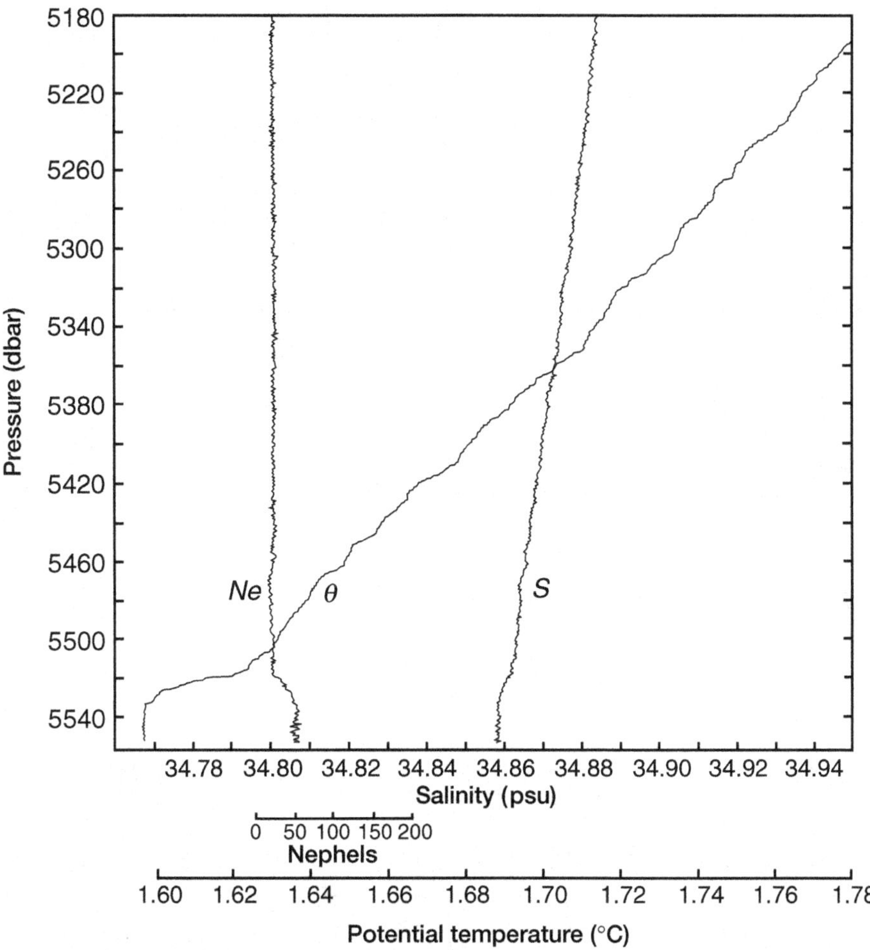

Figure 3.16. A benthic nepheloid layer. Profiles of potential temperature (θ) referred to the sea surface, salinity (*S*) and a measure of the suspended sediment or 'nephel' concentration (*Ne*), at a site in the Hatteras Abyssal Plain where the water depth is about 5550 m. The water properties are relatively uniform in the 25-m-thick benthic boundary layer with higher levels of *Ne*. Small variations in *S* and *Ne* are close to the resolution of the measured values and are probably not real. (From Armi and D'Asaro, 1980.)

northern part becomes strongly stratified in summer. The two regions are separated by a front, typically 1–10 km wide, but many tens of kilometres in length. The mean location of such tidal mixing fronts is determined by a competition between the destruction of stratification through the turbulent mixing caused by seabed-generated turbulence and the production of stratification by solar heating with a surface buoyancy flux $B_0 < 0$. (Recall that B_0 is measured as being positive if there is a positive upward flux of heat.) If, as is generally valid, molecular effects can be ignored, there is only one non-dimensional term on which the location of a front separating stratified and non-stratified water can depend. In a mean tidal flow, $\langle |U| \rangle$, in water of depth h (where both $\langle |U| \rangle$

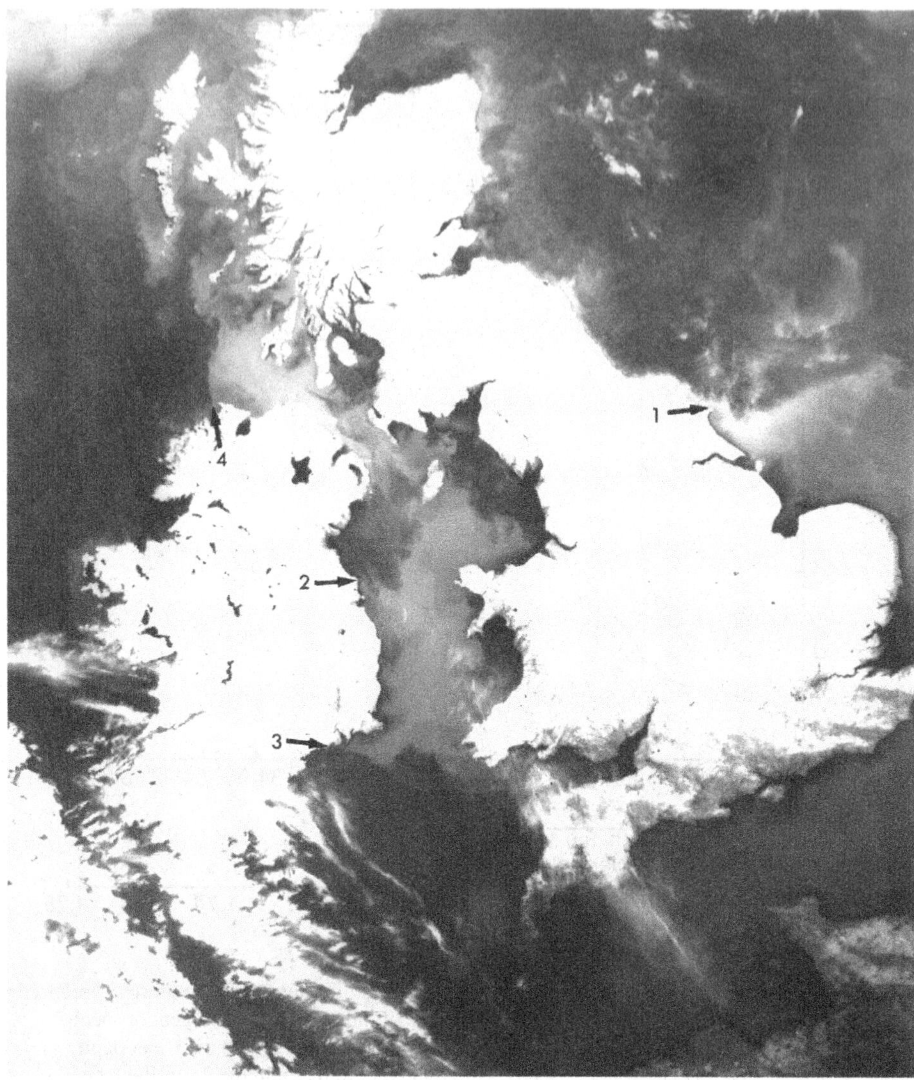

Figure 3.17. Tidal mixing fronts around the British Isles in summertime. Warmer surface (and stratified) water appears relatively dark. Clouds (white) cover the southwestern part of Ireland, the lower left-hand part of the image, and extend from southwest England across the English Channel. The fronts, some marked by arrows, separate regions of stratified waters from waters shown in lighter grey tones that are relatively cold at the surface and that are vertically uniform and unstratified, being mixed by turbulence generated by tidal flows. For example, water north of the front (marked 1) extending across the North Sea from the English to the German coast (not shown) is stratified in temperature whilst that to the south is well mixed and unstratified. Much of the Irish Sea is also well mixed, although there are stratified regions in Cardigan Bay, off west Wales, and further north off the English coastline. There is a front (2) in the western Irish Sea, separating the (lighter-toned) mixed water in the central Irish Sea from the (darker) stratified water near the Irish coastline, and a relatively sharp front (3) is visible in St George's Channel at the southern entrance to the Irish Sea. There is also a front (4) extending from the coast of Northern Ireland towards the Hebrides. (Image courtesy of the University of Dundee, Scotland.)

and h depend on position), this term is $h|B_0|/\langle|U|\rangle^3$. In shallow regions of relatively fast tidal flows where $h|B_0|/\langle|U|\rangle^3$ is relatively small, turbulence generated by shear stress on the bottom reaches the surface and results in mixing throughout the water column, sustaining unstratified conditions. The bottom boundary layer then extends all the way up to the surface and the whole water column is mixed by the turbulence generated at the seabed. Where turbulence is relatively weak or the water depth is large, so that $h|B_0|/\langle|U|\rangle^3$ is large, the input of solar heat in summer produces stratification that prevents complete mixing, and here a thermocline is generally maintained until the onset of winter cooling or severe wind mixing.

- Fronts separating stratified and well-mixed regions form in locations where $h|B_0|/\langle|U|\rangle^3$ reaches a critical value. An empirically determined value of

$$h\,(\mathrm{m})/\langle|U\,(\mathrm{m\,s^{-1}})|\rangle^3 = 500 \qquad (3.11)$$

provides a rough predictor of the location of summertime 'tidal mixing fronts' on the European Continental Shelf, some of which are shown in the infrared image of Fig. 3.17.

The simple formulation $h|B_0|/\langle|U|\rangle^3 = \text{constant}$, commonly referred to as the 'h over u cubed criterion', has been tested successfully in a number of shelf-sea regions around the world, although modifications are needed to account for the effects of seasonal heating, the spring–neap tidal cycle, and the competing input of turbulent energy from the wind.

Because the speed of tidal currents, $|U|$, has generally two maxima during each complete tidal cycle, and the production of turbulent kinetic energy near the seabed is proportional to $U^2|U|$ (Section 2.5.3), the rate of dissipation of turbulent kinetic energy, ε, has a periodicity that is twice that of the tide. Near the seabed, ε is greatest close to the time of maximum flow, but the time at which the maximum rate of turbulent dissipation occurs increases with height above the bed. For example, in tidal currents of 0.7–0.8 $\mathrm{m\,s^{-1}}$ under unstratified conditions in a part of the Irish Sea where the mean water depth is 90 m it is found that the maximum ε occurs at the sea surface about 1.8 h after the maximum at the bed.[13] This time delay in ε appears to be, in part, a consequence of a lag in its local production by the shear-stress rather than an upward diffusion of turbulent energy from the vicinity of the seabed, but exactly how the production and transfer processes may relate to the 'bursts' and ejection processes near the seabed and the arrival of boils at the sea surface has yet to be discovered.

A further process, important particularly in regions of freshwater influence (ROFI), is the periodic stratification resulting from the straining of horizontal density gradients by tidal motions. This tidal straining can partly be understood with reference to Fig. 1.10(a). Suppose that the initial square, ABCD, represents a vertical section down an estuary with fresher, less dense water at AD emerging from a river on the left, and

13 In stratified conditions the lag is greater, the maximum dissipation at a height of 30 m above the seabed being found about 3 h later than that at the seabed.

denser seawater at BC further down the estuary on the right, and that the tide is ebbing, moving to the right in the figure. Friction at the seabed results in a shear flow carrying the less dense water, lying along AD, over the denser water at BC, leading to a distorted shape much as is shown in the figure at time *t*, and resulting in a stable stratification with less dense over more dense water. The reversal of the tidal flow in the following flood tide leads to a reduction in the vertical density gradient as the distorted shape is translated back towards the initial square. The vertical stratification consequently varies through the tidal cycle. In practice, the mean shear is not uniform through the water column, but is greater near the seabed. In the ebb flow some turbulent energy produced by the shear stress near the bed is expended in mixing the developing and increasing overlying stable stratification, and consequently (as in Fig. 3.6(b) (top)) the shear is greater than during flood where stratification is reduced. On the flood tide the shear may even lead to a reversed situation with denser salty seawater being carried over the fresher water coming from the river, possibly resulting in convective instability,[14] so increasing the turbulent kinetic energy near the seabed and further reducing the mean shear, much as illustrated in Fig. 3.6(b) (bottom).

Suggested further reading

Classification of mixing processes and entrainment rates

The idea of classifying mixing processes as external or internal according to their immediate source of energy, with examples of such classification, is developed by Turner (1973; section 4.3.1). He also describes the entrainment assumption (1973; section 6.1.2) first made by Morton *et al.* (1956).

The benthic boundary layer

Chriss and Caldwell's (1982) experiment (results of which are shown in Fig. 3.5) remains one of the most careful and detailed studies of flow close to the seabed.

Wimbush (1970) made very ingenious measurements of the vertical structure of temperature over the floor of the deep ocean. Measurements are elaborated by Wimbush and Munk (1971) and compared with those in the atmosphere. There is no very comprehensive study of the benthic boundary layer in the deep ocean that includes observations of turbulence both very close to the seabed and in the overlying water column, but Armi and D'Asaro's (1980) paper, with its fold-out figure showing their observations of the variations in temperature and currents in the boundary layer over the floor of the Hatteras Abyssal Plain, is very informative.

14 There appear, however, to be no direct observations of the structure of convection (e.g., plumes) in these conditions, and further study is warranted.

The upper ocean mixed layer

Langmuir's short (1938) paper, the first to explain the pattern of circulation known now by his name, provides an instructive, if not amusing, description of his investigations carried out to confirm his notion that circulation cells exist.

Moum and Smyth (2001) give a succinct account of the processes leading to mixing in the upper ocean.

The papers by Shay and Gregg (1984a, corrigendum 1984b, 1986) are important studies of convection in the layer, and the earlier paper includes a comparison with observations of convection in the atmospheric boundary layer.

Schott *et al.* (1996) give an interesting account of their observations of convection in the Gulf of Lions.

Further study

Von Kármán's constant

Von Kármán's constant, k, is known only empirically through laboratory and field observations. Recent experiments to determine k in the atmospheric boundary layer are described by Andreas *et al.* (2006).

Langmuir circulation

Langmuir circulation is a subject that has been given much attention by theoreticians and observationalists. It is generally accepted that a 'vortex force' that depends on the presence of shear and mean particle motion produced by waves (the Stokes drift) drives the circulation. A thorough description of Langmuir circulation and its generation is given by Leibovich (1983). Thorpe's (2004) review provides some information about Langmuir turbulence, but interested readers should refer to original sources, including McWilliams *et al.* (1997).

Laboratory experiments and field observations by Veron and Melville (2001) show that small circulation cells (typically 5 cm wide) are rapidly generated following a sudden increase of wind (Fig. 3.18) but perhaps not by the same mechanism as that which produces the larger cells discovered by Langmuir. These small cells may be of importance in that they exchange water previously in close contact with the air with that in the interior of the near-surface layer, a process of 'subduction' that involves flow separation from the water surface. The transfer process exposes subsurface water (possibly of relatively low gas concentration) to the molecular transfer of gases (as well as heat) from the air, across what is, for the water newly arrived at the sea surface, a large concentration gradient, and may therefore contribute to air–sea gas transfer.

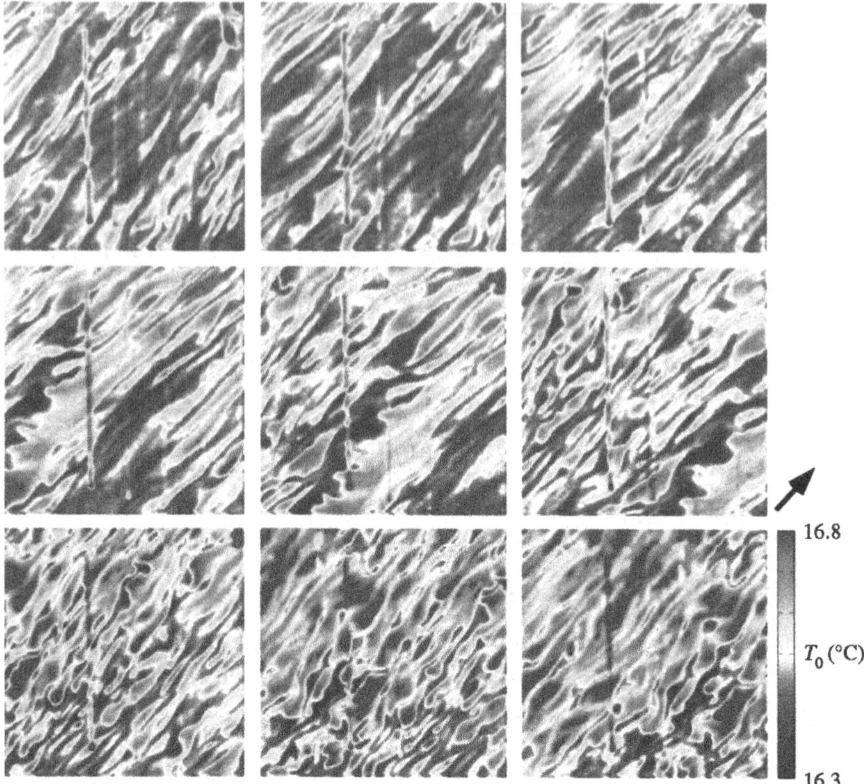

Figure 3.18. Temperature patterns formed as wind increases. Infrared images show the temperature of the water surface at 1-s intervals during an increase in wind speed in a light gust. The sequence is from left to right along the top row, and then similarly along the middle and bottom rows. Warmer linear structures aligned in the wind direction develop as the cool surface skin (described in P.3.9) is replaced by water rising from below. The initial bands rapidly become more intense (warmer) and less regular, although some banding in the wind direction survives. Some advection of the structures from left to right is evident: the apparent temporal evolution may be a result of spatial variations. The mean wind is about 2 m s^{-1} in the direction of the arrow, and the image covers an area 0.53 m square. Vertical bands are artefacts of the instrumentation. (From Veron and Melville, 2001.)

Breaking surface waves

The breaking of surface waves is a topic that, as indicated in Section 3.4.3, has many different facets, beyond their direct generation of turbulence. Breakers form spray, or droplets, and bubbles. The near-surface ocean is therefore a 'two-phase' flow system, and this adds considerable complexity to the flow's dynamics, notably so when droplets or spray are present in high concentration, as is generally the case in hurricane conditions or, with lower wind speeds, near the sea surface at locations where waves are in the process of breaking. The bubbles generated by breakers are important as a conduit for air–sea gas exchange. The gases within subsurface bubbles are at a greater pressure than are those in the atmosphere because of the hydrostatic increase in pressure with depth and because surface tension increases the pressure within bubbles beyond that in their surroundings. The atmospheric gases of which bubbles are composed therefore tend to pass from the bubbles into solution in the surrounding water, increasing its level of gas saturation. Because bubbles oscillate at high frequency for short periods after their formation, they emit sound and this contributes to the general level of sound in the sea. Bubbles also reflect sound (e.g., sounds produced beneath the sea surface by acoustic sonar detection systems or ADCPs) and breakers reflect microwaves (coming from instruments above the sea surface, e.g., radar), factors and instrumental means that are used to detect breaking waves. Further information is given by Melville (1996).

Bottom boundary layer and fronts

Major advances in the understanding of the dynamics of bottom boundary layers in the ocean are likely from the further collection and analysis of data using PIV (see Fig. 2.3).

More about the tidal mixing fronts in shelf seas is found in the reviews by Simpson (1998) and Sharples and Simpson (2001). Simpson *et al.* (2000) describe the *phase lag of dissipation* with height above the seabed in tidal currents. Reference to *tidal straining* and other effects in ROFI can be found in Simpson *et al.* (1990).

The greatest omissions in the topic area of boundary layers are discussions of the relatively large-scale effects of the Earth's rotation, the development of Ekman layers and consequent upwelling or downwelling, and also effects of sloping boundaries in the ocean, particularly those of continental slopes and around islands and ocean ridges. These sloping boundaries are regions both of internal wave generation, through the conversion of energy from the surface tides, and of internal wave breaking. The latter has several important consequences, some of which are similar to those in the familiar surface surf zone described in Section 1.4, whilst several others differ. Crossing isopycnal surfaces, the boundary layers above sloping boundaries, can, for example, lead to diapycnal diffusion. They may also be sites of the generation of pancake-like vortices or 'vortical modes' – see Section 5.4.2 – that carry (or disperse) material away from the sides of the ocean into its interior, a means of forming intermediate nepheloid layers. A related topic, that of flow through topographically confined regions, straits and channels, where the flow is 'hydraulically constrained' and often subject to instability, making these regions sites of relatively intense mixing, is referred to in Section 6.8.

Further reference is to be found in TTO to *convection* (Chapter 4), the *benthic boundary layer* (Chapter 8), the *upper ocean boundary layer* (Chapter 9), *shallow seas* (Chapter 10), *boundary layers on slopes* (Chapter 11) and *flow through passages and channels* (Chapter 12).

Problems for Chapter 3

(E = easy, M = mild, D = difficult, F = fiendish)

P3.1 (M) Geothermally driven buoyancy flux. What is the mean buoyancy flux into the ocean resulting from the average geothermal heat flux through the seabed? Compare this with values of ε typical of the abyssal ocean. Assume that the density of water overlying the seabed is 1030 kg m^{-3}.

P3.2 (D) Spreading of a buoyant plume. Show by dimensional arguments that, at a height z above the seabed, a turbulent buoyant hydrothermal plume of warm water rising in unstratified surroundings of density ρ_0, from a small local source where the buoyancy flux is B_0 and the vertical momentum is negligibly small, must have a mean radius $R = c_1 z$, where c_1 is a non-dimensional constant. The constant, c_1, gives the (constant) angle of spread of the plume. Near the source the plume must therefore spread conically. You may suppose that entrainment occurs by turbulent engulfment of the surrounding water and that molecular processes of heat or momentum transfer are insignificant in increasing the plume's width.

Show also by dimensional arguments that the mean rise speed, W, of the plume at height z is given by $W = c_2 B_0^{1/3} z^{-1/3}$, and that its mean buoyancy (g times the difference in mean density between the plume and its surroundings, divided by ρ_0) at this height must be $c_3 B_0^{2/3} z^{-5/3}$, where c_2 and c_3 are non-dimensional constants.

Show that the mean rate of dissipation of turbulent kinetic energy per unit mass must decrease in height in proportion to z^{-2}.

By examining the rate at which the vertical flux of warm water changes with z, show that the rate of flow into the plume (the entrainment velocity) must be proportional to W, in accord with the entrainment assumption of Morton *et al.* (1956).

What is the angle of inclination of the edge of the plume to the vertical if the entrainment velocity is equal to $0.1W$?

P3.3 (D) Spreading of a jet. A submarine vent emits a turbulent jet of fluid with density ρ equal to that of the uniform overlying seawater, with a flux of momentum $F_{\mathrm{M}} = \int w(\rho w)\mathrm{d}a$, integrated over the small area of the vent, where w is the vertical velocity. Show that, on dimensional grounds, the jet radius at a height z above the vent is proportional to z, so that the jet spreads conically, whilst the mean vertical velocity and average rate of dissipation of turbulent kinetic energy, $\langle\varepsilon\rangle$, at height z are proportional to $F_{\mathrm{M}}^{1/2}\rho^{-1/2}z^{-1}$ and $F_{\mathrm{M}}^{3/2}\rho^{-3/2}z^{-4}$, respectively. Compare these with the corresponding relations for a buoyant plume in P3.2. As in that problem, you may suppose that entrainment occurs by turbulent engulfment of the surrounding water and that molecular processes of heat or momentum transfer are insignificant in increasing the jet's width.

P3.4 (D) Rise of a buoyant plume in stratified water. A turbulent plume is formed by the continuous release of fluid with buoyancy flux B_0 through a small isolated vent on the seabed. If its vertical momentum is negligibly small and if the buoyancy frequency, N, of the overlying water through which it rises is constant and molecular transfers are negligible, show by a dimensional argument that the greatest height above the seabed to which the plume will rise is given by $z_m = q(B_0 N^{-3})^{1/4}$, where q is a dimensionless constant. As in P3.2, you may suppose that entrainment occurs by turbulent engulfment of the surrounding water and that molecular processes of heat or momentum transfer are insignificant in increasing the plume's width.

• The value of q is found empirically to be about 3.8. Because the rising plume has vertical momentum when its density becomes equal to that of its surroundings, the plume 'overshoots' this level, reaching a greater height before its vertical velocity decreases to zero. The plume subsequently collapses and spreads at a height of about $0.8z_m$. The effect of the Earth's rotation is usually important in the subsequent horizontal spread of fluid from hydrothermal plumes.

P3.5 (E) The Kolmogorov scale in a boundary layer. How does the Kolmogorov length scale, l_K, vary with distance from the seabed, with the friction velocity and with the stress within a law-of-the-wall constant-stress layer if $C_D = 2.5 \times 10^{-3}$? Estimate l_K at a height of 1 m from the bed in tidal flows of 0.2 and 1 m s^{-1}.

P3.6 (E) Dissipation in the bottom boundary layer. The mean flow in the section shown in Fig. 2.3 is 0.113 m s^{-1}, as given in the figure caption. What dissipation rate is predicted by the law-of-the-wall relation at the mean height of 0.55 m above the bottom if $C_D = 2.5 \times 10^{-3}$?

The value found is less than the values contoured in Fig. 2.3(b), possibly because only the higher values of ε are contoured and these are not therefore representative of the mean. (Other factors might be (i) the flow is not isotropic, leading to erroneous estimates of ε, (ii) the mean flow was not steady, but modulated by the currents produced by surface waves that induce effects not included in the law-of-the-wall theory, and consequently (iii) because the law of the wall was not valid and because the observations are not in a constant-stress layer. Care is needed in applying the law of the wall to deduce values of turbulent dissipation rates!)

P3.7 (M) The effects of rainfall. Figure 3.19 shows profiles of the buoyancy frequency, N, and the rate of dissipation of turbulent kinetic energy, ε, before and at two times, about 1 and 2.5 h, after a rain squall that, during its passage, increased the wind stress from about 0.06 to 0.5 N m^{-2} and in which the rainfall reached a peak of about 110 mm h^{-1}. Give a qualitative explanation for the changes that occur in N, and also for those in ε above and below the depth of about 20 m, as a result of the squall.

P3.8 (M) Convection under ice. Solar radiation passing through a thin layer of ice formed on the water surface in winter heats the underlying water that is close to freezing point. Explain how this may lead to convective motions in a freshwater lake whilst it does not lead to convection in seawater.

• The convection in ice-covered lakes is a source of turbulent motion that may help, in winter, to maintain diatoms in suspension in the water column even though they are slightly negatively buoyant and so tend to sink to the lake bed.

P3.9 (D) The thermal compensation depth and near-surface temperatures. Incoming short-wave solar radiation penetrates the ocean to a depth that depends on the clarity

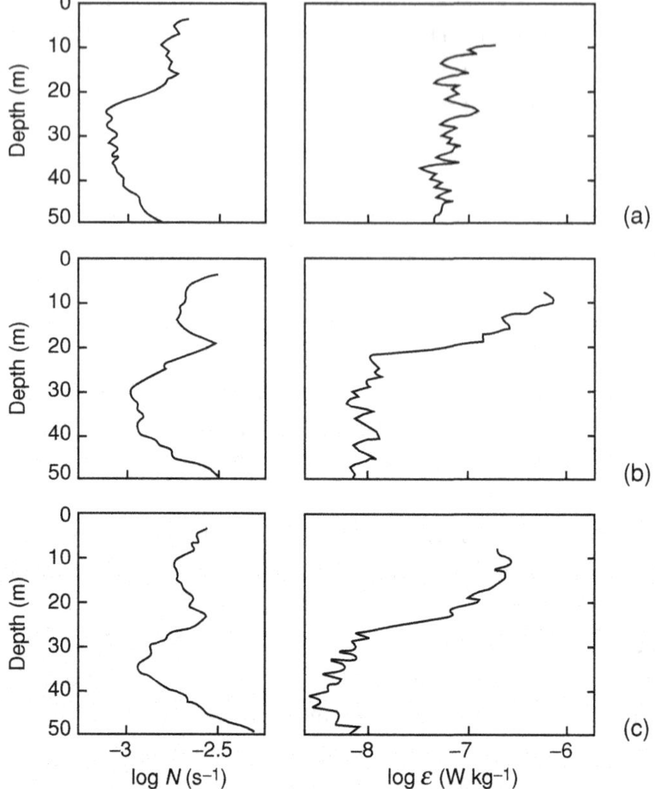

Figure 3.19. Effects of a rainsquall. Profiles of (left) buoyancy frequency, N, and (right) ε, (a) before the onset of rain, and at times (b) about 1 h and (c) about 2.5 h after a rainsquall. (From Smyth *et al.*, 1997.)

of the water (e.g., whether or not it contains suspended sediment or algal blooms). The heat flux decays approximately exponentially with depth over scales, d_e, ranging from about 0.5 to 10 m, the greater values obtaining in the clearer conditions. Heat is transferred into the water at a rate that equals the divergence of the flux. Outgoing long-wave radiation is almost entirely from the water surface, as is heat lost by evaporative cooling. As a result, the temperature in a thin, typically 1 mm thick, layer at the surface is less than that below, sometimes by as much as 0.5 °C. Because of this 'cool skin' of the ocean, temperatures of the sea surface measured by microwave remote sensing, particularly those obtained from satellites, are not identical to those of the mixed layer at depths of 1 m. The skin is, however, disrupted by breaking waves and in conditions of changing wind stress as shown in Fig. 3.18.

• An important consequence of the various processes of heat transport between the ocean and the atmosphere is that the upper layers of the sea may lose heat even though the incoming radiative heat flux exceeds the net outgoing heat flux. The *thermal compensation depth* is the depth at which the vertically integrated heat transfer *into*

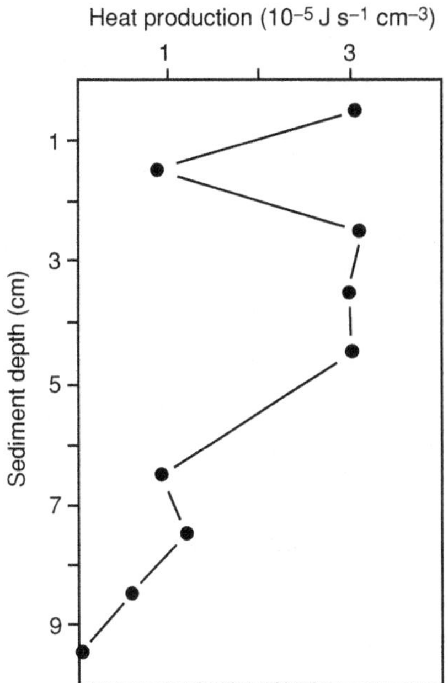

Figure 3.20. Heating by decomposing organic matter: measured rates of heat production by the decomposition of organic matter in the upper layers of the sediment on the sea floor. The units of heat production are 10^{-5} J s^{-1} cm^{-3}. (From Graf, 1989.)

the ocean derived from incoming solar radiation balances the flux outgoing from the ocean at the water surface.

If the net incoming solar radiation at the sea surface is 250 W m^{-2} and $d_e = 1.5$ m, whilst the net outgoing heat flux, the sum of the latent heat flux associated with evaporation, the long-wave heat flux and heat flux carried by molecular transfer (the sensible heat flux), is 100 W m^{-2}, find the thermal compensation depth above which the cooling at the sea surface may lead to unstable stratification. (Make appropriate choices for the values of any parameters required to make the calculation.)

Suppose that rain, consisting of fine drops that do not result in much mixing or turbulence in the surface layer, falls on the sea surface in the Tropics at a rate of 5 mm h^{-1}. If the drops are of the same temperature as the water at the sea surface, but are of density 25 kg m^{-3} less than the density at the sea surface, find the buoyancy flux and from it derive the equivalent flux of heat. What effect does the rainfall have on the stability of the upper layer of the ocean if the values given above for the heat flux (250 W m^{-2} and 100 W m^{-2}) are sustained during rainfall? (See also P3.7.)

P3.10 (D) The absence of geothermally driven convection. Taking the mean geothermal heat flux through the bed of an abyssal plain as 46 mW m^{-2} and the drag coefficient as 2.5×10^{-3}, and supposing that the currents are typically 0.05 m s^{-1}, estimate the Monin–Obukov length scale at the boundary. At what height above the sea floor might free convection be found?

• The well-mixed benthic boundary layer is usually of height less than 60 m, above which the ocean is generally stably stratified. You should conclude that convection resulting from the geophysical heat flux has a negligible role in mixing in the benthic boundary layer.

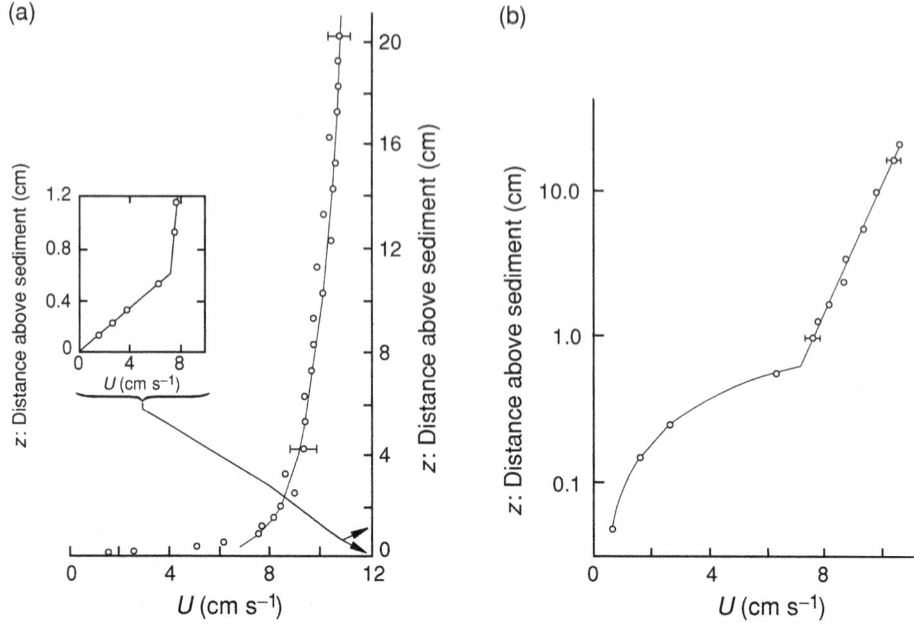

Figure 3.21. Profiles of current speed above the seabed on the Oregon Continental Shelf. (a) A linear plot of average speeds, with an inset showing the speeds in the lowest 1.2 cm of the water column. The same data are shown in (b) but with the height above the seabed plotted on a logarithmic scale. The straight lines represent the best linear fits to data in the inset in (a) within the lower viscous sublayer, where dU/dz is almost constant, and in (b) within the overlying log (or law-of-the-wall) layer, where $U \propto \log_{10} z$. (From Caldwell and Chriss, 1979.)

P3.11 (D) Heating by decomposing detritus. Figure 3.20 shows the heat production in sediments at a water depth of 1430 m on the Vøring Plateau off the Norwegian coastal margin caused by the decomposition of organic matter following a spring bloom of phytoplankton in the upper ocean, the sinking of organic material and its partial burial by benthic organisms. Is the heat produced likely to be important in driving convective mixing in the benthic boundary layer above the sediment? (Use data values given in P3.10 to quantify the answer.)

P3.12 (E) The size of streaks in the viscous sublayer. If the drag coefficient at the seabed is $C_D = 2.5 \times 10^{-3}$, estimate the thickness of the viscous boundary layer and the mean separation and length of streaks within it in a tidal flow of 0.1 m s^{-1}, supposing that the seabed is flat.

Roughness elements of size comparable to the thickness of the sublayer will disrupt the formation of streaks and dominate the nature of the variations in the near-bed flow.

P3.13 (M) Continuity of stress between the viscous sublayer and the overlying turbulent flow. Estimate the friction velocity in the law-of-the-wall layer of Figure 3.21 between 0.01 and 0.1 m from the seabed, and hence the stress exerted by the seabed on the overlying water. Estimate the drag coefficient, C_D, related to the flow at a height of 0.1 m. If the stress in a viscous sublayer near the seabed is $\rho\nu\, \mathrm{d}U/\mathrm{d}z$, where ν is the molecular viscosity, find the near-bed mean velocity shear, $\mathrm{d}U/\mathrm{d}z$, required to transmit the stress at the seabed to the overlying turbulent layer through the viscous sublayer. Compare this shear with that observed. (Following Caldwell and Chriss, 1979, from which Figure 3.21 is taken, you should take $\nu = 1.5 \times 10^{-6}\ \mathrm{m^2\ s^{-1}}$.)

Chapter 4

Turbulence in the ocean pycnocline

4.1 Introduction

4.1.1 Processes of turbulence generation

This chapter is about turbulence within the stratified body of the ocean beyond the direct effects, described in Chapter 3, of its boundaries. The ultimate sources of energy leading to mixing in the ocean are external. The processes causing mixing in the stratified regions of the ocean derive their energy internally, as illustrated in Fig. 3.2, from sources (e.g., radiating internal waves) that may themselves be directly or indirectly driven by external forcing at the boundaries.

Two very different processes usually dominate in the generation of turbulence and diapycnal mixing in the stably stratified ocean. The first is instability resulting from the shear or differential motion of water, i.e., the vertical gradient of the horizontal current, dU/dz, which is often caused by internal waves.[1] This is described in Section 4.2 and some aspects and evidence of the related turbulent motion are presented in Sections 4.3–4.7. The second process is more subtle, a form of convection that results from the different molecular diffusion coefficients of heat and salinity. How these lead to instability is explained in Section 4.8.

1 These internal waves can propagate and transport the energy required to generate turbulence because the ocean is stratified. The shear-generated instability referred to (described in more detail in Section 4.2) is a form of internal wave breaking. Probably less common, and yet to be definitively observed in the ocean, is a form of wave breaking that mimics that of plunging surface waves. It is commonly referred to as 'convective overturn', and involves waves becoming so large that the isopycnal surfaces fold over, resulting in statically unstable regions where density increases upwards and in which convection, and subsequently shear instability, may occur. (The term 'convective overturn' is, however, rather misleading, describing the result rather than the cause of instability: the folding or overturning is a result of the differential advection of particles, more dense over less dense, by the wave.)

4.1.2 The first observations of turbulence in the thermocline

The first published measurements of turbulence within the stratified waters of the thermocline were reported in 1968 by Grant, Moilliet and Vogel. They were made off the west coast of Vancouver Island using hot-film anemometers mounted on a submarine. Grant and his colleagues compared their measurements of turbulence with those made in the mixed layer near the sea surface. In the latter, at a depth of 15 m, turbulence was found to be continuous, but variable in intensity (as shown, for example, in Fig. 2.14). The mean value of the dissipation rate, $\langle \varepsilon \rangle$, was found to be equal to 2.5×10^{-6} W kg^{-1}, which was estimated by fitting the observed one-dimensional spectra of speed determined from the anemometers to the theoretical Kolmogorov spectrum (2.15). The mean rate of loss of temperature variance, $\langle \chi_T \rangle$, was 5.6×10^{-7} K^2 s^{-1}, which was estimated from (2.13) but using the measured fluctuations in horizontal gradient, $\partial T'/\partial x$.

In the stratified water in and below the seasonal thermocline, however, turbulence was patchy, and this has generally proved to be a characteristic of turbulence in the stratified regions of the ocean. Grant *et al.* noted that the mean rates of dissipation in stratified water at a depth of 90 m, $\langle \varepsilon \rangle = 1.5 \times 10^{-8}$ W kg^{-1} and $\langle \chi_T \rangle = 7.2 \times 10^{-8}$ K^2 s^{-1}, were substantially smaller than those they found in the mixed layer.

• The observations draw attention to a fundamental difference between the turbulence in the stratified ocean and that in the ocean's boundary layers or, for example, in ships' wakes and aircraft jets: the latter is generally sustained, being in close proximity to external energy inputs, whereas the former is usually intermittent, being maintained by variable or transient mixing processes, such as internal waves. Rarely are the internal sources of turbulence energy production (e.g., those leading to a transfer of energy to turbulence from the flow field) maintained in a local region of the stratified ocean for times exceeding f^{-1}, known exceptions being where turbulence is sustained in breaking internal waves with frequency near to the inertial frequency f (as illustrated in Fig. 4.11 later) and in double diffusive convection (Section 4.8). Thus, in this chapter, greater attention is paid to transitional processes, those leading from a relatively quiescent flow to turbulent motion, and the subsequent decay of the turbulence to which they lead, than was the case in the discussion of boundary layers in Chapter 3.

At about the same time as the paper by Grant *et al.* appeared, a graphic description of the process leading to turbulence in the thermocline was published. A team of divers led by Woods released dye into the thermocline off Malta, and photographed the disturbances caused by internal waves and turbulence made visible by the dye. The thermocline was found to be layered with internal waves, typically 20 m or more in wavelength, propagating along thin layers of relatively large density gradient (layers made visible by the dye and termed 'sheets' by Woods) between more uniform and weakly turbulent layers in which dye was mixed and more diffuse. This 'fine-structure' of the ocean density, with density gradients varying over vertical scales of 1–50 m, is a common feature of the ocean pycnocline, although the 'layer and sheet' description has

Figure 4.1. Billows in the Mediterranean thermocline: a photograph taken by divers of billows made visible by a sheet of dye within the seasonal thermocline. The dye sheet is viewed obliquely from above. The dye was released into a thin, roughly 0.1 m thick, density interface within the thermocline, a layer where the density increases sharply with depth and along which the dye has spread. In the background an internal wave produces an undulation in the continuous band of the dye. The shear produced at the crests and troughs of the waves results in an instability in the form of bands of billows, of which seven or eight are clearly visible. The billows thicken the dye sheet and are visible in the photograph as bands or as roughly elliptical blobs of dyed fluid. The bands marking the individual billows are normal to the direction of wave propagation and parallel to the vorticity generated by the waves travelling along the interface. (From Woods, 1968.)

not proved to give a useful means of characterizing ocean structure in a quantitative way. Small undulations, about 1 m in length and so much shorter than the internal waves, steepening and rolling up to develop into overturning 'billows', were seen to form on the crests of some of the higher internal waves, as illustrated in Fig. 4.1. The billows generate turbulence and mix water across the sheets, producing a diapycnal flux of density. Laboratory experiments illustrated in Fig. 4.2 have also shown that billows form when waves of sufficiently large amplitude propagate along a stratified layer of thickness very much smaller than the wavelength of the waves. The instability leading to this form of wave breaking and turbulence generation is discussed in the following section.[2]

Woods' observations provide an explanation of the patchy nature of turbulence observed in the thermocline by Grant *et al.* Measurements of the dissipation rates in turbulence produced by the transient and infrequent billows became possible only later, in the 1970s, with the development of the air-foil probes described in Section 2.5.2 and

2 As explained in Section 1.8.1, internal waves do not propagate only along density interfaces. Their upward and downward propagation through the stratified ocean provides a means of energy transport from the seabed and the sea surface, but adds complexity to the problem of predicting the distribution, persistence and nature of mixing caused by their breaking.

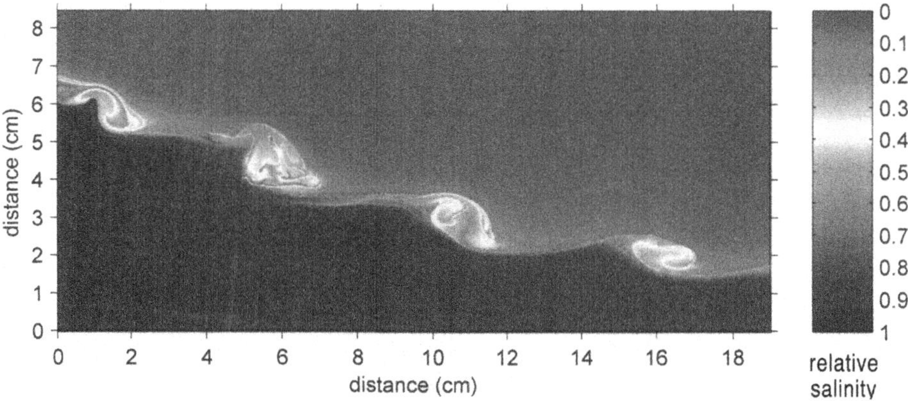

Figure 4.2. Billows produced as a consequence of shear in internal waves on a density interface in a laboratory experiment. The waves are travelling to the right along an interface separating layers of water and denser brine, and the billows have a clockwise rotation consistent with the shear at the wave trough that has moved out of frame to the right. (The shear at the wave crest is in the opposite direction, and billows there are anticlockwise for waves travelling to the right.) The images are produced by adding fluorescent dye to the lower brine layer as it is inserted below the freshwater layer prior to the start of the experiment. Fluorescence is induced by illumination with a thin sheet of light from a laser, and the fluorescence level acts as a surrogate measure of salinity. (From Troy and Koseff, 2005.)

with the design of free-fall instrument packages having very small levels of vibration. These instruments are capable of withstanding huge pressures and, consequently, of obtaining reliable measurements of turbulent fluctuations at depths well beyond those at which divers can work.

4.2 Shear-flow instability and the transition to turbulence

The billows shown in Figs. 4.1 and 4.2 are caused by what is known as Kelvin–Helmholtz instability. The conditions necessary, but not sufficient, for a stably stratified shear flow to be unstable, allowing small disturbances in a finite band of wavenumbers to grow, were discovered by Miles and Howard in 1961.

• Instability of a steady, inviscid, non-diffusive, two-dimensional, parallel horizontal flow (i.e., a steady flow in which ν, κ_T and κ_S are negligible) can occur only when the gradient Richardson number in the flow,

$$Ri = N^2/(\mathrm{d}U/\mathrm{d}z)^2, \tag{4.1}$$

is less than $\frac{1}{4}$ somewhere in the flow, where N is the buoyancy frequency and $\mathrm{d}U/\mathrm{d}z$ (the vertical shear) is the gradient of the horizontal component of current, $U(z)$.

This is known as the Miles–Howard theorem. Flows in which Ri is greater than $\frac{1}{4}$ at all levels, z, are 'stable', in the sense that all small disturbances decay or propagate as

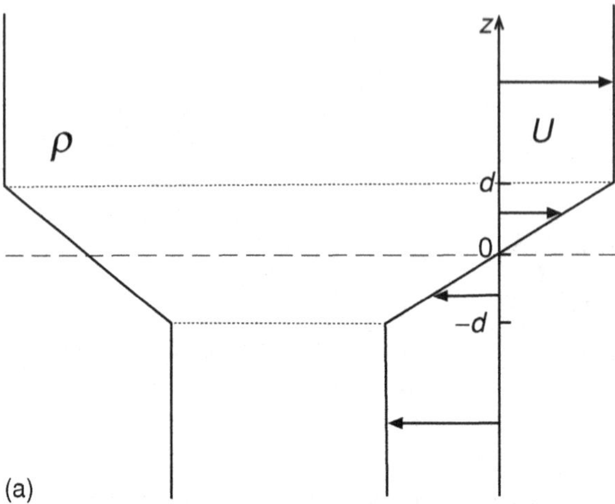

Figure 4.3. Shear-flow stability and instability. (a) The selected profiles of density, ρ, and velocity, U. Two uniform layers are separated by an interfacial region of thickness $2d$, within which the shear and density gradient (or buoyancy frequency) are constant. The stability diagram for this particular flow is shown in (b), the curve dividing the unstable region from that in which the flow is stable. The Richardson number, Ri, of the flow is based on the uniform gradients of density and velocity in the interfacial layer between two uniform layers and, in this case, this is equal to the minimum Richardson number, Ri_{min}. The stability diagram has axes kd, a non-dimensional form of a wavenumber k of disturbances to the flow, and Ri_{min}. In this particular case $Ri_c = \frac{1}{4}$, and, for all values of $Ri_{min} > \frac{1}{4}$, the flow is stable and there are no disturbances of any wavenumbers that grow. For a value of $Ri_{min} < \frac{1}{4}$ there is a range of wavenumbers within which disturbances of the flow will grow and, for such disturbances, the flow is 'unstable'. (c) The growth rate of these unstable disturbances, $q = a^{-1}\,\mathrm{d}a/\mathrm{d}t$, where a is the disturbance amplitude, non-dimensionalized with the shear, $\mathrm{d}U/\mathrm{d}z$, in the interfacial layer, as a function of kd at various values of Ri_{min}. (The shear, $\mathrm{d}U/\mathrm{d}z$, has dimensions T^{-1}.) The growth rate increases as Ri_{min} decreases. Flows with other density and velocity profiles will have different stability diagrams and growth rates. (From Miles and Howard, 1964.)

internal waves without increase in amplitude. The term 'somewhere in the flow' in the theorem gives importance to the size of the smallest Richardson number, Ri_{min}, found as the flow speed and density vary in z. If Ri_{min} exceeds $\frac{1}{4}$, then $Ri > \frac{1}{4}$ everywhere and, according to the theorem, the flow is stable.

For instability to occur, the minimum value, Ri_{min}, must be less than $\frac{1}{4}$, but Ri_{min} $\frac{1}{4}$ does not ensure that instability will occur; the theorem implies a necessary, but not sufficient, condition for instability. For a given flow, the largest value of Ri_{min} for which instability is possible is known as the critical Richardson number, Ri_c, of the flow. The value of Ri_c depends on the shape of the density and velocity profiles, and for some flows is smaller than $\frac{1}{4}$. (One example of a flow with $Ri_c < \frac{1}{4}$ is that of dense water flowing down a slight slope beneath a uniform layer of less dense water.) When, in a given flow, Ri is less than Ri_c at some level z, small waves in a limited band of wavenumbers will grow, usually in the vicinity of z. The stability diagram of Fig. 4.3(b) shows the range of disturbances that are unstable in the particular flow shown in Fig. 4.3(a), and the

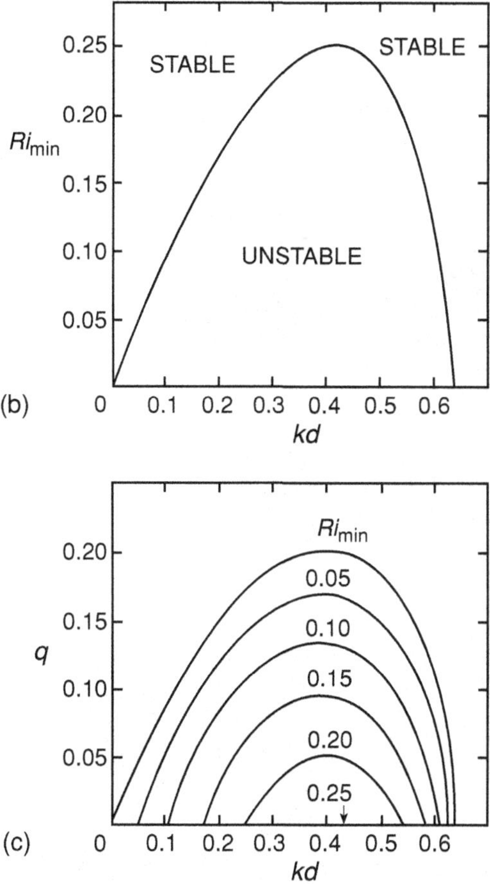

Figure 4.3. (*cont.*)

corresponding growth rates are plotted in Fig. 4.3(c). The occurrence of growing disturbances implies that the flow is unstable: if disturbed (and small disturbances are omnipresent in the ocean), it cannot be sustained in its original state. [**P4.1**]

How the instability develops in a laboratory experiment is shown in Fig. 4.4. Two layers of different densities, the less dense layer above the other, move at the same speed but in opposite directions parallel to the relatively thin interface between them. The Richardson number, Ri, varies with depth and is smallest at the centre of the interface, and for this flow (but, as mentioned above, not all) the critical value, Ri_c, equals $\frac{1}{4}$. No growing disturbances are observed when the minimum Richardson number, $Ri_{\min}$, exceeds $\frac{1}{4}$, and the flow is then stable. When $Ri_{\min}$ is less than $\frac{1}{4}$, however, waves appear spontaneously and grow exponentially with e-folding time scales less than the buoyancy scale, $2\pi/N_{\max}$, where $N_{\max}$ is the maximum buoyancy frequency (that in the centre of the interface). These waves 'roll up' to produce billows (Fig. 4.4(c)), vortices with axes aligned normal to the vertical plane of original shear flow, i.e., in the

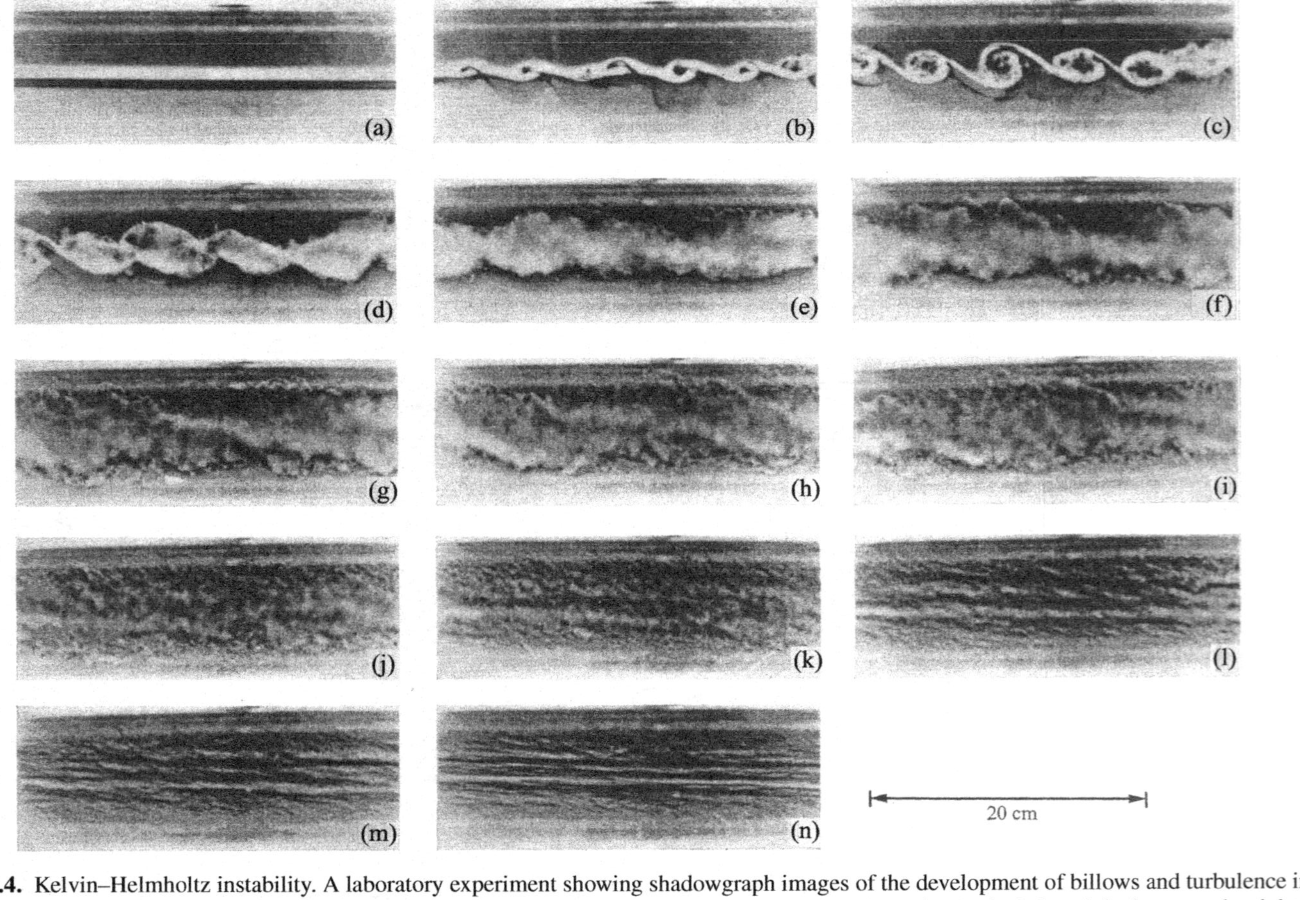

Figure 4.4. Kelvin–Helmholtz instability. A laboratory experiment showing shadowgraph images of the development of billows and turbulence in an initially laminar stratified shear flow between two uniform layers of different densities, the upper (less dense) moving to the left and the lower to the right: (a) the laminar flow; (b) and (c) the growing and amalgamating, and eventually, at (d), turbulent, billows. Turbulent motions collapse, (j)–(m), leaving a striated, near-horizontal, fine-structure within a broadened interface between the upper and lower layers. The wavelength of billows in (b) is close to that of the fastest-growing disturbances predicted by a stability diagram like that of Fig. 4.3(b) but calculated for the error-function profiles of velocity and density profiles of the laboratory experiments. (From Thorpe, 1971.)

direction of the vorticity vector of the shear flow. This is the same relative orientation as have the bands marking billows in Fig. 4.1; they lie 'across the flow'.

The Miles–Howard theorem is valid only within the very limited conditions for which it is stated and in which it can formally be proved. It is strictly applicable only to steady flows, but the conditions (e.g., the value of Ri_c) under which small disturbances may grow in steady shear flows can be applied in flows that are varying relatively slowly with time, for example the flows produced by internal waves.[3] The minimum gradient Richardson number produced in the flow field of internal waves travelling on a relatively thin interface between two layers of different densities occurs near the crests and troughs of the waves. In the temporally varying shear and stratification caused by the waves, the minimum Richardson number must, however, be substantially less than the critical value, $Ri_c = \frac{1}{4}$, of the equivalent steady flow before small disturbances can grow sufficiently to produce overturning billows. **[P4.2]** Like the transition from laminar to turbulent flow investigated by Reynolds (see footnote 3 of Chapter 1), the critical value of the controlling parameter (i.e., Ri_{min} here) may depend on the nature of other disturbances present in the oceanic flow, including residual turbulence from past internal wave-breaking events.

Unlike the rapid transition described by Reynolds, however, the transition from laminar flow to turbulence resulting from Kelvin–Helmholtz instability passes relatively slowly through a set of distinct stages. If Ri_{min} is substantially less than Ri_c, neighbouring billows may rotate around one another and merge, a process known as 'pairing' that transfers energy to larger scales. In general, relatively small-scale internal motions develop within the billows, their early stages being regular and spatially periodic, though rapidly becoming disorganized, three-dimensional and turbulent. The turbulent patches, still retaining the vestiges of periodicity of the billows, become elongated in the flow direction and amalgamate as shown in Fig. 4.4(e), resulting in a layer of turbulence containing small-scale density fluctuations. Turbulence eventually subsides, leaving a layer thicker than the original interfacial layer (Fig. 4.4(n)). The gradient Richardson number in this layer is about 0.32, which is greater than $\frac{1}{4}$, so, according to the Miles–Howard theorem, the flow is now stable with respect to small disturbances.

Although the small-scale structure of the density and flow fields during and following Kelvin–Helmholtz instability has not been measured in laboratory experiments, results from numerical studies shown in Fig. 4.5 indicate that the flow and density field is anisotropic in the early stages of turbulence when motion is dominated by the billows or their remains (Fig. 4.5(b)), but becomes isotropic (i.e., gradients in the flow quantities are independent of direction) during a period of energetic turbulence (Fig. 4.5(c)), before again becoming anisotropic as the turbulence decays and the density field becomes dominated by fine-scale layers (Fig. 4.5(d)). The decay of turbulence is approximately exponential in time, t, with $\varepsilon \propto \exp[-t/(q\tau)]$, with $q = 1.0 \pm 0.1$, where τ is equal to the buoyancy period, $2\pi/N$, and N is the buoyancy frequency in the turbulent layer.

3 'Relatively slowly' means that the time taken for disturbances to grow is very much less than the time over which the mean flow changes (e.g., the wave period in internal wave flows).

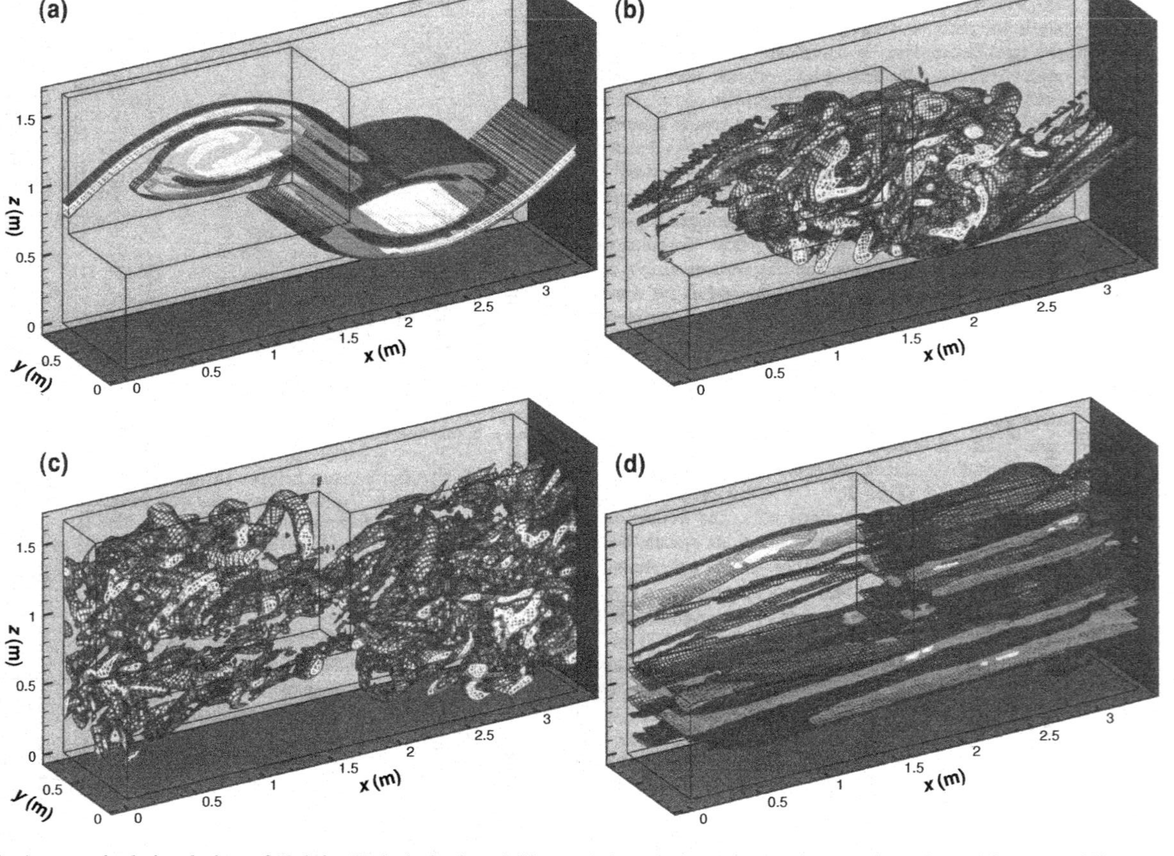

Figure 4.5. A numerical simulation of Kelvin–Helmholtz instability and the subsequent turbulent motion viewed from an oblique angle. The initial stage, (a), shows two billows that are beginning to pair together. (The billow on the left appears twisted. Only its two-dimensional shape is shown, projected onto the back 'wall' of the computed region, rather than – as for the second billow – an oblique view of its three-dimensional structure.) Turbulent motion of relatively small scale sets in at stage (b), spreading vertically, (c), and eventually collapsing at (d) to leave a thickened interface with a laminated structure of fine horizontal layers. (Compare this with Fig. 4.4 at stages (c), (d), (f) and (n), respectively. The parameter shown in the images is half the mean square vorticity of the flow. (From Smyth and Moum, 2000. Reprinted with kind permission of the American Institute of Physics.)

A field of small-scale variations in temperature, salinity or density remaining after the decay of turbulent motion is termed 'fossil turbulence'. Such anisotropic remnants of turbulence may persist long after the turbulence has decayed, the variations in the properties (e.g., temperature) being gradually removed by the action of molecular diffusion and, where the residual field contains regions of statically unstable fluid, by weak convection controlled by viscosity and molecular diffusion.

The energy dissipated by turbulence in the Kelvin–Helmholtz transition from laminar to turbulent flow and back to laminar flow can be quantified by subtracting the potential energy gained by increasing the layer thickness from the kinetic energy lost by the mean flow. The ratio of the potential energy gained to the kinetic energy lost, a measure of the 'efficiency' of the mixing in terms of producing diapycnal mixing, depends on the minimum Richardson number at which instability starts, but is typically in the range 0.1–0.25. Most of the kinetic energy lost by the mean flow is transferred to and dissipated in turbulence. Although, if the surrounding water is stratified, a fraction of the turbulent energy may be radiated as internal waves, this is generally very small and most of the turbulent energy is dissipated through viscosity into heat, and is therefore transferred from the motion and density field into a molecular form. As already seen (e.g., in P2.4), the heating (and therefore changes in density and potential energy produced by the viscosity heating) in the stratified ocean is minuscule and is not available to initiate any significant further motion or mixing.

• Much of the kinetic energy lost in Kelvin–Helmholtz instability is effectively removed from the ocean rather than being used in increasing the potential energy. Whilst the relatively small fraction transferred to potential energy may initiate motions and so feed back kinetic energy, for example though the collapse and spread of a mixed patch as in Fig. 1.16, the energy transferred by turbulence to heat has no significant role in the subsequent dynamics. [**P4.3**]

The time from the onset of turbulence in the Kelvin–Helmholtz billows to its eventual decay in the laboratory experiments, τ_{KH}, is given empirically by

$$\tau_{KH} \approx 15U/(g\Delta\rho/\rho_0), \tag{4.2}$$

where the flow speed varies from $-U$ below the original interface to U above and the density increases from $\rho_0 - \Delta\rho$ above the interface to $\rho_0 + \Delta\rho$ below. [**P4.4**]

4.3 The Richardson number in the ocean

Turbulence in the stratified ocean is 'patchy'. Internal waves produce transient shear, locally reducing the Richardson number and occasionally leading to instability or wave breaking (as in Fig. 4.2) that creates local patches of turbulence that subsequently decays or collapses. Because the instability of a stably stratified shear flow is dependent on the Richardson number, the local value of *Ri* in the ocean may be a useful guide to the factors leading to turbulence and can provide a means to quantify mixing. We therefore turn next to the measurement of *Ri*.

The vertical gradient of the horizontal current, du/dz, can be measured using a free-fall instrument equipped with an air-foil probe as explained in Section 2.5.2, and density may also be determined over small vertical scales, so near-instantaneous values of Ri can, in principle, be found. The resulting values are highly variable and, although they provide a statistical measure of the field of motion, measurements from vertically falling instruments are generally not helpful for describing the temporal or spatial conditions leading to instability in local domains (or depth ranges); billows grow over finite times at depths where Ri is subcritical, $<Ri_c$. Instead, averaged values of Ri at a selected depth have commonly been used to investigate the prevalence of conditions necessary for the onset of Kelvin–Helmholtz instability.[4] Two quantities are required: the mean shear, $d\langle U\rangle/dz$, and the mean buoyancy frequency, $\langle N\rangle$, given by $[(g/\rho_0)d\langle\rho\rangle/dz)]^{1/2}$, both representing average values over some short period of time, $\ll 2\pi/N$, at some location.

A scatter plot based on measurements of the square of the velocity difference and the temperature difference made in the seasonal thermocline using a pair of vector-averaging current meters (VACMs), one 3 m above the other, is shown in Fig. 4.6. The horizontal axis is the square of the velocity difference and the vertical is proportional to the density difference, since density was dominated by temperature. The ratio of the plotted quantities is equal to the Richardson number and curves of constant Ri are therefore straight lines. There appears to be a cut-off in the observed values on a line $Ri \approx \frac{1}{8}$, and most of the points lie above the line $Ri = \frac{1}{4}$ where the flow should be stable to small disturbances. The relatively few values of Ri less than $\frac{1}{4}$ may be for transitional conditions of developing instability, returning Ri to a value $>\frac{1}{4}$ for which the flow is stable. The observations suggest that the main cause of small-scale shear, the oceanic internal wave field, may be close to a 'saturation limit' in which the addition of more wave energy will lead to a greater loss of energy through Kelvin–Helmholtz instability to turbulence, i.e., additional wave shear and strain (defined in (4.15)) will lead to more frequent subcritical values of Ri and consequently to more wave breaking.[5] It is evident, however, that small values of Ri are rare – at least in estimates made

4 Average values are often used simply for convenience or because they are the only values than can be measured with a high degree of accuracy with available equipment. There is a further consideration: whether values of Ri provide much information about the stability of an oceanic flow that is, to some degree, already turbulent. Except perhaps to quantify better the turbulent flow itself, there appears to be little merit in estimating Ri at vertical scales less than those of existing turbulent eddies (e.g., scales less than the Ozmidov scale defined in Section 4.4.1), within which flow is unsteady and non-horizontal and therefore does not satisfy the conditions under which the Miles–Howard theorem applies, and where there will be some negative values of Ri where mixing (temporarily) produces static instability. Although laboratory experiments have not been carried out to study the growth of billows in a shear flow that is already turbulent (e.g., a layer with a mean shear that is, when density is horizontally averaged, statically stable, but in which there is turbulent motion at some scale small compared with the thickness of a layer), it is likely that the growth of billows in such a flow will depend mainly on the Richardson number of the mean flow and only to a small degree on the turbulence, provided, for example, that the time scale over which turbulence removes energy from the mean flow is much larger than that of billow growth in an equivalent non-turbulent flow.

5 The numerical calculation of Fringer and Street (2003) referred to in P4.2 shows that billows will occur in waves travelling on a density interface only when $Ri_{min} < \frac{1}{8}$. This is in accord with the cut-off in the observations, but more study is required in order to establish the generality of the result for internal waves in other, more realistic, density profiles.

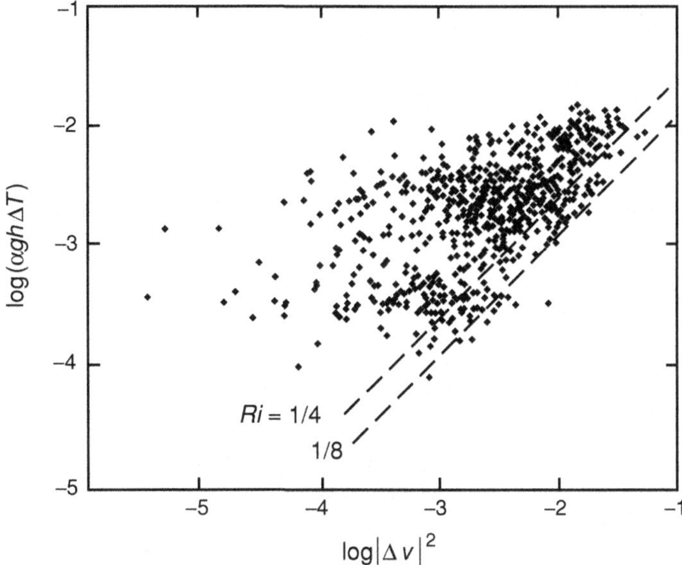

Figure 4.6. Shear, temperature gradient and Richardson number. A scatter plot of values of (horizontal axis) $\log|\Delta v|^2$ and the corresponding values of (vertical axis) $\log(\alpha g h\, \Delta T)$, where both variables are measured in units of $\mathrm{m}^2\ \mathrm{s}^{-2}$. The symbol α denotes the thermal expansion coefficient, and Δv and ΔT are the differences in velocity and temperature, respectively, measured by pairs of vector-averaging current meters located in the seasonal thermocline in the North Pacific Ocean and separated vertically by $h = 3$m. Values are averaged over 112.5 s. The lines show values where the Richardson number is equal to $\frac{1}{8}$ and $\frac{1}{4}$. For most of the measured values, $Ri > \frac{1}{4}$ and the flow is therefore stable with respect to small disturbances. (From Davis *et al.*, 1981, following an earlier study of the scatter of shear squared and N^2 by Eriksen, 1978, off Bermuda at depths of 900 m.)

over 3 m – so that, on average, wave breaking owing to Kelvin–Helmholtz instability is infrequent. **[P4.5]**

The variation of *Ri* represented by the scatter of points in Fig. 4.6 is a reminder of the variable nature of the ocean, and of the fact that, although *Ri* may help to determine the onset of instability in the laminar flow conditions of a laboratory experiment, in the turbulent and wave-affected ocean it is a variable quantity with a statistical distribution. A flow becoming unstable may already contain small-scale turbulent motions, perhaps as a residue of earlier instability (see also footnote 4). A more direct connection between *Ri* and turbulence is needed.

An empirical relationship between the local mean values of *Ri* and the rate of dissipation of turbulent kinetic energy, ε, derived from free-fall instruments has been obtained at depths between 400 and 800 m in the North Atlantic.[6] Averaged values of ε are contoured in N^2–S^2 space in Fig. 4.7, with N^2 and $S\,(=\mathrm{d}U/\mathrm{d}z)$ estimated over vertical scales of 4 m. Contours of ε lie roughly along lines of constant *Ri*, and $\langle\varepsilon\rangle$ decreases as *Ri* increases: $\langle\varepsilon\rangle \sim 5 \times 10^{-10}\ \mathrm{W\ kg}^{-1}$ when $Ri = \frac{1}{2}$ and

6 The data were obtained as part of the NATRE experiment described in Section 4.7.

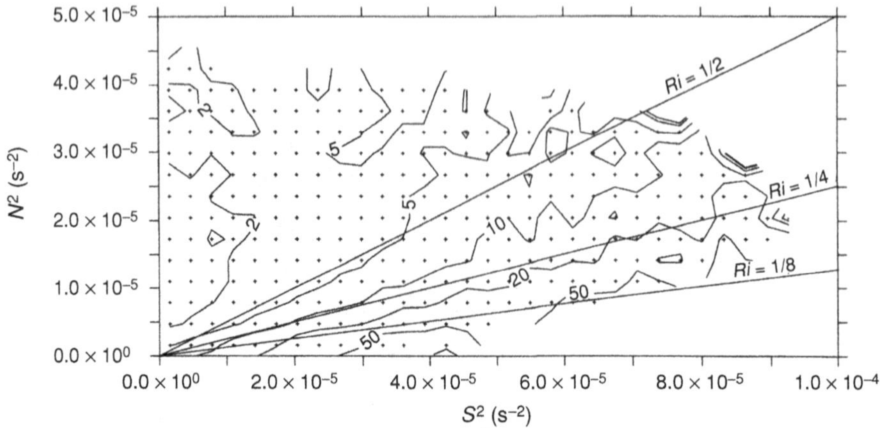

Figure 4.7. Shear, buoyancy frequency and rate of dissipation. Contours of mean dissipation rate ($10^{10}\varepsilon$, with ε in units of W kg^{-1}) in terms of the square of the vertical gradient of the horizontal current, $S = \mathrm{d}U/\mathrm{d}z$, and the square of the buoyancy frequency, N. Values are estimated over vertical distances of 4 m at depths of 400–800 m in the eastern North Atlantic. The straight lines indicate values of Richardson number, Ri, equal to $\frac{1}{8}$, $\frac{1}{4}$ and $\frac{1}{2}$. (From Polzin, 1996.)

$\langle\varepsilon\rangle \sim 50 \times 10^{-10}$ W kg^{-1} when $Ri = \frac{1}{8}$. (It will be noticed that these values are minute in comparison with those measured at shallower depths by Grant, Moilliet and Vogel.) There is a reduction in $\langle\varepsilon\rangle$ as Ri increases but no 'cut-off' as Ri becomes greater than $\frac{1}{4}$ (i.e., where the mean flow is stable with respect to shear instability) or where $Ri = 0.32$, the value to which, in the laboratory experiments, turbulence collapses. The relationship

$$\langle\varepsilon\rangle = f_{\mathrm{r}}(\Delta z)^2\langle N^3(Ri^{-1} - Ri_{\mathrm{c}}^{-1})(Ri^{-1/2} - Ri_{\mathrm{col}}^{-1/2})\rangle/96 \tag{4.3}$$

provides a parametric representation and estimate of $\langle\varepsilon\rangle$ correct to within a factor of about 2 (i.e., with an uncertainty of about 50%). Here f_{r} is the fraction of the record within which Ri is less than its critical value (usually taken as 0.25), Δz is the vertical scale (here 4 m) over which S and N^2 are estimated, Ri_{col} is the Richardson number, about 0.32, to which turbulence will collapse, and $\langle\ldots\rangle$ implies that average values are to be taken over the period or horizontal distance within which ε is estimated.

Similar statistical relationships relating $\langle\varepsilon\rangle$ to the mean shear generated by internal waves have been found. These indicate that internal waves are a major contributor to turbulent mixing in the stratified ocean. The pattern of variation of $\langle\varepsilon\rangle$ with N^2 and S^2 found in deep water and shown in Fig. 4.7 is not replicated in the relatively shallow shelf seas, suggesting the presence of different processes of turbulence generation (see Further study). The absence of a cut-off in $\langle\varepsilon\rangle$ is, however, a reminder of the statistical nature of the observations, the failure to resolve shears and density gradients at the relevant scales (footnote 4), the non-applicability of the Miles–Howard theorem in unsteady flows, and the fact that turbulence, once generated, can be self-sustaining provided that the turbulence Reynolds stress leads to a transfer of energy from the mean flow to turbulence through the shear-stress term in (2.16) at a rate exceeding that of its loss.

Whilst the statistical relationships indicate that *Ri* is related to turbulence, they do not provide information about the physical structure of the ocean, in particular about the size and presence of billows and overturning eddies or the associated production of regions of static instability, which are the characteristic features of the consequences of Kelvin–Helmholtz instability. The presence and vertical extent of regions of static instability provide another means of quantifying turbulence motion, as described in the following section.

4.4 Further turbulence parameters derived from microstructure measurements

4.4.1 Estimation of ε

• The size of eddies provides a way of estimating ε, using the Ozmidov length scale, L_{O}. This is defined as

$$L_{\mathrm{O}} = \langle\varepsilon\rangle^{1/2} N^{-3/2}, \tag{4.4}$$

and is proportional to the vertical size of the largest eddies which can overturn in stably stratified but turbulent water.

(The relationship expressed in 4.4 can be derived on dimensional grounds by supposing that the scale of the inertial motion is large and that consequently viscosity is negligible. The only dimensional quantities characterizing the turbulent eddies in stratified water are then ε and N, the latter providing a measure of the mean density structure within which the eddies overturn. The shear, and therefore *Ri*, is neglected as a contributor on the basis of an assumption made by Ozmidov in 1965 that the Richardson number adopts a universal critical value of order unity in turbulent flow. More generally, however, the expression on the rhs of (4.4) might be multiplied by a function of the dimensionless *Ri*.) Although data exhibit considerable scatter, there is evidence that the scale does provide a measure of the eddies that produce regions of static instability; the Ozmidov scale is found to be proportional to an easily measured quantity, the root mean square displacement, L_{T} (sometimes known as the Thorpe length scale), defined in Section 2.4.2. A comparison of the two length scales is shown in Fig. 4.8.

• The proportionality of L_{O} and L_{T} leads to a relation often used to estimate the mean rate of dissipation:

$$\langle\varepsilon\rangle = c_1 L_{\mathrm{T}}^2 N^3, \tag{4.5}$$

where c_1 is a constant equal to $(L_{\mathrm{O}}/L_{\mathrm{T}})^2$, with mean values found empirically to be in the range 0.64–0.91. The scatter in L_{O} versus L_{T} data illustrated in Fig. 4.8 adds additional uncertainty to the estimates of ε obtained using (4.5). Approximate values

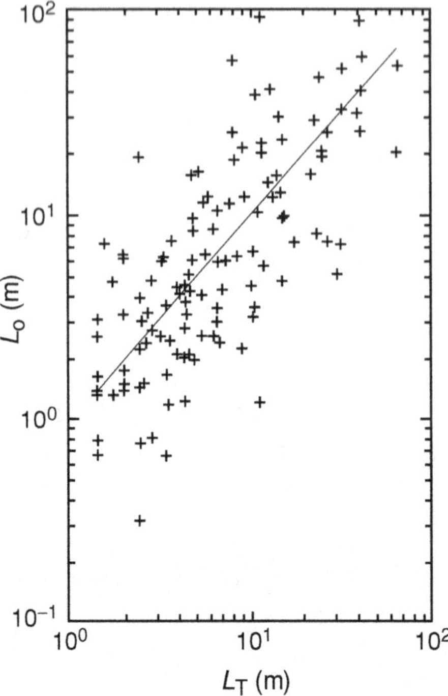

Figure 4.8. Comparison of the Ozmidov length scale, L_O, with the Thorpe scale, L_T. Data in this example come from observations in deep water in the Romanche Fracture Zone. Values of ε and N (and thence L_O) were found using a high-resolution profiler (HRP; shown in Fig. 2.13(b)) and L_T estimated from CTD data. The fitted line is $L_O = 0.95L_T$. (From Ferron *et al.*, 1998.)

of the mean dissipation rate, $\langle\varepsilon\rangle$, can, however, be estimated from vertical profiles of density, without the demand for the resolution of the micro-scale measurements of velocity shear required in using (2.9) or (2.11). (It is often assumed that temperature, which is easier to measure than density, can be used as a surrogate to infer density; the vertical scale of density overturns is commonly determined from those observed in temperature profiles, justifiably so for temperature overturns in freshwater lakes with temperatures in excess of 4 °C, where density is maximum. The vertical scale of density overturns can also be determined from those of temperature in the ocean, where salinity as well as temperature affects density, provided that the relation of temperature to density is monotonic.) Studies have shown that the root mean square value of displacements, L_T, is dominated by the larger, more accurately determined displacements, η, found in a reordered density profile as in Section 2.4.2. The value, N, in (4.5) is derived from the mean of the reordered density profiles, which is notionally the mean profile disturbed by overturning turbulent eddies. The mean dissipation rate, $\langle\varepsilon\rangle$, may consequently be estimated approximately using (4.5) from conventional CDT profiles lowered at about 1 m s^{-1} from a ship and recording at rates usually exceeding 30 Hz, provided that the heaving of the ship in rough weather does not produce vertical motions of the slowly lowered CTD that carry it back into water it has previously disturbed.

The Ozmidov scale characterizing the largest eddies that can overturn also represents an upper bound of the vertical scale at which the motion and density in turbulent eddies may be isotropic. At larger scales, eddies (if they exist) will be flattened, with greater horizontal than vertical scale. The ratio of the Ozmidov scale to the Kolmogorov scale (at which viscous dissipation becomes significant) is $L_O/l_K = [\langle\varepsilon\rangle/(\nu N^2)]^{3/4}$ or $I^{3/4}$, where I is the isotropy parameter defined in (2.14). If this is very large, which is a requirement identified in Section 2.3.5 as a necessary condition for isotropy at some small eddy scale, there is an appreciable range of scales between those of the largest eddies that can overturn and those of smaller eddies within which viscosity becomes significant. The greater the separation of the two scales, the more likely it is that an inertial subrange exists and that, in the inertial cascade of energy by eddy interactions from large to small scales, any anisotropy that may exist at the larger scales will be lost, resulting in isotropy at sufficiently small scales.

It should be recalled, however, that the Ozmidov scale, like L_T, is a statistical 'ensemble' mean quantity. It does not provide a measure of the vertical dimension of the single largest eddy to overturn in a given period of time or region of space. Isotropy is also a statistical description of the mean state of motion, and $I > 200$ will not ensure the directional uniformity of any particular structure within the turbulent flow at any particular moment.

4.4.2 Estimation of eddy diffusion coefficients

Two relationships discovered in the 1970s have proved valuable in quantifying the effects of ocean turbulence from measurements of ocean microstructure using free-fall instruments. They are important in relating turbulence to temperature microstructure and mixing.

The first is an equation for the eddy diffusion coefficient of heat:

• $$K_T = \kappa_T C, \tag{4.6}$$

where C is known as the Cox number,

$$C = \langle(\partial T'/\partial x)^2 + (\partial T'/\partial y)^2 + (\partial T'/\partial z)^2\rangle/(\mathrm{d}\langle T\rangle/\mathrm{d}z)^2, \tag{4.7}$$

the ratio of the variance of the gradient of temperature fluctuations, T', averaged over all three coordinate directions, x, y and z, to the square of the mean vertical temperature gradient, $\mathrm{d}\langle T\rangle/\mathrm{d}z$.[7] **[P4.6]** If, as often assumed, the turbulent temperature field is isotropic at scales that contribute most substantially to the terms in (4.7), it follows that

$$C = 3\langle(\partial T'/\partial z)^2\rangle/(\mathrm{d}\langle T\rangle/\mathrm{d}z)^2. \tag{4.8}$$

Whilst in principle (4.6) and (4.8) allow K_T to be estimated using free-fall instruments that measure the vertical variations of ocean temperature, in practice it places

7 Equation (4.6), discovered by Osborn and Cox (1972), is derived from an equation for the molecular transfer of heat in a turbulent flow.

severe instrumental demands on the resolution of very small variations in temperature gradient that dominate the average.[8]

The second relation[9] relates the eddy diffusion coefficient for density (or mass) to the rate of dissipation of turbulent kinetic energy per unit mass:

• $$K_\rho = \Gamma\varepsilon/N^2, \tag{4.9}$$

where Γ is an 'efficiency factor' and N is the mean buoyancy frequency.

This widely used relation is derived from the assumption that the three terms on the rhs of the turbulent energy equation (2.16) sum to zero. This supposes that, in the local region of study, the ocean is in a steady state (which is sometimes an unjustified assumption) and that other terms contributing to the change in turbulent kinetic energy are negligible. The balance of terms gives

$$\langle uw\rangle \mathrm{d}U/\mathrm{d}z + \varepsilon + g\langle w\rho'\rangle/\rho_0 = 0. \tag{4.10}$$

The 'flux Richardson number', R_f, is a non-dimensional value that is quite distinct from the gradient Richardson number, Ri. It relates to energy transfers in a turbulent flow rather than to conditions in which a flow may become turbulent, and is defined as the ratio of the rate of removal of kinetic energy by buoyancy forces to the production of turbulent kinetic energy by the shear,

$$R_\mathrm{f} = g(\langle w\rho'\rangle\rho_0)/(-\langle uw\rangle \mathrm{d}U/\mathrm{d}z). \tag{4.11}$$

From its definition in Section 2.2.2, the eddy diffusion coefficient of density is given by $K_\rho = -\langle w\rho'\rangle/(\mathrm{d}\langle\rho\rangle/\mathrm{d}z)$. Using the definition of the buoyancy frequency, $K_\rho = g\langle w\rho'\rangle/(\rho_0 N^2)$, and from (4.10) and (4.11), the eddy diffusivity K_ρ can be written in terms of R_f as

$$K_\rho = [R_\mathrm{f}/(1 - R_\mathrm{f})]\varepsilon/N^2, \tag{4.12}$$

or as (4.9) if

$$\Gamma = R_\mathrm{f}/(1 - R_\mathrm{f}). \tag{4.13}$$

(The efficiency factor, Γ, is then equal to the rate of increase of potential energy in a turbulent flow divided by its rate of loss of kinetic energy.)

• Estimates of $R_\mathrm{f} \approx 0.17$, obtained from laboratory experiments and from measurements in the ocean (e.g., see Section 4.4.3), give $\Gamma \approx 0.2$, and this is the commonly adopted value. The precise value of Γ is, however, subject to continuing debate, and values exceeding 0.2 are found in conditions where double diffusive convection is possible, as explained in Section 4.8. [**P4.7**]

The eddy diffusion coefficients, K_ρ and K_T, are equal when density is determined by temperature alone or when the contributions of salinity to density are negligible, and

8 As for ε in Section 2.5.2, semi-empirical methods of interpolation of spectra have been devised to determine the contribution of small unresolved scales.

9 Equation (4.9) was first proposed by Osborn (1980).

(4.9) then allows estimates of K_T to be obtained from microstructure measurements of ε and ρ (and hence N) made with a free-fall instrument. Measured values are commonly of order 10^{-5} m^2 s^{-1} in the ocean pycnocline except within a few hundred metres of rough topography, e.g., the continental slopes, mid-ocean ridges and seamounts, where higher values are found (Section 4.7).

An example of the observed variations in ε and in estimates of K_ρ obtained using (4.9) is shown in Fig. 4.9.

4.4.3 R_f and the ratio of the eddy coefficients of mass and momentum

Using the definitions $K_\nu = -\langle uw \rangle/(\mathrm{d}U/\mathrm{d}z)$ given by (2.2) and $K_\rho = -\langle w\rho' \rangle/(\mathrm{d}\langle\rho\rangle/\mathrm{d}z)$ as in the preceding section, with the equation (4.11) for R_f, the ratio of the eddy coefficients of mass and momentum (or eddy viscosity) is found to be related to the flux and gradient Richardson numbers by

$$K_\rho/K_\nu = R_\mathrm{f}/Ri. \tag{4.14}$$

In 1931 Taylor examined measurements of velocity and density made by J. P. Jacobsen between 1913 and 1919 at pairs of stations in the Schultz Grund and Randers Fjord of the tidal Kattegat, and derived the ratio K_ρ/K_ν and (using 4.14) the flux Richardson number, R_f. (Estimates of the gradient Richardson number, Ri, were obtained from measurements separated vertically by about 2.5 m in the Schultz Grund and 1 m in the Randers Fjord.) The mean ratio, K_ρ/K_ν, is in the range 0.05–0.03 for Ri in the range 4–10,[10] and appears to decrease with increasing Ri. The flux Richardson number, R_f, estimated from (4.14) is significantly less than unity and the rate of removal of kinetic energy by buoyancy forces is therefore substantially less than the production of turbulent kinetic energy by the shear, the difference being the contribution of turbulent kinetic energy to ε. Although the data are scattered, Taylor was able to draw three additional conclusions important in the development of the understanding of turbulent motion in stratified flows: (i) that K_ν is more than 100 times greater than the kinematic viscosity, ν, so that friction (or stress) in the water column must be achieved by turbulent motion, not molecular viscosity alone; (ii) that turbulence is found (but is now known generally to be patchy) even when the mean value of Ri exceeds unity, and (iii) that the data in stratified waters are not consistent with an earlier assumption by Reynolds and Prandtl that $K_\rho = K_\nu$ in turbulent flow.

Subsequent laboratory studies confirmed that the ratio K_ρ/K_ν decreases with increasing stability or Ri, and that R_f is substantially less than unity and, although typically about 0.17 (as mentioned in Section 4.2.2), may depend on Ri.

10 Values quoted by Ellison and Turner (1959).

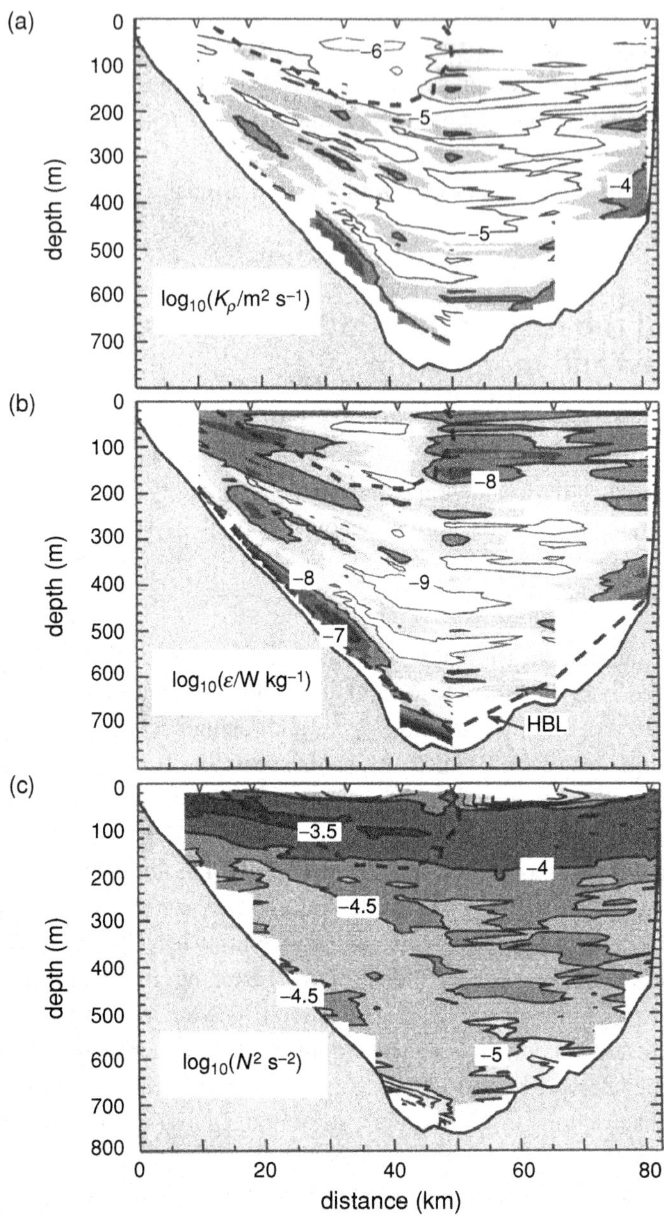

Figure 4.9. Diffusion, dissipation and buoyancy frequency in a section across the Straits of Florida. The variations of $\log_{10}$ values of (a) K_ρ, (b) ε and (c) N^2 are shown. Measured values of N^2 based on density differences measured over 10 m in the vertical, and of dissipation, ε, averaged over 10 m and measured by the MSP free-fall probe, are used to determine K_ρ using (4.9) with $\Gamma = 0.2$. The dissipation, ε, varies over two orders of magnitude, and K_ρ ranges from 10^{-6} to 10^{-4} m^2 s^{-1}, with greatest values close to the sloping sides of the Strait. (From Winkel *et al.*, 2002.)

4.5 Entrainment into the surface mixed layer

The near-surface mixed layer deepens through the action of processes that are not fully understood.

The assumption that entrainment is proportional to the velocity difference at the edge of a vertically rising plume mentioned in Section 3.2.2 may be extended to conditions in which entrainment into the mixed layer is impeded by buoyancy forces at its base. Work must be done in transporting the denser pycnocline water upwards into the less dense mixed layer. A further parameter, a bulk Richardson number, $Ri_B = g\,\Delta\rho\,D/[\rho(\Delta u)^2]$, a non-dimensional measure of the dynamical effect of the mixed layer flow, is available to determine the entrainment rate, where $\Delta\rho$ and Δu are, respectively, the density and velocity changes across the lower boundary of the mixed layer of thickness D. (Although the length scale, D, does not characterize the pycnocline thickness, it does limit the vertical dimensions of turbulent eddies in the mixed layer and is therefore relevant to the mixing at its lower boundary.)

Numerous laboratory experiments have been performed to simulate the oceanic conditions and to determine the non-dimensionalized rate of entrainment of water, $w_c/\Delta u$, across a density interface separating turbulent and laminar layers as a function of the bulk Richardson number. These include experiments in which turbulence is generated by a grid of bars oscillating rapidly and mixing a layer within a relatively deep stratified fluid. The mixed layer is bounded by interfaces across which the density changes and beyond which it is uniform or of uniform gradient. (The characteristic velocity scale, Δu, may be taken as the rms velocity in the turbulence near the interface.) In other experiments a homogeneous turbulent layer flows over (or under) a static denser layer, generating a shear flow at the interface. The non-dimensionalized entrainment rate, $w_c/\Delta u$, generally decreases as Ri_B increases (from a value of about 0.1 in the absence of stratification when $Ri_B = 0$), and there appears to be general agreement that $w_c/\Delta u \propto Ri_B^{-3/2}$, at least in some range of $Ri_B > 1$.

Laboratory experiments with shear flows find that both Kelvin–Helmholtz instability and Holmboe[11] instability are involved in the processes of entrainment, but are dominant in different ranges of Ri_B. Precisely how the conclusions of laboratory experiments relate to the oceanic mixed layer where turbulence is generated by shear, convection, breaking waves and Langmuir circulation has yet to be resolved.

4.6 Observations of mid-water mixing processes

In recent years there have been several observations of turbulence associated with internal waves. Figure 1.13, for example, shows an internal wave detected using a high-frequency acoustic sonar. The acoustic scattering is from particles or organisms in the water column and from temperature microstructure resulting from shear-induced

11 See Further study and Fig. 4.20.

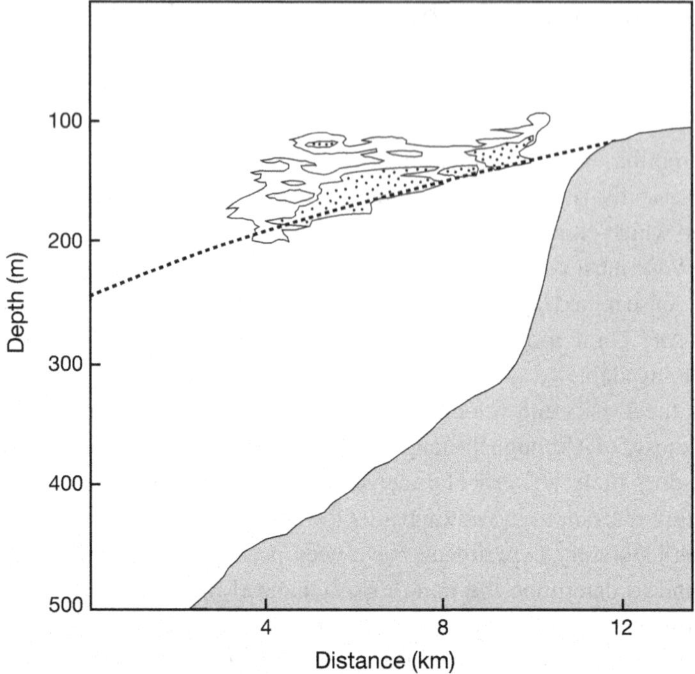

Figure 4.10. Dissipation in an internal-wave beam. Relatively high values of dissipation, ε, are found along the path of a beam of internal tidal energy originating from a region near the shelf break in Monterey Bay, California. The horizontal-to-vertical scales are distorted. The dashed line shows the calculated path of an M_2 tidal ray generated at the shelf break. Two contours of ε, at about 2×10^{-7} W kg^{-1} and 1×10^{-6} W kg^{-1}, are shown. Values greater than 1×10^{-6} W kg^{-1} are stippled. (From Lien and Gregg, 2001.)

turbulence and mixing caused by the wave. The wave is asymmetrical, with a wave trough that is more pronounced than the wave crest, the result of the pycnocline on which the wave travels being much closer to the sea surface than to the sea floor. As a consequence, it is at its trough that the wave induces the greatest shear, initiating instability and mixing, and leaving the trail of acoustically detected microstructure.

Three further examples of turbulence apparently caused by internal waves are shown in Figs. 4.10–4.12. The first of these shows a band of turbulence reaching some 6 km into deep water from the continental shelf break (the edge of the shelf or the top of the slope). Its inclination is consistent with the downward propagation of an internal wave beam of M_2 tidal frequency, probably generated near the shelf break by motions caused by the surface tides. Whether the internal tide is itself breaking or, perhaps by producing an increased level of shear, is causing waves in the background internal wave field to break is not revealed by the observations.

Figure 4.11 shows variations in shear, strain (see (4.15) below), ε and K_ρ resulting from near-inertial internal waves (perhaps generated by an earlier passing storm) radiating downwards from the surface. The regions of high and persistent dissipation, ε,

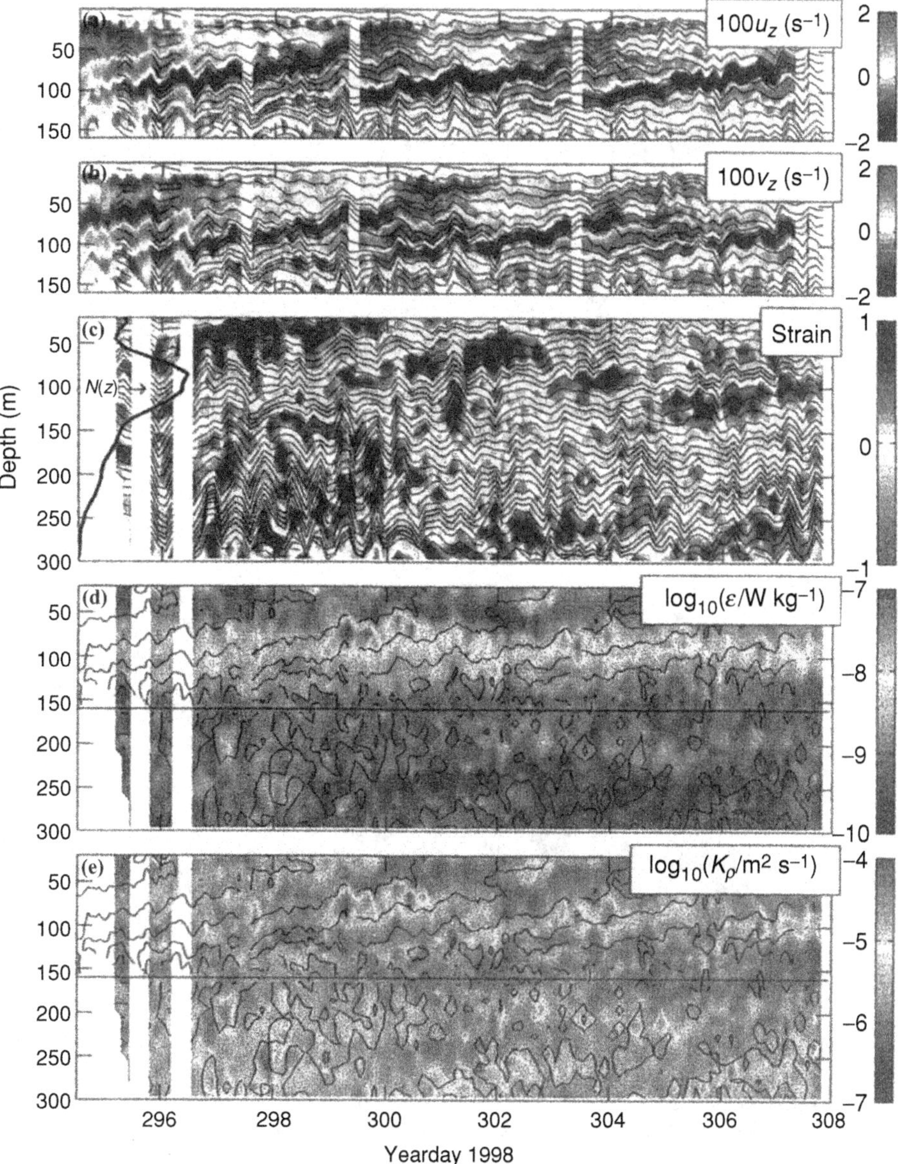

Figure 4.11. Observations of mixing in the Banda Sea made over a period of 14 days. From top to bottom: the vertical derivatives of the east (u) and north (v) components of current, the strain given by (4.15), $\log_{10} \varepsilon$ and $\log_{10} K_\rho$, the latter determined from (4.9) with $\Gamma = 0.2$. The vertical derivatives of the velocity components (shear) show bands of high vertical gradients that become shallower with increasing time. Their period (the time between their successive arrivals at a given depth) shows that they are caused by waves of inertial frequency. Although the phase propagates upwards, the energy of the inertial waves propagates downwards in accordance with Fig. 1.14. The mean profile of N is shown to the left of the strain. High values of ε and K_ρ are visible as lighter tones in the region of large density gradient (large N) or pycnocline near 100 m depth as the rising layers of high shear pass through it. (From Alford and Gregg, 2001.)

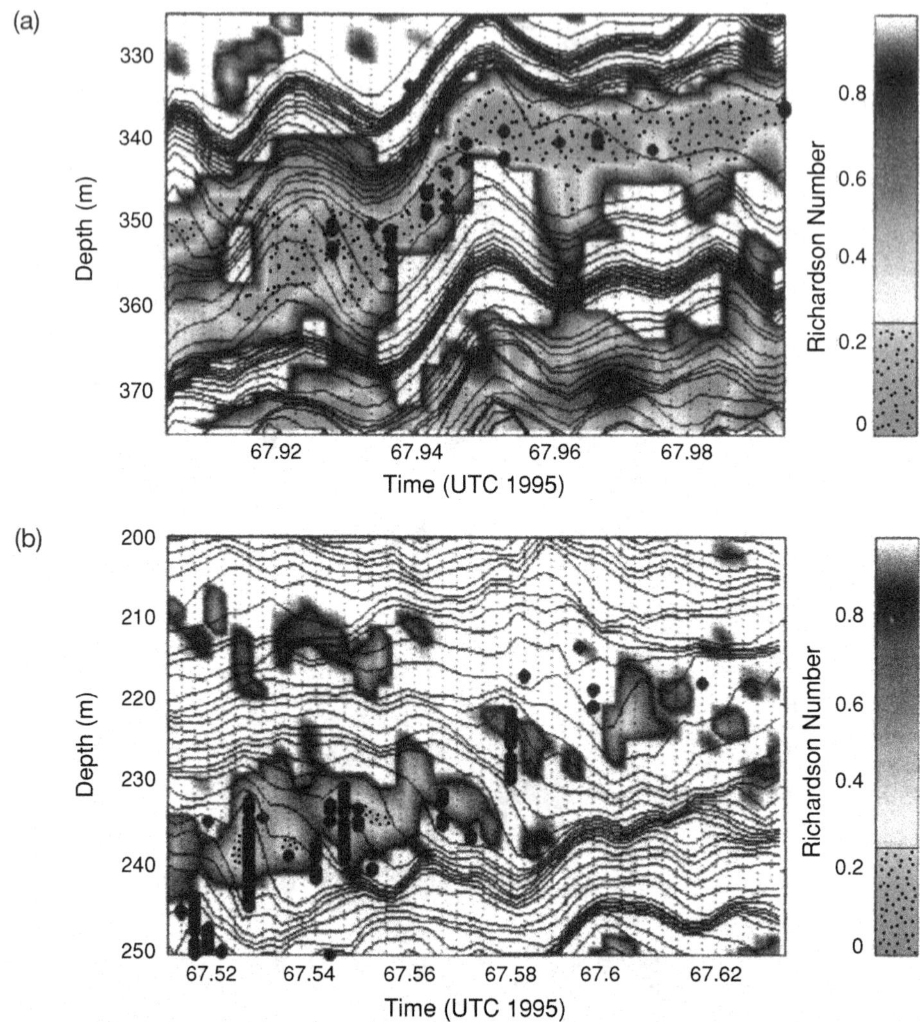

Figure 4.12. The location of density overturns within isopycnals. In the two examples shown, the thin lines are isopycnal surfaces, the large dots are regions where the displacement scale exceeds 2 m and the stippled regions are where *Ri* (measured over 6.4 m) is less than 0.25. (From Alford and Pinkel, 2000.)

appear to be related to the upward propagating phase of the waves but, as in Fig. 4.10, it is not clear whether they lead directly to turbulence by increasing the shear and strain, or only indirectly by affecting the ambient waves.

Figure 4.12 shows a time series of observations made from the Floating Instrument Platform, FLIP, using an acoustic Doppler sonar and a lowered CTD with a vertical resolution of 0.15 m. The lines shown are isopycnal surfaces, their vertical changes marking the presence of internal waves propagating or being advected past FLIP by a relative flow. The stippled regions are where *Ri* (estimated over a vertical scale of 6.4 m) is less than 0.25. The large dots are locations where the displacement scale, L_T, exceeds 2 m, indicating regions of unstable stratification and revealing the very patchy nature of statically unstable and potentially active zones of mixing in the pycnocline. Patches of overturns sometimes follow, and sometimes cross, the isopycnals, and not all overturns are located where the 6.4-m *Ri* is small.

Whilst, however, there is, on average, a significant correlation between the locations of these overturns and those of low *Ri*, the distortion of the density field by the internal waves, sometimes known as the strain, γ, appears to provide a better index of overturning. The strain of the wave field can be written in terms of the buoyancy frequency (or, equivalently, using (1.5), in terms of the vertical density gradient) as

$$\gamma = [(\langle N^2 \rangle / N^2) - 1], \tag{4.15}$$

and is positive where the vertical density gradient is reduced below the mean value at a given level by the presence of the waves and negative where it is increased. It appears that shear is not the key factor describing the relation between internal waves and the presence of unstable stratification; the waves produce shear but also modulate the stratification. (As in the studies on which Figs. 4.6 and 4.7 are based, much finestructure may be poorly or inadequately resolved by the coarse 6.4-m resolution of *Ri*.)

Kelvin–Helmholtz billows have been detected by sonar from acoustic scattering off layers of particles (sometimes zooplankton) on the continental shelf and in regions of strong shear in sea straits. (They have also been detected by radar in the atmosphere.) An example from the Strait of Gibraltar is shown in Fig. 4.13. The Mediterranean is a region of high evaporation rates and, consequently, high salinity. Water from the North Atlantic flows eastward into the Mediterranean though the Strait (partly to compensate for the loss of volume through evaporation) above a layer of warmer and saltier water that flows westwards as an undercurrent out of the Mediterranean, subsequently forming a 'tongue' of high-salinity 'Mediterranean Water' at a depth of about 1200 m in the eastern North Atlantic as shown in Fig. 5.1 later. Shear between the two layers in the Strait is particularly high where the outflow accelerates in passing over the constrictions imposed by sills. Figure 4.13 shows the resulting mixing, which is marked by billows with a height of about 50 m over the Camarinal Sill. Tidal flows through the Strait result in a modulation of the flow over the sills and the generation of internal bores or hydraulic jumps in which dissipation rates reach 10^{-4} W kg^{-1}.

Kelvin–Helmholtz billows have also been observed using arrays of thermistors in the ocean and in lakes and are probably of common occurrence, but there appear to be no observations that follow the onset, decay and changing structure of the related turbulence.

4.7 The rate of diapycnal mixing

Observations of ocean microstructure with free-fall instruments have proved invaluable in establishing the general variations in diapycnal diffusivity in the ocean. By the

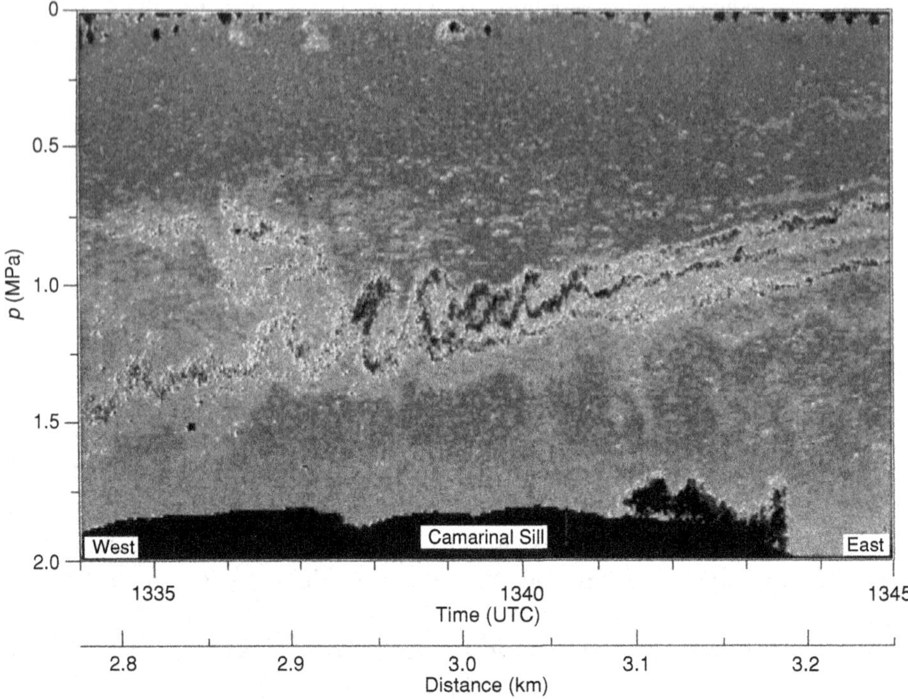

Figure 4.13. Billows in the Strait of Gibraltar: an acoustic (sonar) image of billows over the Camarinal Sill in the Strait during a period of 11 min as the vessel carrying the acoustic transducers passed over the sill from west to east. Reflections from the seabed are visible at depths corresponding to a pressure of 1.8 MPa (about 180 m). The larger billows (at times 1338–1340) are about 25 m high and occur at or near the boundary between the overlying eastward-going Atlantic water and the lower westward-going Mediterranean water (see also Fig. 2.4(b)). The billows are visible because the scatterers of sound, e.g., fish swim bladders and zooplankton, are advected by water, or because the growing billows create temperature microstructure that reflects high-frequency sound. Comparison of the shape of the billows with that of Fig. 4.4 indicates that the water above them is indeed moving to the east relative to that below. (From Wesson and Gregg, 1994.)

late 1980s, however, there was some doubt about the magnitudes of the mean eddy diffusion coefficient of heat estimated from microstructure measurements based on (4.9) in conditions under which it is appropriate to take $K_\rho = K_T$. In 1966 Munk had carried out a calculation that concluded that the mean value of K_T in the abyssal Pacific Ocean is about 10^{-4} m^2 s^{-1} (now usually referred to as Munk's canonical value of eddy diffusivity), but the microstructure measurements gave a value about an order of magnitude less, 10^{-5} m^2 s^{-1}.

• Munk's conclusion that $K_T \approx 10^{-4}$ m^2 s^{-1} is based on a so-called advection–diffusion balance. A reduction in temperature at a given level resulting from the slow upward movement of deep cold water, water that has been cooled and has sunk in the Antarctic and spread towards the Equator (as described in Section 6.4.3), is countered by heat diffused downwards from the warmer upper ocean, so keeping the net

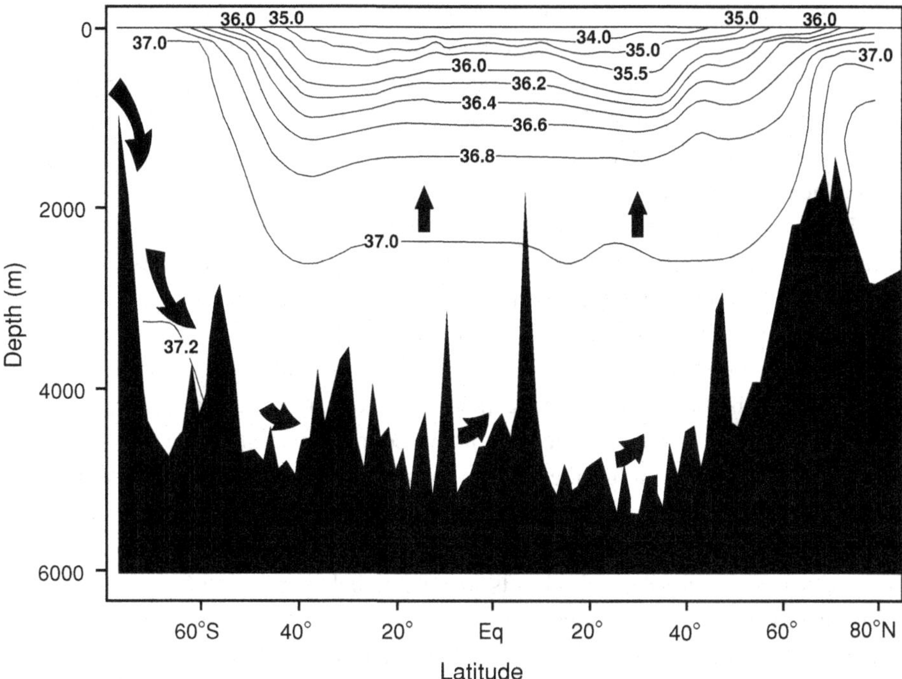

Figure 4.14. A north–south section of potential density, σ_θ, referenced to 2000 m and zonally averaged across the Atlantic. Cold Antarctic Bottom Water (AABW) originating in the Antarctic spreads northwards on the bottom of the Atlantic Ocean (as well as in the Indian and Pacific Oceans), leading to a slow rise in the overlying water as indicated by the arrows. The temperature of this water is maintained constant by a downward diffusion of heat. The balance of advection and diffusion was used by Munk (1966) to obtain an estimate of the vertical eddy diffusion coefficient, K_T. The zonally averaged topography fails to represent adequately the true depth of the Greenland–Iceland sill, which is about 600 m. (From Toggweiler and Key, 2001.)

temperature, and density, approximately constant (Fig. 4.14). **[P4.8]** The rate of production of the colder deep water can be estimated, and therefore so can its upward speed, which is found to be about 4 m per year. Munk's calculation effectively provided a long-term average value of K_T for the whole water mass below 1500 m at moderate latitudes in the Pacific.[12] Values of eddy diffusivity similar to Munk's have been found using an advection–diffusion-balance method in deep basins in other parts of the ocean. [**P4.9**]

The significant discrepancy between Munk's value and the estimates based on data from free-fall instruments made it appear possible that the values derived from the

12 Munk considered four processes that might produce the required mixing: breaking internal waves, the effects of mixing at the lateral boundaries of the ocean, double diffusive convection (Section 4.8) and the diurnal vertical migration of swimming zooplankton. He was able to discount the third and fourth because of their spatial limits. Ganachaud and Wunsch (2000) used methods similar to Munk's to derive the diapycnal diffusivity in the Atlantic, Indian and Pacific Oceans at depths of about 2000–3800 m, and also below about 3800 m. Values of K_T range from $(3 \pm 1.5) \times 10^{-4}$ m^2 s^{-1} in the intermediate deep Atlantic to $(12 \pm 7) \times 10^{-4}$ m^2 s^{-1} in the deep Indian Ocean.

latter were somehow in error. If the measurements of shear from which ε is derived are accurate, the problem might be to do with under-sampling. This could occur in two ways. Because only a limited number of estimates of ε (and therefore K_T) are obtained from a set of free-fall microstructure profiles, and because ε is known to be approximately log-normal, too few of the relatively high values of ε might be sampled, so that the estimate of the overall mean value is too low. Alternatively, it was possible that the rather small number of profiles gathered by the late 1980s had not sampled the ocean sufficiently well in both space and time (perhaps missing regions and periods of large K_T) to obtain a reliable estimate of the mean diffusivity.

An ingenious method based on the use of an injected tracer was devised to obtain a direct measure of a 'long-term' and large spatial average value of the vertical diffusivity. Although this was incidental to the main objectives of the experiment, it provided a test of the reliability of the estimates of K_T derived from microstructure measurements. A chemical, sulphur hexafluoride, SF_6, is found to be almost neutral in the ocean in that it is chemically stable and does not associate with particles. It can, moreover, be detected at concentrations of about one part in 10^{17}, concentrations far lower than for the tracers used previously, and at levels at which it makes an insignificant contribution to the density of seawater. (For comparison, the fluorescein dye used in earlier dispersion studies can be detected at best down to concentrations of about one part in 10^{12}.) Methods were devised to release the SF_6 tracer into the sea, and, in May 1992, a patch containing about 140 kg of the tracer was inserted into the pycnocline at a depth of 300 m in the northeastern Atlantic in an experiment led by Ledwell and Watson called the North Atlantic Tracer Release Experiment (NATRE). The patch was followed using acoustically tracked floats, and its subsequent vertical, or diapycnal, diffusion was monitored by ship surveys over a period of 30 months, during which time the original 20-km-wide patch had spread to reach a horizontal size of more than 1000 km. (This spread is shown later in Fig. 5.13 and the horizontal dispersion of the tracer is discussed in Section 5.4.3.) The vertical diffusion of the patch, from about 20 m to 150 m in 30 months, is shown in Fig. 4.15, and from the increased patch thickness a value of the vertical diffusivity of the tracer can be determined: $(1.7 \pm 0.2) \times 10^{-5}$ m^2 s^{-1}. This is consistent with estimates of K_ρ derived using free-fall instruments in the same area and depth range.

• Although the experiment raised questions about whether the diapycnal diffusivity of SF_6 is exactly the same as that of density or temperature, the results demonstrate that the measurements of K_ρ by the free-fall instruments are reliable to within a factor of about 2 (or 50% uncertainty) and vindicated the estimation of K_T using free-fall instruments.

Later experiments using microstructure measurements and SF_6 have shown that the diapycnal diffusivity is often much greater than 10^{-5} m^2 s^{-1}, reaching 10^{-3} m^2 s^{-1} or more in regions above rough topography, possibly because of the local breaking of internal waves, notably those of tidal frequency (Section 6.3.2). Much higher values of ε and K_T are, for example, found over the mid-Atlantic Ridge than over nearby abyssal plains. As shown in Fig. 4.16, there is evidence that flows within canyons on the flanks

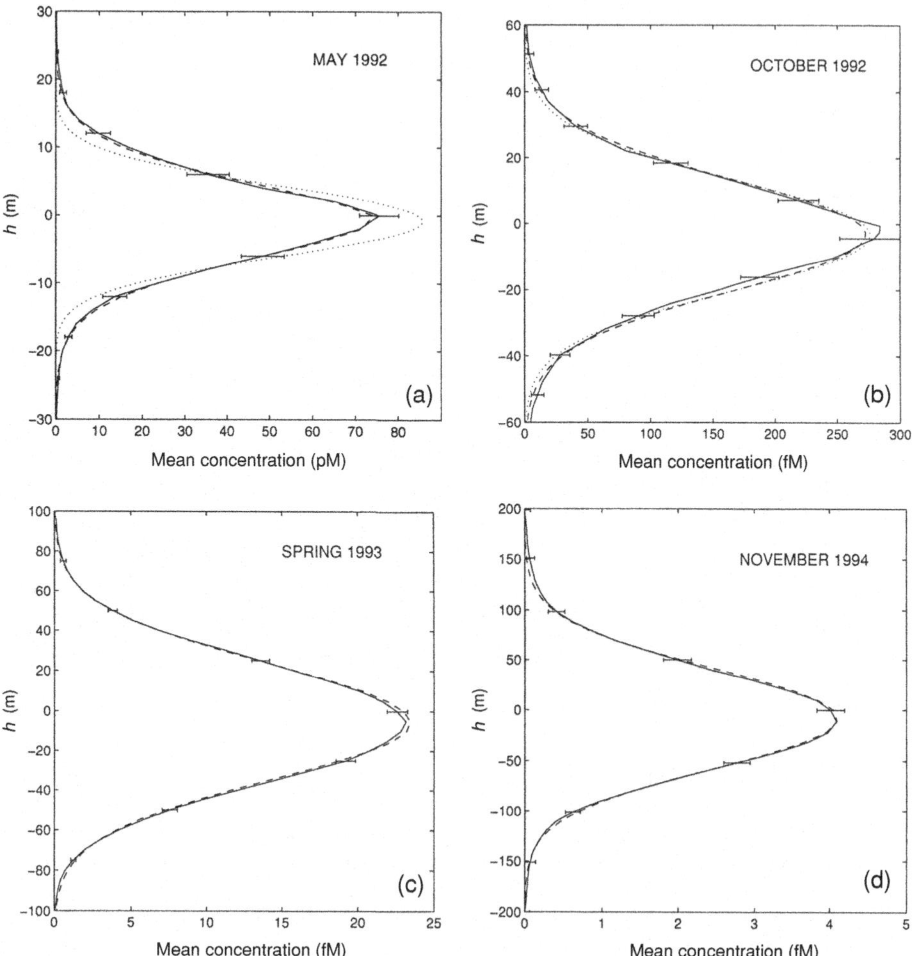

Figure 4.15. The vertical diffusion of the SF_6 tracer during NATRE. The mean measured concentrations are shown as a function of height above their mean level during four different surveys of the diffusing and dispersing patch made over a period of 30 months. The four months in which the measured values were obtained are shown. The horizontal and vertical scales vary so that, for example, the maximum concentration decreases with time whilst the thickness of the patch increases (1 pM = 10^{-12} moles per litre, i.e., 10^{-9} mol m^{-3}, and 1 fM = 10^{-15} moles per litre.) The horizontal dispersion of the SF_6 patch is shown in Fig. 5.13. (From Ledwell *et al.*, 1998.)

of the Ridge may contribute to enhanced mixing. These relatively recent observations, and the lack of earlier measurements in such regions, provide an explanation for the apparent discrepancy from Munk's estimate: the free-fall microstructure probes were not under-sampling values of ε, but the spatial distribution of the microstructure observations made by the late 1980s was inadequate to account for the regions in which mixing is greatest. Few (if any) estimates of K_T had been obtained in the relatively small geographical regions, especially those close to the mid-ocean ridges, where

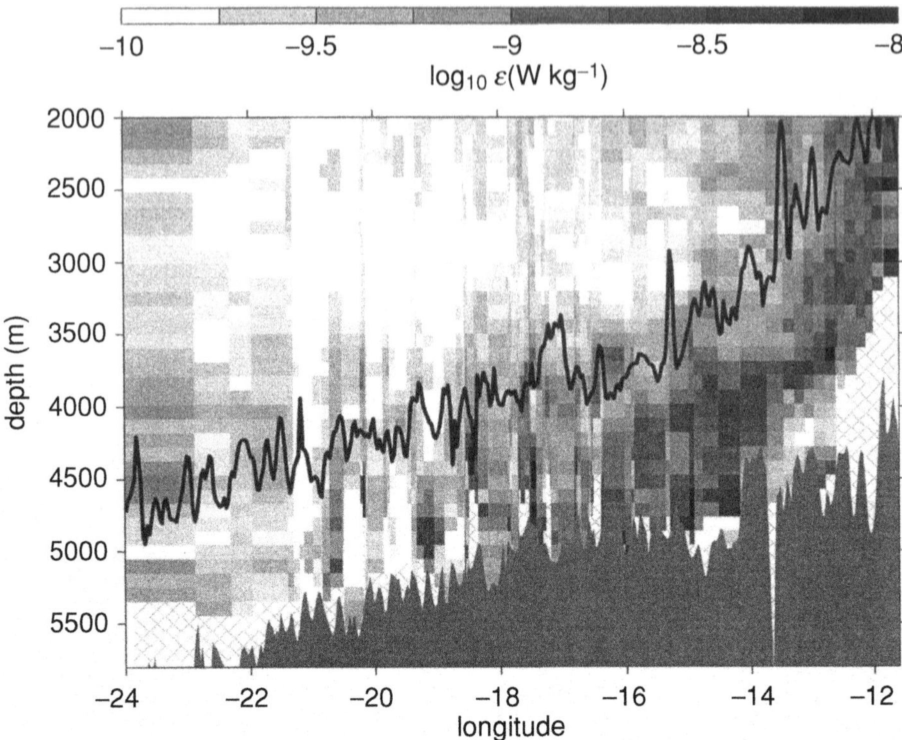

Figure 4.16. Enhanced dissipation over rough topography: rates of dissipation, ε, measured by recording a series of free-fall microstructure profiles in an east–west section over the Mid-Atlantic Ridge at the edge of the Brazil Basin. (The topography is shown in Fig. 5.14.) Individual profiles are shown as columns of rates of dissipation averaged over 100-m-depth intervals. The topography underlying the profiles is shown in grey, and the black line above it marks the level of the crests of the topography surrounding the position of the profiles. Most of the high values of dissipation lie between the bottom and the nearby ridge crests, and thus within canyons. Within 500 m of the bottom the dissipation is often two orders of magnitude (i.e., 100 times) greater than at mid-depth or at similar heights above the nearby abyssal plains. (From St. Laurent *et al.*, 2001.)

dissipation is now known to be relatively large, so the mean spatial average value of K_T derived from the then existing microstructure measurements was much lower than Munk's value.

• It is now apparent not only that turbulent mixing is temporally varying and patchy in the pycnocline, but also that there are geographical regions, 'hot-spots', where mixing is generally much more intense than average. The sources of the energy required to support diapycnal diffusion in the ocean are discussed in Chapter 6. [**P4.10**]

4.8 Double diffusive convection

Whilst internal wave breaking contributes substantially to the mixing of the stratified ocean, a further process is evident in regions where such mixing is relatively weak and in which the vertical gradients of temperature and salinity have the same sign, both positive or both negative. A form of convective motion known as double diffusive convection is then possible.

An extraordinary, if hypothetical, means of producing perpetual motion was suggested by Stommel, Arons and Blanchard in 1956. They supposed that a long thin-walled heat-conducting tube is lowered vertically into the ocean at a location where the water near the surface is warmer and more salty than it is at depth, and that the water within the tube is moved upwards, perhaps initially by a pump. Since the temperature within the ocean decreases with depth, the water driven upwards within the tube will be at a lower temperature than that outside and, all along the length of the tube, it will be heated by heat conduction through the tube walls. Its salinity cannot change by conduction through the impermeable tube walls but, being raised, is less within the tube than it is outside. This effect, and the heating through the walls, makes the water within the tube less dense than that outside. The water in the tube has everywhere above it fluid of lower density than that of the surrounding water, and the pressure at the bottom of the tube (and at every level inside it) is therefore less than that in the water nearby. There is consequently a pressure force that can drive the water in the tube upwards. The motion will be sustained so long as the positive upward salinity gradient remains. (Equally, if the water is first driven downwards, the downward motion will also continue.) This phenomenon is known as 'the perpetual salt fountain'. It works (at least in theory) because heat, but not salinity, is passed through the tube walls.

Although the idea did not lead to a pragmatic design of an efficient engine, friction and bio-fouling being insurmountable problems, it led Stern in 1960 to the idea that, because the molecular diffusivity of salt, κ_S (about 1×10^{-9} m^2 s^{-1}) is very much less than the thermal conductivity, κ_T (about 1.4×10^{-7} m^2 s^{-1}), the tube walls were really unnecessary for motion to be continuously driven. Seawater in which the temperature and salinity both increase upwards – for example as in Fig. 1.12 above 900 m – can be unstable even if the vertical density gradient is stable ($\mathrm{d}\rho/\mathrm{d}z < 0$). A small volume of (non-turbulent) water moved upwards in such conditions receives heat by thermal conduction from its warmer surroundings and so its temperature rises. Although its salinity will also rise through molecular transfer from its relatively saline surroundings, the rate of salt transfer is very much less than that of heat, and consequently there are conditions in which the decrease in density of the displaced fluid volume through rise in temperature exceeds the increase through salinity. On becoming less dense, fluid continues to rise under buoyancy forces developing in a form of convective motion, but one that differs in the way it is driven from that described in Section 3.2. In terms of the classification suggested in Fig. 3.2, double diffusive convection is an internal process. Its scale of size or velocity does not depend directly on the fluxes of energy through the boundaries of the ocean or on distance from a boundary.[13]

13 A further distinct type of convective instability known as 'cabelling' results from the non-linear dependence of density on temperature and salinity and from the mixing of water masses of equal density but different temperatures and salinities. **[P4.11]**

- There are two types of double diffusive convection, the first known as the 'finger regime', when, as above, less dense but warmer and saltier water lies over colder fresher water. It takes the form of ascending and descending convective columns or fingers of water, typically 1–6 cm wide, 'salt fingers', as in Fig. 4.17. Because the molecular conductivity of heat is greater than the molecular diffusivity of salt, the density of relatively cold and fresh ascending fingers becomes less, as described in the preceding paragraph, being reduced by the more rapid transfer of heat than of salinity from the surrounding descending fingers. The latter, losing heat more rapidly than salinity, become denser but remain more saline than their ascending surroundings and tend to continue their descent.
- The second type of double diffusive convection is the 'diffusive regime', in which relatively warm and salty water lies beneath less dense, colder and fresher water. If it is displaced downwards, the colder water becomes less dense as a consequence of the molecular transfers with its relatively warm but salty surroundings, and rises buoyantly to recover its original position, but, having achieved a lower density than when it started, overshoots its original position and continues to rise, but now losing heat and becoming denser. Its subsequent re-sinking and rising lead to a growing oscillation. Fluid initially moved upwards undergoes an increase in density, with subsequent sinking and overshoot, leading to a similar growing oscillation. Such growing oscillations are termed 'overstability'. The diffusive regime of convection can be active in the Arctic, for example above a relatively warm and salty intrusion of water of Atlantic origin.

The development and motions of both types of instability are impeded by the frictional effects of viscosity, and the width and growth rate of the fastest-growing salt fingers depend on the kinematic viscosity.

We have so far described what happens when both temperature and salinity increase or decrease upwards. There are two other possible classes of temperature and salinity variation. If salinity decreases upwards and temperature increases upwards, the contributions of both quantities lead to decreasing density and so produce a density gradient that is negative (i.e., $d\rho/dz < 0$, with density decreasing upwards). This is a regime of static stability that is termed 'doubly stable'. In the opposite case, when the salinity increases upwards and temperature downwards, both contribute to a density that increases upwards. Consequently the density gradient is positive and the water column is 'statically unstable'.

- A parameter determining the nature of double diffusive instability is the density gradient ratio,

$$R_\rho = (\alpha\, dT/dz)/(\beta\, dS/dz), \tag{4.16}$$

where $\alpha\, dT/dz$ is the contribution of the mean vertical temperature gradient, dT/dz, to the vertical density gradient and $\beta\, dS/dz$ is the corresponding contribution of the mean vertical salinity gradient, dS/dz. The ratio, R_ρ, is less than zero in the doubly

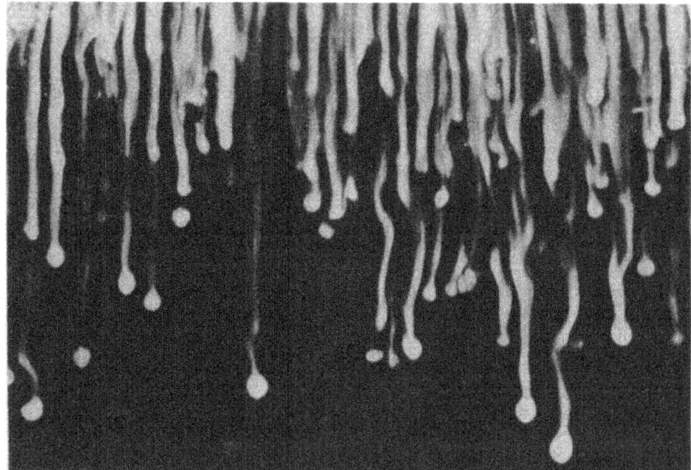

Figure 4.17. Salt fingers. These are formed in the laboratory by carefully allowing a warm weak brine solution containing a fluorescent dye to spread over a cold fresh layer of water. A vertical sheet of light is used to cause the dye to emit light and so make the developing fingers visible in this horizontal view. The 'fingers' are typically 2 mm in width in the laboratory experiment, but in the ocean fingers may be some few centimetres wide. (From Huppert and Turner, 1981.)

stable regime ($\mathrm{d}T/\mathrm{d}z > 0$, $\mathrm{d}S/\mathrm{d}z < 0$) and in the statically unstable regime ($\mathrm{d}T/\mathrm{d}z < 0$, $\mathrm{d}S/\mathrm{d}z > 0$), but $R_\rho > 0$ in both the double diffusive regimes. From (1.4) the vertical density gradient is

$$\mathrm{d}\rho/\mathrm{d}z = \rho_0(-\alpha\,\mathrm{d}T/\mathrm{d}z + \beta\,\mathrm{d}S/\mathrm{d}z), \tag{4.17}$$

and for overall static stability this must be <0. Using (4.16), (4.17) can be written

$$\mathrm{d}\rho/\mathrm{d}z = \rho_0\beta\,(\mathrm{d}S/\mathrm{d}z)(1 - R_\rho), \tag{4.18}$$

which is <0 if $\mathrm{d}S/\mathrm{d}z > 0$ (as in the finger regime), provided that $R_\rho > 1$. Similarly, $\mathrm{d}\rho/\mathrm{d}z < 0$ if $\mathrm{d}S/\mathrm{d}z < 0$ (in the diffusive regime), provided that $R_\rho < 1$. In the finger regime $R_\rho > 1$; in the diffusive regime $0 < R_\rho < 1$.

The convection described so far does not involve turbulence; the motions driven by instability are laminar. Provided that the contribution to turbulence from other sources such as breaking internal waves is relatively weak, the initial form of instability in double diffusive convection (i.e., the salt fingers in the finger regime) is found to evolve into a spatially coherent structure containing layers of uniform temperature and salinity. The layers appear as a 'staircase' in the temperature and salinity profiles, a '*T*–*S* staircase' as illustrated in Fig. 4.18. The weakly turbulent layers are typically 2–50 m thick, stacked one above the other. In the finger regime, the initial finger form of instability with laminar flow persists within the relatively thin (0.1–1 m thick) interfaces between the layers, interactions between fingers leading to their disintegration at the edges of the uniform layers. The layers, a form of temperature and salinity

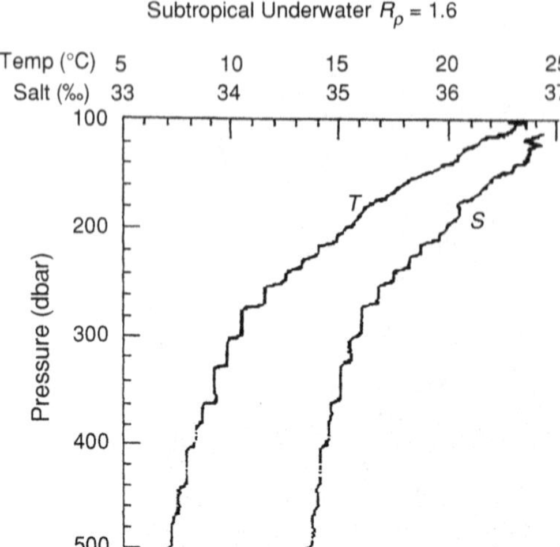

Figure 4.18. A *T–S* staircase at depths between 200 and 400 m in the subtropics. Both temperature and salinity increase upwards, favouring the formation of salt fingers, and $R_\rho = 1.6$. (From Schmitt, 1981.)

fine-structure, are sustained by a convective buoyancy flux through the salt fingers, and are observed when $1 < R_\rho < 1.7$. The first observation of such staircases was in the late 1960s at depths of some 1200 m, beneath the relatively warm and saline Mediterranean outflow spreading westwards from the Strait of Gibraltar into the eastern North Atlantic.

There is evidence that the magnitude of the efficiency factor, Γ, that appears in (4.9) can be affected by double diffusive convection even in conditions under which the presence of turbulence prevents the formation of double diffusive layers. Data obtained from a free-fall microstructure probe during NATRE, in which there was no clear evidence of staircase-like structures, have been analysed to determine the statistical variation of Γ as a function of the Richardson number, *Ri*, and the density gradient ratio, R_ρ, as shown in Fig. 4.19.[14] In the doubly stable regime shown in Fig. 4.19(a), where $R_\rho < 0$, values of Γ are near 0.2 across the ranges of sampled *Ri* ($0.01 < Ri < 100$), in accordance with the value commonly adopted. Similar values are found in the finger regime when R_ρ is greater than about 2, and also in this regime when $1 < R_\rho < 2$ and the shear is large so that $Ri < 1$ (Fig. 4.19(b)). When the shear is relatively small, however, so that $Ri > 1$, and when the stability parameter is moderate (in the finger regime when R_ρ is greater than unity but less than about 2), Γ is found to have values between 0.4 and 1 (the top left quadrant of Fig. 4.19(b)). The greater efficiency is equivalent to an increase in the flux Richardson number, R_f, or a greater fraction of energy being transferred to potential energy from turbulent kinetic energy.

14 Supposing, as earlier when salinity plays no part, that $K_T = K_\rho$, and using (4.6), (4.9) and the relation between χ_T and C (P4.6), an expression is found for the efficiency factor: $\Gamma = N^2\chi_T/[2\varepsilon(\mathrm{d}\langle T\rangle/\mathrm{d}z)^2]$. Measured values of ε, χ_T (or C), N and the vertical temperature gradient are used to find Γ. Its difference from the value 0.2, the value observed in regions where double diffusive convection plays no part (as in Fig. 4.19(a)), may be taken to imply that the effect of salinity is dynamically important in some parts of the ocean.

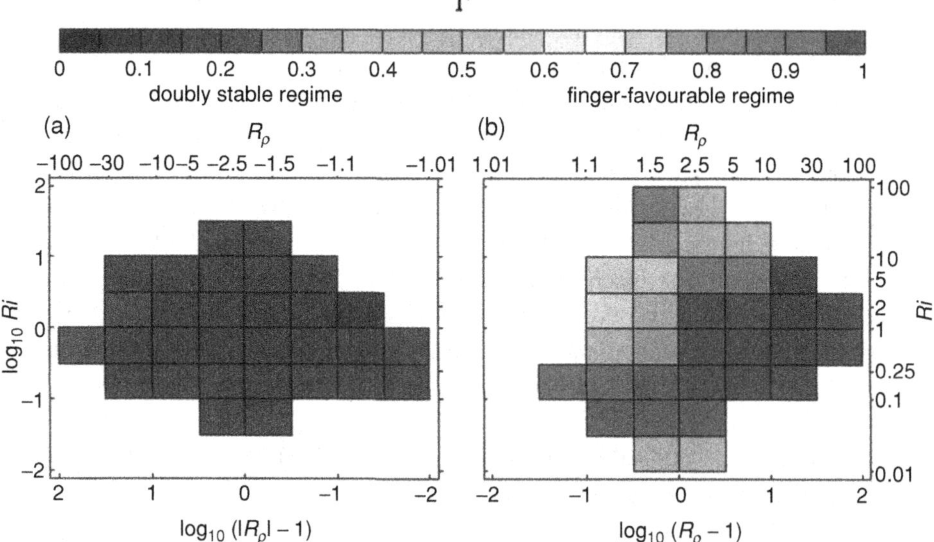

Figure 4.19. The variation of the efficiency parameter, Γ, with the Richardson number, *Ri*, and the density gradient ratio, R_ρ. In the oceanic regime in which double diffusive convection cannot occur, shown in (a), values of Γ (defined in 4.9) are generally close to 0.2. Larger values are found when convection is produced as a consequence of diffusion in the finger regime shown in (b), provided that $1 < R_\rho < 2$ and *Ri* is not small; i.e., $\Gamma > 0.2$ in relatively low shear when turbulence is unlikely to be generated by shear-flow instability and conditions favour salt-fingering. (After St. Laurent and Schmitt, 1999.)

• In the finger regime when the shear is weak ($Ri > 1$) and when there is moderate stability ($1 < R_\rho < 2$), double diffusive convection may contribute substantially to the mixing process, even in conditions under which thermohaline staircases are not produced. No observations of turbulence and its relation to *Ri* and R_ρ seem to be available yet for the less common conditions favouring the diffusive regime.

Suggested further reading

Diapycnal mixing

Taylor's (1931) paper is a classic example of how important conclusions may be drawn from relatively simple measurements. The methods used by Jacobsen to derive K_ρ and K_ν are described, with some modifications, by Proudman (1953).

Turner's (1973) book gives an excellent introduction to the theoretical and laboratory investigations of K_ρ/K_ν or R_f, and includes discussion of laboratory experiments on turbulent entrainment in stratified shear flows.

Woods' (1968) paper provides an excellent and unique photographic description of internal wave breaking and diapycnal mixing in the ocean. Other than shadowgraph images of ocean fine-structure and microstructure, notably of salt fingers, e.g., see Kunze *et al.* (1987), there are few underwater images that demonstrate the turbulent nature of the ocean or its sometimes coherent structure.

Gregg (1987) provides a fine and detailed review of the knowledge of ocean mixing in the 1980s obtained from the early, and still limited, use of microstructure profilers.

Munk's (1966) paper entitled 'Abyssal recipes', in which the canonical value $K_T = 10^{-4}$ m^2 s^{-1} is derived, was of great originality at the time and is worth reading alongside its successor, Munk and Wunsch (1998), published 32 years later, if only to appreciate how much (and how slowly) ideas have developed and thinking has changed. The discovery of relatively high turbulence dissipation rates over rough topography over the mid-Atlantic Ridge at the eastern boundary of the Brazil Basin by Polzin *et al.* (1997) gave particular stimulus to the study of mixing in the abyssal ocean.

Ledwell *et al.*'s (1998) account of the results of NATRE relating to diapycnal diffusion is a benchmark paper in the subject.

Double diffusive convection

The note on the perpetual salt fountain by Stommel *et al.* (1956) still bears its hallmark of originality and is a reminder of the continuing search for means of extracting energy from the ocean. The discovery of *T–S* staircases is described by Tait and Howe (1971).

Turner's (1973) book gives a still highly relevant account of the basic processes involved in double diffusive convection, but the relevant chapter (Chapter 8) should be read in conjunction with a more recent account, for example Schmitt's (2001) article that gives a concise account of more recent discoveries and the effect of the convection in the ocean.

Further study

Flow instability

The discovery of the theorem now known by their names is described in separate papers by Miles (1961) and Howard (1961). Drazin and Reid's (1981) textbook provides a comprehensive account of the mathematical theory of the stability of fluids in motion, and is useful as a basic reference.

Only a cursory description of the transition from laminar flow to turbulence following Kelvin–Helmholtz instability has been given here. Other types of shear-flow instability are known. One that may be common at the foot of a surface mixed layer as a means of erosion of the thermocline is Holmboe instability (Fig. 4.20). In this the velocity shear extends beyond the region of large density gradient. The instability takes the form of billows that detach stratified water from the region of high density gradient, rather than leading directly to a decrease in the high gradient. Strang and Fernando

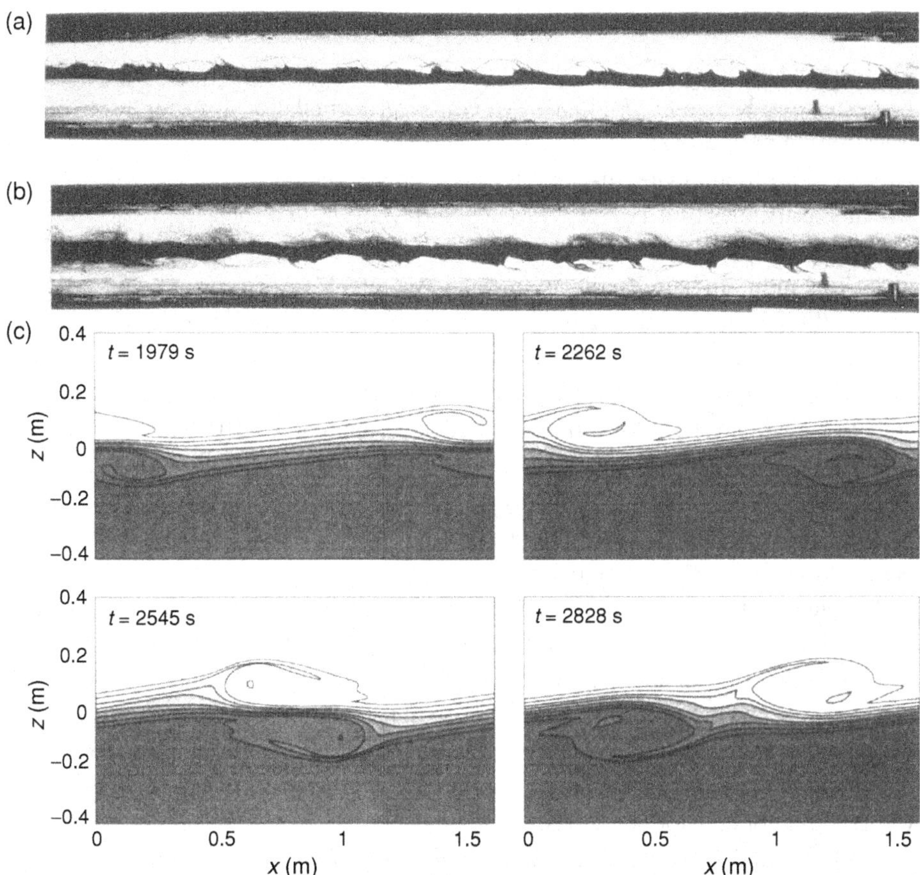

Figure 4.20. Holmboe instability. The instability occurring (a) on the upper side and, (b), at a later stage, on both sides of a dyed interface between layers of different densities in a laboratory experiment. (c) A numerical simulation of the instability at four sequential stages of development, but with flows opposite to those in (a) and (b) The lines represent isopycnal surfaces. Relative times are marked at the top on the left. (Parts (a) and (b) from Thorpe, 1968, and (c) from Smyth and Winters, 2003.)

(2001) describe laboratory experiments on the processes and rates of entrainment into a turbulent flow, and provide references to earlier studies of entrainment.

Some of the processes leading to internal wave breaking are described by Staquet and Sommeria (2002).

Mixing in straits

Wesson and Gregg (1994) give a detailed account of their measurements of turbulence over the Camarinal Sill in the Strait of Gibraltar, background to which is provided by Armi and Farmer (1988).

Parametric representation

The parameterized relation for ε (4.3) was derived by Polzin (1996). It is also well worth reading Kunze *et al.* (1990). An empirical formulation for ε in terms of the properties of the internal wave field was devised by Gregg (1989), but MacKinnon and Gregg (2003) found that a different formulation applied on the New England Shelf, with highest values of ε at large buoyancy frequency, N, and shear, S.

A comprehensive and instructive analysis of data obtained from lowered ADCPs and CTDs in the Indian, Pacific and Atlantic Oceans has been carried out by Kunze *et al.* (2006) to determine ε and K_T using a parameterized relation that depends on internal-wave shear and strain. Weak eddy diffusivities of order 10^{-5} m^2 s^{-1} are found in most of the upper ocean, throughout the water column near the Equator, and over smooth topography, but eddy diffusion coefficients approaching 10^{-4} m^2 s^{-1} are often found in the bottom 1000 m, and may extend into the main pycnocline. Sections made through the oceans show that vertically integrated values of ε are very heterogeneous. Readers may also find the observations and discussion of reduced mixing in equatorial waters by Gregg (2003) of interest.

Measurements by Peters *et al.* (1988) of velocity, density, ε and χ_T in the equatorial Pacific have been analysed to derive empirical relations between K_ρ/K_ν and Ri. As in Section 4.4.3, these are consistent with a decrease in K_ρ/K_ν with increasing Ri. (The authors, however, caution against the use of their formula to parameterize mixing in locations where the effects of a large mean shear and appreciable internal waves driven by the diurnal convection are absent or less substantial than in the Equatorial Undercurrent or equatorial regions.)

Further information about *Kelvin–Helmholtz instability*, the *breaking of internal waves* and *mixing in mid-water* is to be found in Chapters 3, 5 and 7, respectively, of TTO.

Problems for Chapter 4

(E = easy, M = mild, D = difficult, F = fiendish)

P4.1 (M) The gradient Richardson number. The steady horizontal flow with speed $U = U_0 + U_1 \tanh(az)$ and density $\rho = \rho_0[1 + \Delta \tanh(az)]$ has critical Richardson number, Ri_c, equal to $\frac{1}{4}$. The terms U_0 and U_1 are the mean and fluctuating speeds, respectively, ρ_0 is a reference density, Δ is a fractional density change and a is an inverse length scale such that $2a^{-1}$ characterizes the thickness of the velocity and density interface at $z = 0$.

Find the gradient Richardson number, Ri, expressed as a function of z, U_1, Δ, a and of g, the acceleration due to gravity.

What is the minimum Richardson number, $Ri_{\min}$, expressed in similar terms?

If the interface scale $2a^{-1} = 1$ m and $\Delta = 3 \times 10^{-5}$, what is the smallest velocity difference, $2U_1$, across the interface that would be required in order to make the flow unstable with respect to small disturbances?

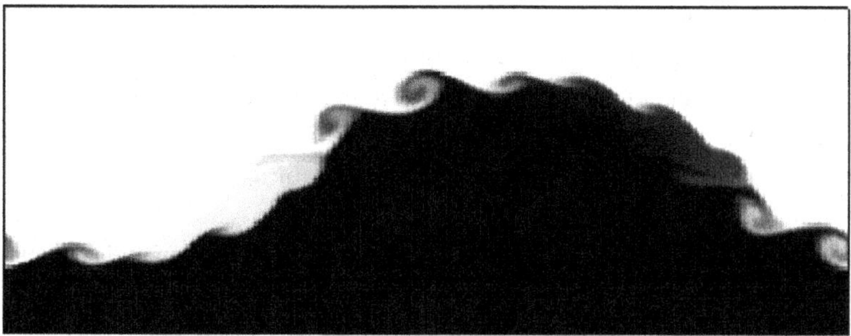

Figure 4.21. Billows growing on an interfacial wave: a numerical simulation of billows growing at the crest and trough of an internal gravity wave travelling along a density interface. (From Fringer and Street, 2003.)

P4.2 (M) Billows produced by internal waves. Figure 4.21 shows billows formed by Kelvin–Helmholtz instability at the crests and troughs of internal waves travelling on a thin interface in a numerical simulation made by Fringer and Street. Are the waves travelling to the left or to the right? If the velocity difference across the interface at the wave crests and troughs is $2a\sigma$, where a is the wave amplitude and σ is the wave frequency, estimate the smallest wave slope, ak, of internal waves of wavenumber k travelling along an interface between the two relatively deep layers that is required in order to reduce the Richardson number at the crests and troughs to $\frac{1}{4}$ when the wavelength of the interfacial waves is 20π times the interface thickness. You may suppose that the dispersion relation is $\sigma^2 = g'k$, where $g' = g\,\Delta\rho/\rho_0$, and the densities of the upper and lower layers are $\rho_0 - \Delta\rho$ and $\rho_0 + \Delta\rho$, respectively.

• Small instabilities require time to grow into billows. It is found that, in the periodic shear induced by the internal waves, billows grow to an overturning state, causing mixing, only if the shear is maintained and Ri is less than the flow's critical value of $\frac{1}{4}$ for a sufficiently long time. Fringer and Street find that the time required demands that the Richardson number is reduced to $\frac{1}{8}$ at the crests and troughs for billows to overturn.

P4.3 (F) Energy transfers and efficiency of Kelvin–Helmholtz instability. As a consequence of Kelvin–Helmholtz instability, the thickness of an interface between two uniform deep layers is increased from $2d$ to $2D$. The initial laminar flow and density are continuous and vary linearly with vertical coordinate, z, in the interfacial region from uniform horizontal x-directed speed, U, and density $\rho_0 - \Delta\rho$ in the upper deep layer to values of $-U$ and $\rho_0 + \Delta\rho$, respectively, in a lower layer, as illustrated in Fig. 4.3(a). The ratio $\Delta\rho/\rho_0 \ll 1$. (The velocity and density profiles are approximations to the error-function profiles of the laboratory experiments shown in Fig. 4.4.) After the transition caused by the Kelvin–Helmholtz instability, the velocity and density in the interface are still continuous with linear variations in z, as in Fig. 4.3(a). The initial gradient Richardson number, Ri, in the interfacial region has a value $<\frac{1}{4}$. After the turbulent motion resulting from instability has dissipated, the Richardson number in this region is 0.32, as observed in laboratory experiments.

In the model transition described above, find expressions (in terms of U, ρ_0, $\Delta\rho$, h and H) for (a) the change in kinetic energy of the mean flow per unit horizontal area, ΔKE, during the transition; (b) the corresponding change in potential energy per unit horizontal area, ΔPE; and (c) the energy per unit area dissipated by turbulence in the transition. Derive an equation for the *efficiency* of the transition, $\Delta\mathrm{PE}/\Delta\mathrm{KE}$, as a function of Ri. What is its maximum value?

• Most of the kinetic energy lost is dissipated in turbulent motion and a relatively small fraction converted into potential energy.

P4.4 (M) The time scale of turbulence collapse. How long is the period of collapse of turbulent motion following Kelvin–Helmholtz instability in terms of the buoyancy period in the resulting thickened layer? Compare this period with that of the approximately exponential decay of $\varepsilon \propto \exp[-t/(q\tau)]$, where $q = 1.0 \pm 0.1$ and τ is equal to the buoyancy period.

P4.5 (F) Richardson numbers derived from measurements made at vertically separated points. If the mean density gradient is stable, show that the Richardson number, $Ri(2h)$, estimated by taking differences of density and horizontal velocity at two points, $z = -h$ and $z = h$, separated by a finite vertical distance, $2h$, must be greater than or equal to the smallest gradient Richardson number in $-h \leq z \leq h$.

• This means that the Richardson numbers observed by taking differences in density and velocity at two vertically separated points on, say, a mooring, will generally exceed, and thus overestimate, the smallest gradient Richardson number, Ri. In particular, the minimum Richardson numbers, Ri_{min}, on which the Miles–Howard criterion for the stability of a flow is based, and which are required to establish the stability of a flow, will be overestimated.

P4.6 (E) The Cox number and the rate of loss of temperature variance. How are the Cox number, C, and the rate of loss of temperature variance, χ_T, related?

P4.7 (E) Estimation of turbulent dissipation rates. Show that, in conditions under which the eddy diffusion coefficients of heat, K_T, and of density, K_ρ, are equal, (4.6) and (4.9) provide a further way to determine ε from measurements.

P4.8 (D) The advection–diffusion balance. A thin layer of water of thickness δz lies at position z (measured upwards) where the mean vertical temperature gradient in the ocean is $\mathrm{d}T/\mathrm{d}z$. This gradient is positive, but varies with depth. If the water moves steadily upwards at speed w, by how much does the temperature at level z decrease after a small time δt? What is the rate of change of heat per unit horizontal area in the layer?

If the vertical downward flux of heat per unit volume in the water resulting from turbulent mixing is $\rho_0 c_p K_T\, \mathrm{d}T/\mathrm{d}z$, where K_T is the eddy diffusion coefficient of heat (which we suppose to be independent of z), find the downward heat flux at the boundaries of the layer at z and $z + \delta z$, neglecting w, and hence determine the rate of change of heat per unit horizontal area in the layer due to transfers by turbulence, ignoring the heating caused by turbulent dissipation and the steady vertical rise of water.

If the temperature (and heat content) within the layer is constant, obtain an equality based on the balance of heat change caused by steady vertical movement, or advection, and the vertical diffusion of heat by turbulence.

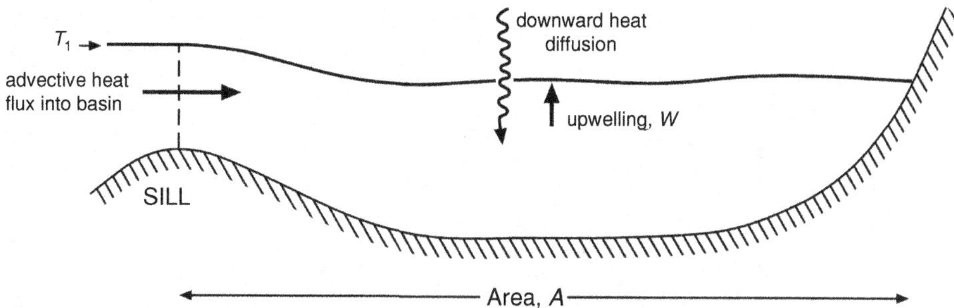

Figure 4.22. Heat balance in a deep-ocean basin. Flow enters a deep-ocean basin over a sill on the left. The basin is closed on the right. There is a downward flux of heat by diffusion from water of higher temperature above the T_1 isotherm and upward (upwelling) motion at speed, w, within the basin at the level of the T_1 isotherm, maintaining a steady state.

Use this equality to find the eddy diffusion coefficient of heat, K_T, if the temperature variation in the ocean is steady and is given by $T = T_0 \exp(z/d)$, where $d = 1$ km and T_0 is a constant reference temperature, when the vertical upwelling speed, w, is 4 m yr^{-1}.

P4.9 (D) Estimates of diffusion in deep ocean basins. Hogg *et al.* (1982) and Saunders (1987)[15] calculated K_T in deep ocean basins that are surrounded by topography and closed except for influx through channels (the Vema Channel into the Brazil Basin and the Discovery Gap in the eastern North Atlantic, respectively) within which measurements of the temperature and currents allow estimation of the net flux of heat entering the basins below an isothermal surface of temperature T_1 that intersects the basin sides (Fig. 4.22). In a steady state in which there are no long-term changes to the position of the isothermal layer or to the inflow, and if the mean inflow is Q (measured in m^3 s^{-1}) and the area of the isothermal surface within the basin is A, what is the mean upward speed, w, of the water rising across the isothermal surface?

An expression for the flux of heat entering the basin through the channel is $\rho c_p \int uT\, \mathrm{d}a$, where T is the water temperature, u the horizontal flow speed into the basin and da an element of the area of a vertical section across which the integral is taken. What is the flux of heat carried by the advection of the rising water across the area, A, of the T_1 isothermal surface?

If there is no geothermal heat flux through the seabed within the basin and the mean temperature of water in the basin does not change, the difference between the flux of heat through the channel into the basin and that leaving by advection through the isothermal surface must be balanced by a diffusive flux through the surface of area A. Supposing that the mean temperature gradient across the surface, $\mathrm{d}T/\mathrm{d}z$, is measured, what is the net downward diffusive flux of heat if the eddy diffusion coefficient is K_T?

Write down the equation of heat balance and use it to derive an equation for K_T.

15 Morris *et al.* (2001) give a comprehensive review of more recent estimates of K_T using the same methodology in the Brazil Basin, and make a comparison with estimates derived from dissipation measurements.

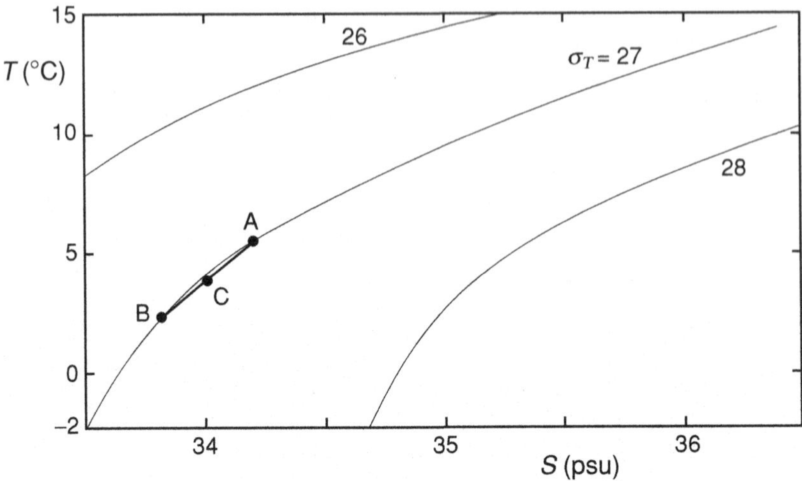

Figure 4.23. Cabelling. A *T–S* diagram, a graph showing how temperature, *T*, and salinity, *S*, vary on curves of constant density, or σ_T. The mixing of equal volumes of the water masses represented by A and B on the same σ_T curve results in water of temperature and salinity shown by point C lying on the line connecting A and B, but below the curve of constant density. Since the density below the constant-density curve is greater than that on it, the mixed water mass at C is of greater density than those at A and B and will tend to sink, an instability known as 'cabelling'.

If the flow from the Madeira Abyssal Plain through the Discovery Gap into the closed basin, the Iberian Abyssal Plain, is 0.21 Sv, the heat flux, estimated as $\rho c_p \int u(T_1 - T)da$ below the $T_1 = 2.05$ °C potential temperature isothermal surface, is 17.3 GW, the area, A, of the 2.05-°C isothermal surface within the Iberian Abyssal Plain is 1.2×10^5 km^2 and the temperature gradient across the T_1 isothermal surface is 1.0×10^{-4} °C m^{-1}, derive an estimate for K_T. (Assume that the density is 1030 kg m^{-3} and that $c_p = 3.99 \times 10^3$ J kg^{-1} °C^{-1}.)

(Saunders also takes into account the geothermal heat flux through the bed of the Iberian Plain.)

P4.10 (D) The effect of marine snow. 'Marine fluff' is an accumulation largely composed of loosely bound organic particles or flocs that have fallen (as 'marine snow') onto the seabed. The deposit grows rapidly following 'spring blooms' of phytoplankton, to a thickness of some 0.05 m in about 2 weeks. (See also P3.11.)

Supposing that the flocs of which the snow is comprised each capture a volume of water equal to E times their own volume whilst falling (E is an efficiency) and carry it on average a distance L downwards through the temperature-stratified water column before it is exchanged with, and passed into, the surrounding water, but ignoring the heat carried by the lattice-like material of which the flocs are composed, derive an expression for the effective rate of downward transport of buoyancy. Hence find an effective vertical diffusion coefficient, K_p, associated with the particles' sinking. You should explain any assumptions made. If $E = 1$ and $L = 10$ m, what is K_p?

Is marine snow likely to be an important contributor to the vertical diffusion of heat in the ocean?

P4.11 (D) Cabelling. Equation (1.4) is a local linear approximation to the equation of state. More precisely, density has a non-linear relation to temperature, salinity and pressure, e.g., see Gill (1982; Appendix 3). The *T–S* diagram (temperature T is plotted vertically and salinity, S, horizontally) sketched in Fig. 4.23 shows lines of constant σ_T (or density) as functions of T and S. The lines of constant σ_T have negative curvature (concave downwards). Points A, $(S + \delta S, T + \delta T)$, and B, $(S - \delta S, T - \delta T)$, represent two masses or bodies of water of equal density. Sigma-T, σ_T, increases as S increases or T decreases, so that the density of water at salinities or temperatures below the line of constant σ_T on which points A and B lie is *greater* than that at A and B. If mixed together in equal amounts, water masses at A and B would combine to give a body of water of salinity S and temperature T, shown by point C, halfway between A and B. This lies below the line of constant σ_T and is therefore of greater density than the water masses A and B. (Mixing in unequal proportions would lead to water with temperatures and salinities on the line joining A and B.) Mixing between water masses of equal density but unequal temperatures and salinities may therefore result in water of greater density that may consequently sink.

Supposing that the density of the *T–S* curve on which points A and B lie is given by

$$\begin{aligned}(1000 + \sigma_T)\ \text{kg m}^{-3} &= 1027.0\ \text{kg m}^{-3} \\ &= 10^3(1 - 8.052 \times 10^{-5}T \\ &\quad + 8.046 \times 10^{-4}S - 3.017 \times 10^{-6}T^2)\end{aligned}$$

with $T\ (> 0)$ in °C and S in psu, and that the salinities at the locations A and B are 34.2 and 33.8 psu, respectively, find the corresponding temperatures at A and B. Hence find the salinity and temperature of the equal-volume mixture represented by point C. How much colder is this than the point on the *T–S* curve on which A and B lie which corresponds to the same salinity as the mixture, C?

If mixing between the water masses A and B occurs over a time period of a year with an associated rate of dissipation of turbulent kinetic energy, ε, of $2 \times 10^{-8}\ \text{W kg}^{-1}$, can the heat generated by the turbulence be sufficient to prevent the increase in density through cabelling?

Chapter 5

Turbulent dispersion

5.1 Introduction

5.1.1 The properties of dispersants

The objective of this chapter is to describe some of the ideas and observations that have been devised to assess and quantify rates of turbulence dispersion in the ocean.

There are several reasons why dispersion, introduced in Section 1.5.1, is of importance in the ocean. It is dispersion that determines the distribution of the naturally occurring 'tracers', such as salinity, for example the area within the North Atlantic that is affected by the high salinity emanating from the Mediterranean through the Straits of Gibraltar (Fig. 5.1). The volume of the water column in the Pacific affected by the plume containing helium 3 (^{3}He) coming from hydrothermal vents in the East Pacific Rise (Fig. 5.2) is a consequence of several processes of dispersion, notably in the initial buoyant ascent of the plume, including the entrainment of surrounding water, and the subsequent advection and spread in the stratified ocean of the water 'labelled' by the ^{3}He. In many cases, especially those relating to the accidental discharge of toxic chemicals or oil into the sea or the development and spread of harmful algal blooms (HABs; Fig. 5.3), the dispersion of solutes or particles by turbulent motion may have dire consequences as the pollutants spread to sensitive regions, especially those near shore where there can be a detrimental effect on mariculture, human health and recreation. Prediction is therefore of great practical importance.

The material – the solutes, particles or organisms (such as, for example, the salinity, ^{3}He, oil drops or films and algae) – carried in seawater and spread by its turbulent motion is referred to as the 'dispersant'. As the examples mentioned above suggest, dispersion frequently relates to stirring motions that exceed those at which turbulence is

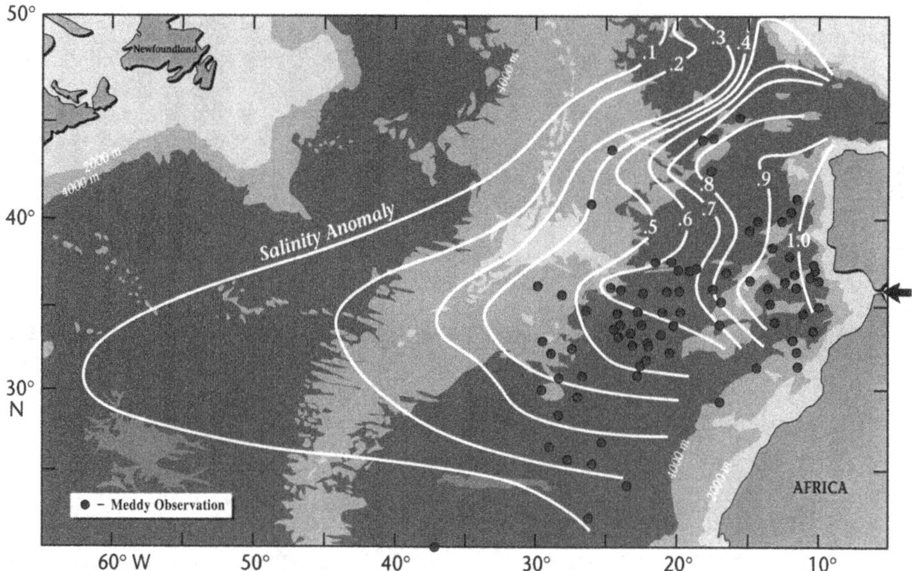

Figure 5.1. Spread of the Mediterranean Water at a depth of 1100 m in the northeast Atlantic. The Mediterranean Water is relatively salty, a result of high evaporation and relatively low precipitation in the Mediterranean and connected seas, and enters the North Atlantic through the Straits of Gibraltar (marked by an arrow). The contours are of the excess of mean salinity relative to a reference value of 35.01 psu. Circles represent the reported locations of Meddies (see Fig. 5.16 later). The underlying topography is shown in light to dark shades changing at 2000 and 4000 m depth. (From Richardson *et al.*, 2000.)

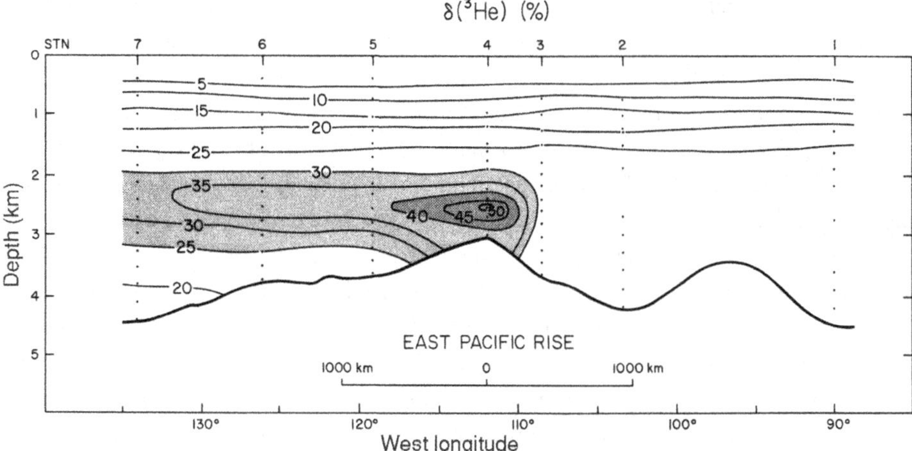

Figure 5.2. A plume of ^{3}He originating from hydrothermal activity on the East Pacific Rise and advected to the west by the mean currents. The hydrothermal plumes rising over the East Pacific Rise (e.g., see Fig. 3.4) are not resolved in this section but will contain even higher concentrations of ^{3}He than those contoured. (From Lupton, 1995.)

Figure 5.3. A harmful algal bloom (HAB). The bloom of toxic algae, cyanobacteria, in the Baltic Sea has a filamentary structure aligned downwind (towards the lower left), perhaps affected by Langmuir circulation, and has been broken by the wake of the research vessel. (Photograph by the Finnish Institute of Marine Research. With permission of the Scientific Commission on Oceanic Research.)

isotropic; the relevant stirring motions are often highly anisotropic, with vertical scales constrained by buoyancy and with motions and spread of dispersants sometimes also differing in two horizontal directions.

• Dispersion depends, however, not just on the turbulence processes leading to dispersion, but on the nature of the dispersant, for example whether it is a solute in water (like a dye), or consists of discrete particles or tends to form clusters (flocs) of particles or, in the case of heavy oils, 'blobs'. Dispersion depends also on the buoyancy of the dispersant, whether it floats to the surface (as do light oils) or sinks (as do sedimenting particles or algal detritus). The dispersant may consist of a dye that has little dynamical effect on the seawater into which it is dissolved or, like salinity, may have a buoyancy effect or even lead to double diffusive convection. Other dispersants, at least at high concentrations, may change the turbulence properties of the water into which they spread: an oil film can reduce the frequency of wave breaking and hence the generation of near-surface turbulence, whilst high concentrations of sediment particles eroded and dispersed by a rapid current over a mobile seabed can significantly increase the density of the sediment–water mixture and affect the energetics of the turbulent motion.

Figures 5.1 and 5.2 illustrate an important property of fluids in general and the ocean in particular. Generally, a maximum in the concentration of a solute (such as salinity or ^{3}He) cannot be created by natural processes within the body of the ocean without the presence of an internal source. Quantities like salinity cannot be produced or increased within the water, but only at its boundaries (e.g., by layers of salt at the seabed dissolving into the overlying water or through evaporation at the sea surface). If a maximum concentration is observed *within* some finite volume of water, in the absence of an internal source it can only be in water that has come from a boundary source; maxima, if they occur, are a result of fluid being transferred from a source at the boundaries of the ocean. (The one-way transfer of water from near the sea surface to greater depth is commonly termed 'subduction'. Fronts, for example, are regions in which subduction often occurs. On a smaller scale, the vertical transport in convection may also result in subduction.) The source and subsequent transfer may be relatively continuous, leading to a gradual reduction in peak concentration with distance from the source as the quantity is diffused and mixed with the low-concentration surroundings (e.g., as suggested by Fig. 5.2) or otherwise removed from the water.[1] Sources and transfer may also be discontinuous – or made so by the presence of eddies, resulting in local maxima of concentration – but the concentration must decrease as the quantity is carried away from its source.[2]

Vertical profiles in which a maximum concentration (or measure of a quantity) is found in mid-water therefore indicate that water has been advected (usually horizontally but, in convective conditions, perhaps vertically), and can be a valuable clue to processes of intrusion or relative motion. Maximum values occur naturally near the source of quantities on the ocean boundaries, e.g., of ^{3}He at the source of the plume on the East Pacific Rise. Care must be taken, however, in interpreting profiles, particularly when temperature is the observed fluid property. Surface cooling, or the cool skin of the ocean and the radiation of heat from the atmosphere, may lead to vertical profiles in which the temperature has a maximum value below the water surface. Unlike other transfer processes, radiation can transfer heat through the water and raise its temperature. But this is, effectively, a process of creation from an internal source of heat energy (where there is a divergence in the radiative heat flux), not one at a boundary.

The value of these comments about concentration maxima is in their application in tracing soluble dispersants as they spread and in discovering their source.[3] In contrast to solutes, particles may, however, cluster and increase in local (but not large-volume average) concentration as they disperse.

More complex interactive factors affect dispersion. Combinations of particles and buoyant fluid or solutes occur. The hydrothermal plumes from 'black smokers' (Fig. 3.4) contain mineral particles which, being denser than the water surrounding them and hence negatively buoyant, will eventually sink from the spreading plume, leaving a surface deposit lying on the seabed beneath the plume and around its source.

1 Removal of gases from solution, for example, may be by their being taken up by marine organisms, adsorbed onto the surfaces of falling particles and diffused into rising bubbles, or through their undergoing chemical interactions.

2 An example is that of salinity in Meddies in the North Atlantic; see Fig. 5.16 later. The existence of a maximum salinity in a Meddy signals that the origin of the water (if not the Meddy) is at a boundary, and in this case the source may be traced to the Strait of Gibraltar.

3 We have here considered maximum concentrations. The same ideas can be applied to minima, with the proviso that care must be taken when there are biologically or chemically related processes that act to decrease the local concentration of a dispersant or fluid tracer.

A similar 'loss' of dispersant from the water column may occur if the dispersant evaporates on reaching the sea surface. Dispersants may also dissolve into seawater or change their form under pressure. An example is that of the bubbles of radius less than 100 μm that are dispersed downwards from breaking waves by turbulence in the upper ocean. Gases pass from the bubbles into solution in the seawater surrounding them. The volume of bubbles carried downwards consequently decreases both because they lose gas into solution and because of the bubbles' compression by increasing hydrostatic pressure. Both the bubbles themselves and the several atmospheric gases that they contain may separately be regarded as the dispersants of interest, rendering the problem one of simultaneously related, 'multiple' dispersion.

The complexity introduced by the varied properties of dispersants makes their spread in the ocean difficult to address in general terms. We shall focus attention on the dispersion of just two types of dispersants, the first being those that float at the sea surface or those which, like neutrally buoyant particles or floats, remain on a given isopycnal surface; and the second, dyes or solutes at concentrations too low to affect the turbulent motion that, unimpeded by their own buoyancy, follow the motion and vertical diffusion of the turbulent water.

• The dispersion of particles that float and are therefore constrained to remain on the sea surface is effectively *two-dimensional*. Floats that are neutrally buoyant on some subsurface isopycnal within the pycnocline, being constrained to remain on the isopycnal surface in mid-water once they have reached it, may also disperse two-dimensionally on this surface. Floating particles of different densities are constrained to follow the mean motion of different isopycnal surfaces that, in general, move at different speeds, so that neighbouring particles on different isopycnals will move apart in the vertical mean shear as illustrated in Fig. 1.10(a). Both these examples contrast with the dispersion of solutes that, not retaining their density, are free to be transferred across isopycnal surfaces by diapycnal mixing or may be spread vertically by turbulent motions within the mixed layer, rendering the problem '*three-dimensional*'. (Floats that are neutrally buoyant within the mixed layer present an intermediate problem of dispersion, one that is three-dimensional but in which the floats are constrained to remain within a limited depth range.)

When the source of a dispersant is horizontally extensive and does not vary significantly in position, for example the (almost two-dimensional) sea surface or a uniform seabed, or the (one-dimensional) outer edge of a surf zone, the problem of dispersion may simplify to one with variation in only a single direction (i.e., depth, height above the seabed, or distance from the edge of the surf zone). The mean dispersant concentration, for example, will then vary only in a direction normal to the source. Such problems are sometimes addressed as being of diffusion, for example of bubbles from the sea surface or of sediment from the seabed, but are more properly regarded as the dispersion of particles or solute by turbulent motion, giving appropriate emphasis to the processes involved and the interaction of turbulence and dispersant.

But first, methods by which to quantify dispersion are required.

5.1.2 Appropriate measures

The measure of a dispersant that is required in observations or in a predictive model will often depend on the 'impact' that the dispersant has, perhaps on marine life or on the recreational use of affected beaches. This 'impact' of a dispersant may relate either to its mean concentration or to the greatest concentration that occurs. For example, it is the mean concentration that relates to the total amount of the dispersant ingested by organisms or, in the case of oil, that may coat the organisms unfortunate enough to be enveloped within it for an appreciable period of time. It may be this accumulation that determines whether organisms survive or not. Alternatively, however, it may be the maximum concentration or a specified concentration level that is important if, at some particular concentration, the dispersant becomes lethal to an organism.

Some measure of the spread of a dispersant is required in predictive models, often in a parametric form that is related to the properties of the environment and the dispersant at the (sometimes relatively large) scales resolved by the model. The measure will depend on whether the dispersant is introduced into the ocean continuously from a location (Figs. 5.4 and 5.5(a)), like oil leaking from a stricken ship or particles in a hydrothermal plume, when a statistically steady state may be achieved; or the dispersant is released at some particular time as a patch or group of particles, an 'instantaneous release' as illustrated in Fig. 5.5(b) and in Fig. 1.7 (to which might be added a steady advective flow). In the former case, the mean or maximum concentration of dispersant and its width at some distance from the source may provide useful measures, whilst in the latter the size of the patch and its mean or maximum concentration as functions of time may be more appropriate, depending on the nature of the impact of the concentration on the environment.

• In cases in which the turbulent field of motion is homogeneous, it might be supposed that the spread from a continuous release can be represented as a series of releases of groups or clusters of particles or patches of dye that are advected by the mean flow from the fixed release point, thus effectively reducing the prediction problem to that of instantaneous release. This assumption has to be treated with considerable caution. Eddies of scale greater than the plume dimensions pass the source and carry the plume to and fro, causing it to meander as shown in Fig. 5.4(a). Near the source these large eddies do not cause particles in the plume to move significantly apart from one another, but lead to temporal variations in the location of the plume. The meanders result in variations of the measured values of the dispersant concentration at a fixed point downstream of the continuous source. Concentration is zero if the plume is totally absent or has temporarily moved away from the point, and non-zero when it is present (as illustrated in Fig. 5.5(a) by the section of concentration through the plume). The mean width of the area affected by a plume at some distance downstream of the source is greater than the instantaneous plume width. The time-averaged concentration at any point downstream of a discontinuous release will depend on the spread of the clusters of particles or instantaneous clouds of dispersant carried to the location of the point,

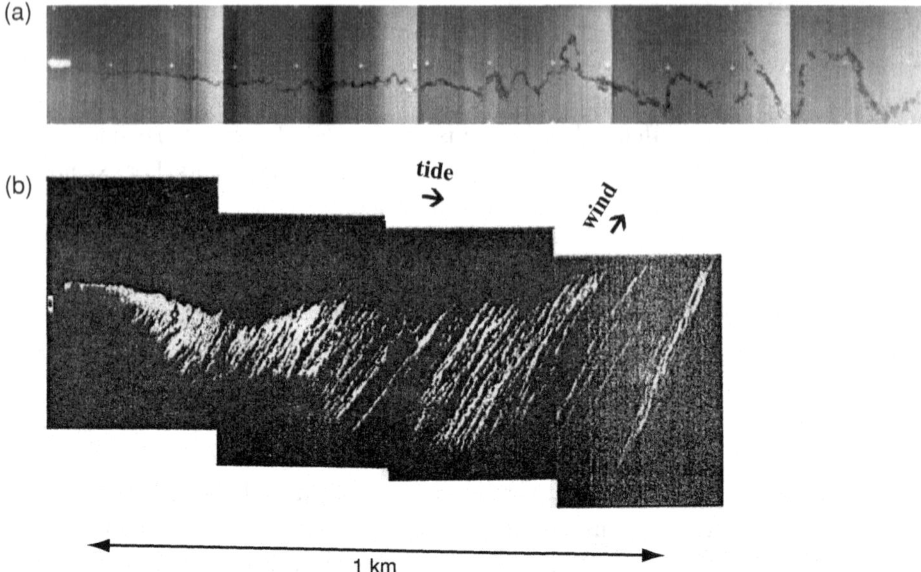

Figure 5.4. Plumes of floating oil. These composite images of two plumes of oil released continuously at a constant rate from fixed locations (to the left) were taken from an aircraft in authorized experimental studies of dispersion in the mixed southern North Sea. (a) Infrared images show growing meanders in the plume of oil carried to the right in a tidal current. The crosses are 200 m apart and the length of the meanders is 100–300 m. (Image kindly provided by Professor A. J. Elliott from work conducted in collaboration with the UK Warren Spring Laboratory.) (b) A composite photograph shows a meandering plume of oil, floating on the sea surface, advected to the right from the release point in a mean tidal current and mean surface wind drift. The wavelength of the meanders appears to be about 400–600 m. The plume breaks up into filaments about 10–30 m apart, aligned roughly in the direction of the wind, probably as a consequence of Langmuir circulation as suggested by the numerical model shown in Fig. 5.20 later. (From Thorpe, 1995.)

how these overlap, and how often large meanders cause relatively small clusters to be carried across the location of the point. The maximum concentration in a set of patches at a given distance from release may depend on patch size, as discussed in the following section, and differ from the maximum concentration found at the same distance in a continuous plume.

5.1.3 Effects of relative eddy and patch sizes

It is suggested by the sketch of the distortion of Welander's checkerboard pattern in Fig. 1.7 that dispersion may depend on the size of a dispersing patch of dye or particles. The particles at points A, C, D and E, belonging to a single eddy, remain close together whereas B, in a neighbouring eddy, is carried in its flow field away from the other particles. The checkerboard patch extends across two neighbouring eddies and consequently becomes elongated.

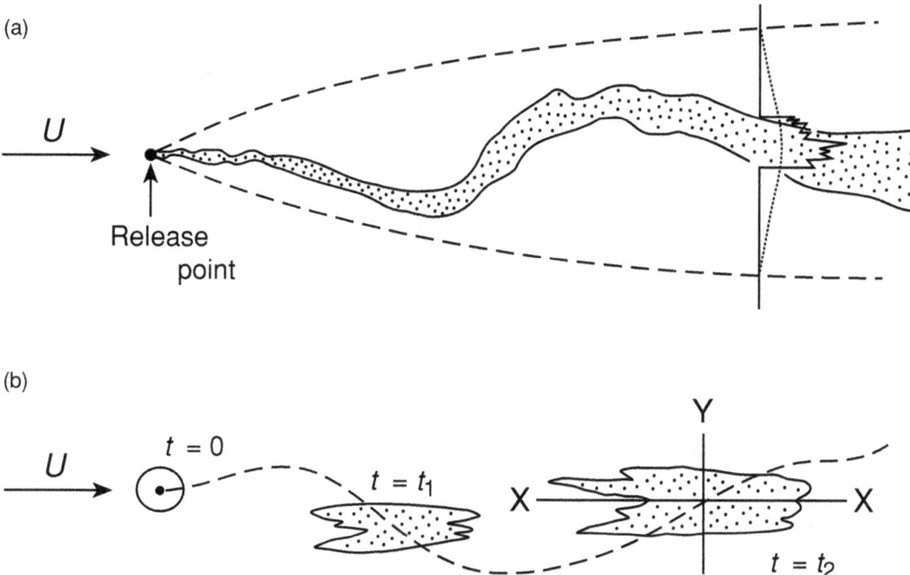

Figure 5.5. Dispersion in a mean current, U, at moderate times after release. (a) Dispersion of a plume of dye emitted continuously from a fixed point and shown at a time after a statistically steady state has been reached, well after the dye was first carried past the vertical line where (dotted) time-averaged and (full line) instantaneous dye concentrations are shown. The dashed line shows the outer limits reached by the dye plume, increasing in width in the downstream direction. (b) A patch of dye at times $t = 0$, $t_1 > 0$ and $t_2 > t_1$. The dashed line indicates the path of the centre of the patch. The distributions of measured concentration in directions along and across the mean direction of flow (e.g., XX and YY at $t = t_2$) will generally differ. In both cases the path of the dye is affected by eddies of scale larger than the instantaneously dyed region and will meander or fold as shown in Fig. 5.4(a). In shallow water, holes may develop in patches of dye or foam at the sea surface where water carried upwards by boils replaces the water surface (Figs. 1.4 and 1.6). Diffusion of a dyed patch will eventually differ in the zonal and meridional directions (see Fig. 5.13(b) later).

The effects of the relative patch and eddy sizes on dispersion are illustrated in Fig. 5.6. Whilst, in general, these effects depend on the statistics of the eddy sizes, Fig. 5.6(a) illustrates how eddies much larger than patches of dye or particles advect the patches, which are distorted by the shear or convergence field within the boundaries of the large eddies and are only slowly dispersed (or increased in area or volume), mainly by the motions of smaller eddies. Once the patch size is of scale comparable to that of turbulent eddies (Fig. 5.6(b)), it is (on average) distorted in the field of shear and convergence between eddies, and drawn out into the thin filaments or streaks sketched in Fig. 1.7. Examples are shown in Fig. 1.6(a), the observed floating filaments of oil drawn out by the presence of boils and, at a much larger scale, in Fig. 5.13(a) later, the filaments of a chemical tracer, SF_6, in the thermocline. When eventually the patch reaches a size much greater than the turbulent eddies (Fig. 5.6(c)), their effect is to

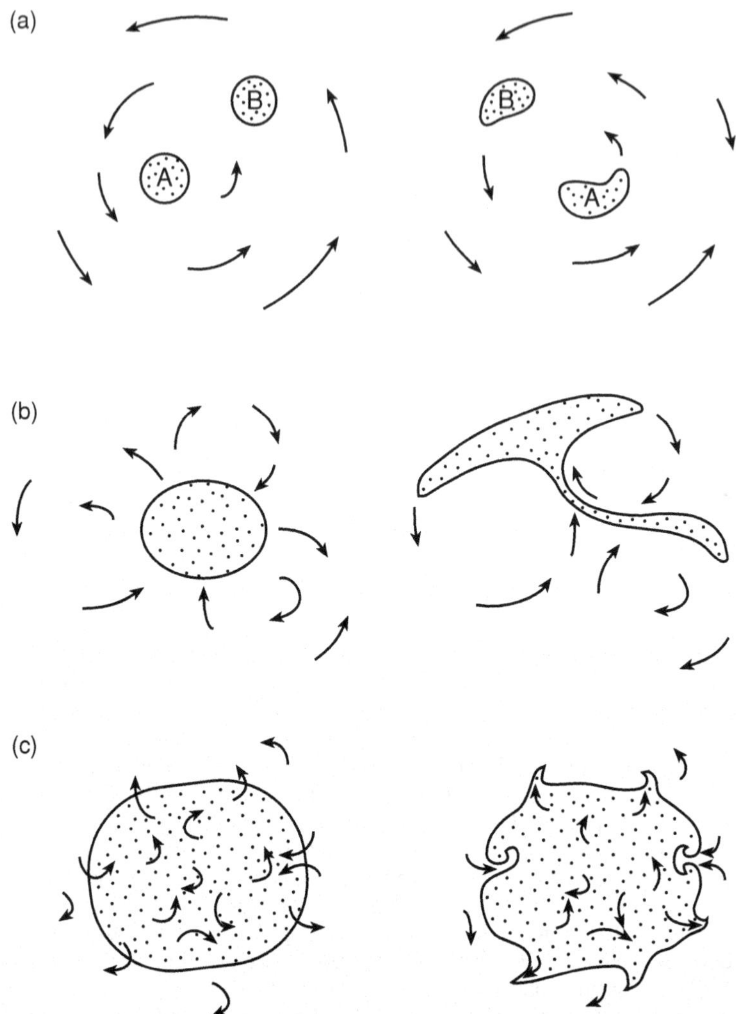

Figure 5.6. Dependence of dispersion on patch and eddy size. Patches or particles are marked by stippled regions and are shown initially (left) and at some later time (right) as they are distorted in fields of eddies indicated by the arrowed flow lines. (a) Eddy scale much greater than patch size. There is relatively little growth in the mean dimensions of patches, except for that caused by shear within the eddy. (b) Similar patch and eddy sizes. The patch is drawn into filaments or streaks by the eddies. (c) Patch size much greater than eddies'. The eddies mix within the patch and extend the patch boundaries relatively slowly.

mix fluid within the patch itself and slowly to continue the process of spreading of the patch at its outer boundaries. (Eventually the filaments of SF_6 diffuse into one another, producing the large, but still inhomogeneous, patch shown in Fig. 5.13(b).)

There are several ways in which time can enter the problem of dispersion. As shown in Fig. 5.6, the rate of dispersion of a patch is related to the size of the patch and, since this increases with time, to the time over which it has previously spread.

Dispersion of a patch will be relatively slow whilst fluid remains (on average) within an eddy (Fig. 5.6(a)). This introduces a concept of eddy lifetime,[4] the period during which an eddy retains a coherent structure before it decays or disintegrates, leaving the patch of dye or particles that was within it to be invaded by further eddies. These may distribute the material originally within the eddy amongst several others and expose material to the process of dispersion illustrated in Fig. 5.6(b). Furthermore, the nature of eddy disintegration may be important, ranging from a process involving the pairing of neighbouring eddies and the formation of a single eddy (an increase in eddy size) to the fragmentation of a dominant eddy into smaller eddies. In stratified regions the form of eddies and their effect on dispersion may change with time, originally three-dimensional eddies collapsing to form horizontal vortices (see Fig. 1.16).

Other temporal processes that force motion, but which are not described as 'eddy-like', may contribute to the dispersion of a patch of dispersant in the ocean as the time after its release increases. These form part of the non-steady and anisotropic conditions that typify ocean turbulence. Over sufficiently short periods of time the speed and direction of the wind may be relatively uniform and statistically steady, with means that are changing very slowly, but over a period of a few hours the wind speed and direction can change substantially, and so will the motions forced by the wind within the upper ocean. Such variations in forcing will, for example, change the orientation of the streaks of oil shown in Fig. 5.4(b) and affect the spreading rate of the oil on the sea surface. The processes leading to turbulence in the mixed layer often change over a diurnal cycle (Section 3.4.2). These external effects, appearing as an unsteadiness of the mean flow or stratification over times of order a day, may, over much longer periods (e.g., months), be regarded as contributions to the spectrum of turbulent motions: such natural variability induced by externally forced processes of mixing may be contained within the terminology 'turbulent motion'. In practice great care is required in making estimates (or in deriving coefficients) of dispersion in quasi-steady or highly variable conditions, in properly assessing average values in the presence of the inherently variable state of the ocean and in assigning the dispersion rates that are appropriate in particular conditions where prediction of dispersion is required.

As well as temporal variability, spatial non-uniformity in dispersion may also be encountered, for example as a dispersing patch or plume reaches the boundary of a region confined by tidal fronts (Fig. 3.17) or a region of enhanced mixing caused by topography (Fig. 4.16), and due account of the presence of such features and of localized processes of dispersion is required in predictive modelling.

4 The lifetime of an eddy, or of other coherent structures characteristic of turbulent motion, is often brief. In many cases the lifetime of eddies recognized in shadowgraph images or by PIV is comparable to the time required for water particles to travel once around a typical eddy's circumference. If the lifetime is much shorter, the structure of eddies will be transient, lacking coherence and possibly making them unrecognizable, beyond any means of detection because of their short duration. In such circumstances a description of turbulence as a collection of eddies may be inappropriate and an alternative descriptive term is needed.

5.2 The dispersion of particles

5.2.1 Autocorrelation and integral scales

Dispersion of particles or markers such as floats, or, in the atmosphere, balloons, by turbulent motion was one of the first topics related to turbulence to be addressed by theoreticians, notably L. F. Richardson and G. I. Taylor (see footnotes 8 and 10, respectively, in Chapter 2), both of whom were also able and skilled at performing experiments to test their predictions, and did so.

Dispersion is related to the spreading or relative position of particles released as a group at some instant of time, $t = 0$, but we begin by considering the tracks of single particles moving with the turbulent water. If the speed of a particle at time t in direction x (for the sake of argument, supposed horizontal) is given by $U_o + u(t)$, where U_o is the mean speed of a group of particles, then the location of the particle relative to its starting or release point is $X(t) = U_o t + x(t)$, where $x(t) = \int_0^t u(\tau)\mathrm{d}\tau$. The relative spread of particles depends on whether or not their motions are similar, and consequently on whether their locations remain similar. This is related to the idea, introduced in Section 5.1.3 and taken up again in Section 5.2.2, that eddies that are large relative to the separation of particles will move all the particles in a similar way, and so contribute little to increasing the size of a cluster, whereas eddies of size comparable to or smaller than the distance between particles will move them apart.

A measure of the time scale over which the speed of a particle or float carried by the motion remains similar during its encounter with eddies in a turbulent flow is provided by the 'velocity autocorrelation function', $R(\tau)$, the mean value of the product of the speeds, u, of a single particle or float released into the ocean measured at two times that are separated by a time interval, τ, and normalized and non-dimensionalized by u', the standard deviation of the particle speed, u:

- $$R(\tau) = \langle u(t)u(t+\tau)\rangle/u'^2 \equiv \left\{\mathrm{Lim}_{T\to\infty}\left[\frac{1}{T}\int_0^T u(t)u(t+\tau)\mathrm{d}t\right]\right\}\Big/u'^2, \qquad (5.1)$$

where $u'^2 = \mathrm{Lim}_{T\to\infty}[(1/T)\int_0^T u^2(t)\mathrm{d}t]$. The non-dimensional function R is equal to 1 when $\tau = 0$, but $R(\tau)$ tends to zero as τ tends to infinity because a particle's (or a float's) speeds become incoherent and uncorrelated at large separation times, τ.[5]

In practice R is calculated from averages over many (i.e., an 'ensemble' of) particles or floats in the ocean, and assumptions are often made about the spatial and temporal uniformity, or homogeneity, of the turbulent velocity field. This places limitations on

5 In estimating R, it is supposed that floats accurately follow the motion of the water and can be used to find its speed, so that R represents a measure of relative motion, not just of floats, but also of water particles. This is not necessarily true; often there will be some 'slippage' that must be accounted for. Cross-correlation functions, of which R is one, are discussed in footnote 11 in Chapter 2. Being relatively easy to calculate, they were once often used to describe the structure of varying flows and the variation of quantities such as temperature or dispersant concentration as well as velocity components. They are sometimes derived from measured values at points separated in space rather than in time. They are related to spectra as noted in Section 2.3.6. Instead of cross-correlation functions, it is now more common to use spectra or structure functions to describe the structure of turbulent flows.

the durations of time or extents of the ocean locations characterized by measurements of dispersion. Measured values are often representative only of a particular period of time or position in the variable ocean: hence the requirement for care in their use that is noted in the last section.

• The smallest time, τ, at which R becomes zero provides an approximate measure of the time scale over which motion remains coherent. As we shall explain, however, a more useful measure, related to the rate of separation of particles and hence to the size of a dispersing patch, is obtained by integrating R over separation times, τ. This provides a single measure, a time scale, called the 'Lagrangian integral time scale',

$$T_{\mathrm{L}} = \int_0^{\infty} R(\tau)\mathrm{d}\tau, \tag{5.2}$$

that represents the time for which a particle's speed remains strongly coherent or self-correlated. Using u' to characterize the speed of particles, a corresponding 'Lagrangian integral length scale',

$$L_{\mathrm{L}} = u'T_{\mathrm{L}}, \tag{5.3}$$

can be defined, providing a measure of the scales of eddies in the x direction of a flow field. (Eulerian time and length scales are commonly used to describe dispersion; they are found simply by calculating R from the velocities, u, measured at a fixed point instead of from those of flow-following floats or particles.) The scales T_{L} and L_{L} provide measures of the turbulence but not, immediately, of its dispersive effects.

A theorem that is central to the study of dispersion is that derived by Taylor in 1921. It provides the link from measures of the movement of single particles to the dispersion of many. It shows how turbulence velocities are connected to particle displacements, and may be stated as follows.

• In a homogeneous turbulent flow, the rate of change of the variance of particle positions, $\langle x^2 \rangle$, which we suppose to be determined from an ensemble of releases or from a large number of particles at the same time after release, is related to R by

$$(\mathrm{d}/\mathrm{dt})[\langle x^2(t)\rangle] = 2u'^2 \int_0^{t} R(\tau)\mathrm{d}\tau. \tag{5.4}$$

This provides a means of determining an 'eddy dispersion coefficient', K_{H}, given by

$$K_{\mathrm{H}} = \frac{1}{2}(\mathrm{d}/\mathrm{d}t)[\langle x^2\rangle], \tag{5.5}$$

from the measured autocorrelation function of the velocities of a set of particles or floats. The coefficient K_{H} is positive for a growing patch of particles or floats, but can be (temporarily) negative in regions of convergent flow. The coefficient is a measure of the mean rate of separation of fluid particles in the x direction in a turbulent flow. A coefficient related to particle separation in the y or z directions can be defined similarly.

For an ensemble of particles released from a fixed location, $\langle x^2(t)\rangle \approx u'^2t^2$ for $t \ll T_{\mathrm{L}}$, a result found by integrating (5.4) and recalling that $R(\tau) \approx 1$ for small τ.

Since, at time t, particles have been carried by the mean current through a distance $U_0 t$, this implies that the width of a patch of neutrally buoyant particles (given by $\langle x^2(t)\rangle^{1/2}$) grows linearly with distance. (This result may be compared to the prediction of a linear spread of a turbulent *buoyant* plume or a turbulent jet, a result that can be derived by means suggested in P3.2 and P3.3.)

• On substituting $\langle x^2(t)\rangle \approx u'^2 t^2$ into (5.5), the dispersion coefficient, $K_H \approx u'^2 t$, is found to increase linearly with time $t \ll T_L$ and is *not* constant; it depends on time or on the mean separation, $\langle x^2(t)\rangle^{1/2}$.

At large times, $t \gg T_L$, (5.2) and (5.4) give $\langle x^2(t)\rangle \approx 2u'^2 T_L t$, so a patch of particles of size proportional to $\langle x^2(t)\rangle^{1/2}$ will tend to grow with a parabolic relation to time (i.e., $\langle x^2(t)\rangle^{1/2} \sim t^{1/2}$) at large distances from the source.

• It follows from (5.5) that, at large times $t \gg T_L$, $K_H \approx u'^2 T_L$: the dispersion coefficient tends towards a constant value, $K_{H\infty}$, which, using (5.3), can be written

$$K_{H\infty} \approx u' L_L, \tag{5.6}$$

where L_L may be interpreted as an effective 'mixing length' of the turbulent diffusive field of motion. The time T_L provides information about how long it will be before the rate of dispersion of floats, $(d/dt)\langle x^2(t)\rangle$, becomes constant and no longer depends on the time after their release.

5.2.2 Richardson's four-thirds power law

Richardson drew attention to the importance of relative scale. As explained in Section 5.1.3 and Fig. 5.6(a), small patches – and hence pairs of particles that are relatively close together – are simply advected as a patch, or together moved as a pair, by eddies of relatively larger scales. Pairs of particles are, however, moved apart by eddies of scales comparable to or less than the distance between them, as illustrated in Fig. 5.6(b). Small clouds of particles are advected by the motions of the large eddies (those of size much greater than the cloud) and dispersed mainly by eddies that produce motions which diverge on scales comparable to the cloud size. The rates of dispersion of particles therefore depend on the presence of eddies with the appropriate scale to cause their dispersion, and the coefficient of dispersion will vary according to how effective eddies of this scale are at causing dispersion. Just as the energy of eddies depends on size (see Section 2.3.6), it is argued that their dispersive action through shear or convergence (as in Fig. 1.10) will be scale-dependent. If turbulent motion is produced by interacting eddies with a broad range of scales, the rate of dispersion of a cloud of particles should depend on the size of the cloud.

• This notion of scale-related dispersion is encapsulated in Richardson's law of diffusion:

$$\partial q/\partial t = (\partial/\partial l)[F(l)\partial q/\partial l], \tag{5.7}$$

where q is a concentration and $q\,dl$ is the number of particles that are separated from their neighbours by a distance between l and $l + dl$. Largely on the basis of available atmospheric data, Richardson proposed that the scale-dependent dispersion coefficient, $F(l)$, should be equal to $c_l l^{4/3}$, where c_l is a constant with dimensions $L^{2/3}T^{-1}$, a relation now known as Richardson's four-thirds power law. (Tests of Richardson's predicted dispersion are described in Section 5.3.1. The fact that c_l is a dimensional constant suggests that some physical process, characterized by dimensional quantities and affecting dispersion, has been overlooked in the formulation of $F(l)$. This is referred to again in Section 5.4.2.) [**P5.1**]

5.2.3 Dispersion of pairs of particles

The diffusivity, commonly called the relative diffusivity, of pairs of particles separated by rms distances D (not necessarily in one coordinate direction) is defined as $\frac{1}{2}\,\mathrm{d}D^2/\mathrm{d}t$. This is the value most commonly determined or estimated. For diffusion described by Richardson, the diffusivity is proportional to $D^{4/3}$, so, by integration,

$$D \propto t^{3/2}, \tag{5.8}$$

after time t. (The times, or distances, over which this law applies will be discussed when observations are described in Section 5.3.1.) At large times, t, and when the particle separation distances are large, long after the release of particles,

$$D^2 \propto K_{\mathrm{H}\infty} t. \tag{5.9}$$

Dispersion coefficients can be derived from measurements of the distances by which particles become separated in each of the three coordinate directions. These are often found to be unequal. As suggested in Section 5.1.1, buoyancy forces may constrain vertical dispersion and the presence of a mean shear can lead to the separation of particles even in the absence of dispersion caused by turbulence. Its effect is evident, for example, in the dispersion of floats released on the sea surface, the generally east–west currents in the Tropics leading at large scales to an evident difference in estimates of dispersion coefficients in the north–south (meridional) and east–west (zonal) directions.

5.2.4 Effects of closed vertical circulations on buoyant particles

It is appropriate here to mention some effects that may occur when the dispersed particles are positively or negatively buoyant.

The distribution of sediment particles sinking within a closed circulation, like that in Langmuir cells described in Section 3.4.3, provides an example of non-neutral tracers and an illustration of how dispersion of buoyant particles may differ from that of neutral particles or a dissolved solute. Sinking particles, perhaps entering the ocean as dust following its erosion in desert storms as shown in Fig. 5.7, may be trapped for some time and prevented from sinking in Langmuir circulation cells, as illustrated in Fig. 5.8.

Figure 5.7. A plume of dust carried by easterly winds from the Sahara Desert over the eastern North Atlantic. Such dust storms occur generally in late spring or summer, but this was imaged by satellite on 26 February 2000. By 11 March 2000, the dust plume had reached the northern coast of South America. As well as scattering solar radiation and reducing that reaching the ocean surface, dust transports iron that may be taken up by surface phytoplankton, resulting in increased rates of CO_2 assimilation. (A SeaWiFS image provided by NASA/GSFC and ORBIMAGE, courtesy of Dr Petra Stegmann, University of Rhode Island, Graduate School of Oceanography.)

If the fall speed of particles through still water exceeds the maximum upward flow speed in the circulating cells, the circulation affects the paths of the particles but they always have a positive downward velocity and sink through the cells (Fig. 5.8(b)). If, however, the maximum upward vertical speed in the circulation exceeds the particle fall speed, then there is a region within which particles circulate and may be trapped (Fig. 5.8(c)). For a specified circulation speed, the volume of the region in which particles may be recirculated depends on the fall speed, being greater for the smaller particles that fall more slowly. Positively buoyant, rising particles (e.g., bubbles produced by breaking waves) may also be trapped if their rise speeds are less than the maximum downward speed in the circulation.

In practice, however, Langmuir circulation is not as drawn in the simplified Fig. 5.8(a) – the circulation appears to be more intense near the surface than at depth, as in Fig. 3.12(b), and is unsteady. Relatively small-scale turbulent motions may eject particles from one cell into another or the circulation may, on average, break down

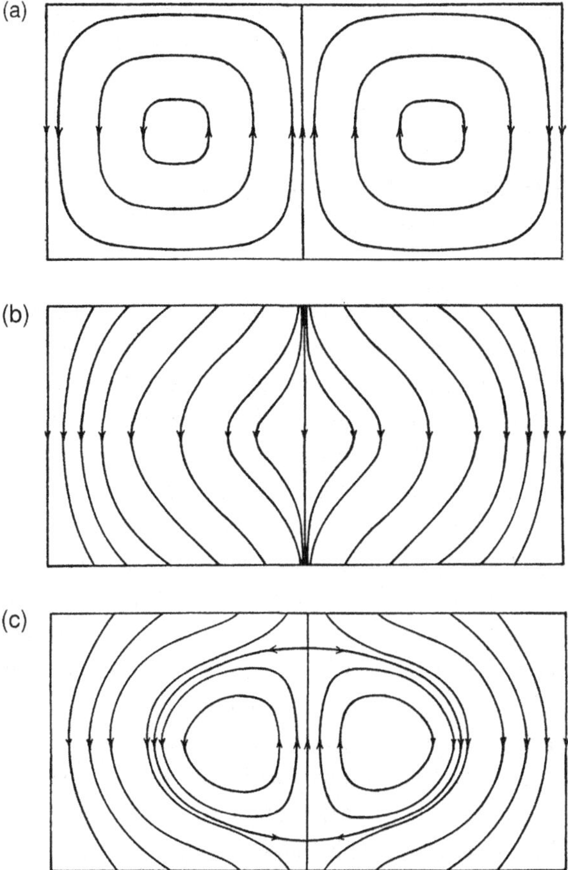

Figure 5.8. The trapping of small sinking particles by closed circulation cells, e.g., Langmuir circulation cells. (a) The assumed flow circulation pattern. (b) Paths of negatively buoyant particles that fall at speeds faster than the maximum upward speed of flow in the circulation. (c) Paths of particles falling at speeds less than the maximum speed in the circulation. Particles released from the upper horizontal line (e.g., the sea surface) will not enter the closed region of particle recirculation, but any that do will remain trapped if the circulation is steady. An example of the inverted situation of rising buoyant particles is that of bubbles produced by breaking waves. If their rise speed is less than the downward motions in Langmuir circulation, small bubbles will be temporally trapped when transported by transient turbulent motions into the region of closed circulation. (From Stommel, 1949b, who first described the phenomenon.)

before neutrally buoyant particles can be carried around a cell. The time for which individual windrows are observed to persist indicates, however, that a local convergence near the windrows is sustained for periods of time roughly corresponding to that for a neutral particle to be carried around the periphery of a cell (see also footnote 4). The presence of intense bands of subsurface bubbles detected by upward-looking sonar in the regions of near-surface convergence suggests that, even in their transient condition, the cells trap slowly rising bubbles. [**P5.2**]

Other patterns of closed circulation occurring, albeit temporarily, in turbulent flows (e.g., rising vortex pairs, Fig. 1.9) may trap and transport particles in a similar way, even those that are not neutrally buoyant.[6]

5.3 Observations of the dispersion of floats

5.3.1 Surface floats

The first test of Richardson's four-thirds power law (Section 5.2.2) for the dispersion of pairs of floating particles on the sea surface – and indeed the first quantified study of dispersion in the ocean – was by Richardson and Stommel.[7] Measurements were made in 1948 from a pier in Loch Long, Scotland, near to Richardson's house in Kilmun where he lived in retirement. Stommel was on a visit to Imperial College, London, at the time and had written to ask whether he might come to meet Richardson. Richardson's reply was, "Come, but bring some golf balls". Stommel, not a golfer, was perplexed by this but, fortunately, before he could make the necessary purchase of the balls (which was difficult, because of their being in very short supply so soon after the end of World War Two), he received a telegram: "Forget the golf balls, they all sink". On reaching Kilmun, he found that Richardson wanted him to help in an experiment on the dispersion of floating particles. Richardson had found that pieces of parsnip floated, but low in the water and thus were little affected by wind, and they could be seen fairly easily. The experiment was made with parsnips dug from Richardson's garden, cut into pieces and dropped from the pier. Their separation was measured using a sighting device cleverly designed by Richardson. Although the evidence was not strong, it was concluded that the variation in separation was consistent with the proposed four-thirds power law.

After further studies, Stommel concluded that a Fickian description of the diffusion of particles, one having a constant, scale-independent, coefficient of dispersion, is not valid other than for a closely constrained ranges of scales. Stommel's results, obtained from measurements covering a range of particle separation scales, $\langle l \rangle$, from 0.1 to 100 m, support Richardson's four-thirds power law with the dispersion coefficient, which Stommel defined as $\frac{1}{2}(\mathrm{d}/\mathrm{d}t)[\langle l^2 \rangle]$, equal to $c_l l^{4/3}$. The constant, c_l, was found to be about $2.3 \times 10^{-3}\ \mathrm{m}^{2/3}\ \mathrm{s}^{-1}$.

Developments in acoustic and satellite methods of float tracking in the 1970s and 1980s led to increasingly reliable measurements of the long-term diffusion

6 Some particles in the ocean coalesce, forming flocs, mineral particles being held together by organic coatings or mucus material. It is often assumed that the turbulent motion is unaffected by the presence of particles and that the particle distribution is not biased towards regions of larger or lesser rates of dispersion. In reality this might not be the case. Even when particles cannot coalesce if they come into contact with one another, turbulent motion may lead to the congregation of solid particles because their inertia produces a bias in their trajectories that carries them towards regions of flow convergence, where the strain rate is high or the vorticity is low.

7 Richardson's name and the parameters associated with him have been mentioned earlier. Henry Stommel (1920–1992) was a gifted and highly distinguished oceanographer at the US Woods Hole Oceanographic Institution who contributed much to the studies of dispersion and the circulation of the ocean. Further background to the parsnip experiment is given by Ashford (1985).

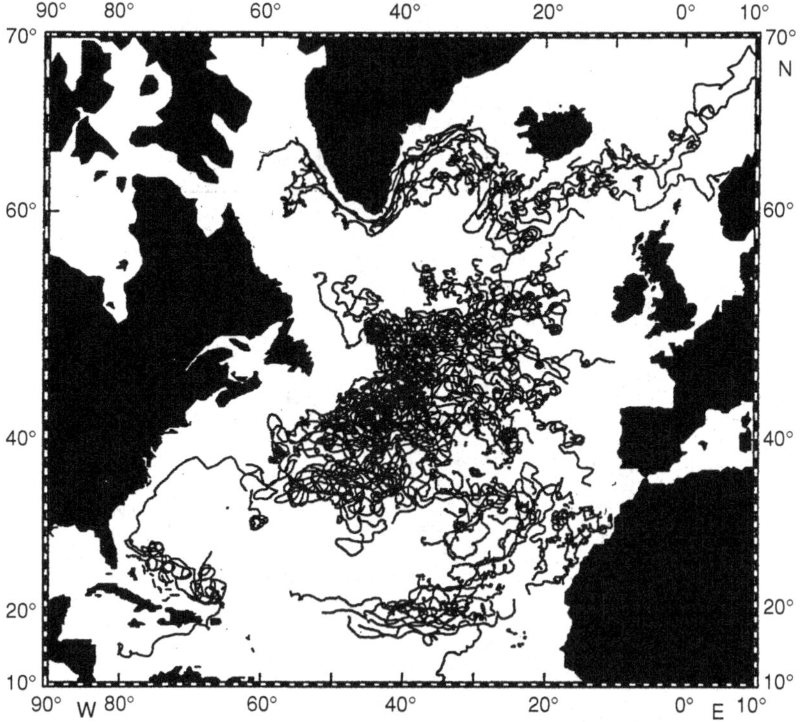

Figure 5.9. Spaghetti diagrams. Tracks of floats drogued to follow the water motion at a depth of 100 m in the North Atlantic showing the motions caused by mean currents and mesoscale eddies. Eddies are found almost everywhere, although the floats have been released not uniformly but according to the requirements of specific studies, so sampling is not uniform. There is only one, for example, that follows the path of the Gulf Stream up the east coast of the USA. (From Brügge, 1995.)

characteristics by following the tracks of a large number of drifters and floats released mainly into the North Atlantic and North Pacific Oceans. The 'spaghetti diagrams', the convoluted tracks of floats (Fig. 5.9), demonstrated the validity of earlier conclusions based on the use of subsurface Swallow floats,[8] namely that the currents in 'mesoscale' eddying and stirring motions of some 50–300 km diameter exceed the mean flows, showing that, at these scales, the water motion is more analogous to a randomly alternating electric current than to a direct current. At the large scales of the mesoscale eddies, the Earth's rotation is important and motion is approximately geostrophic. Mesoscale eddies are detectable from satellite measurements of the sea-surface temperature as illustrated in Fig. 5.10 here and Fig. 5.15 later.

Near-surface currents can be followed with a satellite-linked surface float attached by a cable to a subsurface high-drag drogue. (In some early experiments the drogues

8 Swallow floats are named after their inventor, John Swallow (1923–1994). They work on the principle that, a Swallow float being basically a structure (an aluminium tube) that is less compressible than water, it is stable when floating at a depth at which its density matches that of water. Downward (or upward) displacements cause its density to become less (or greater) than that of the more (or less) compressed surrounding water, so it will rise (or sink) towards its initial depth.

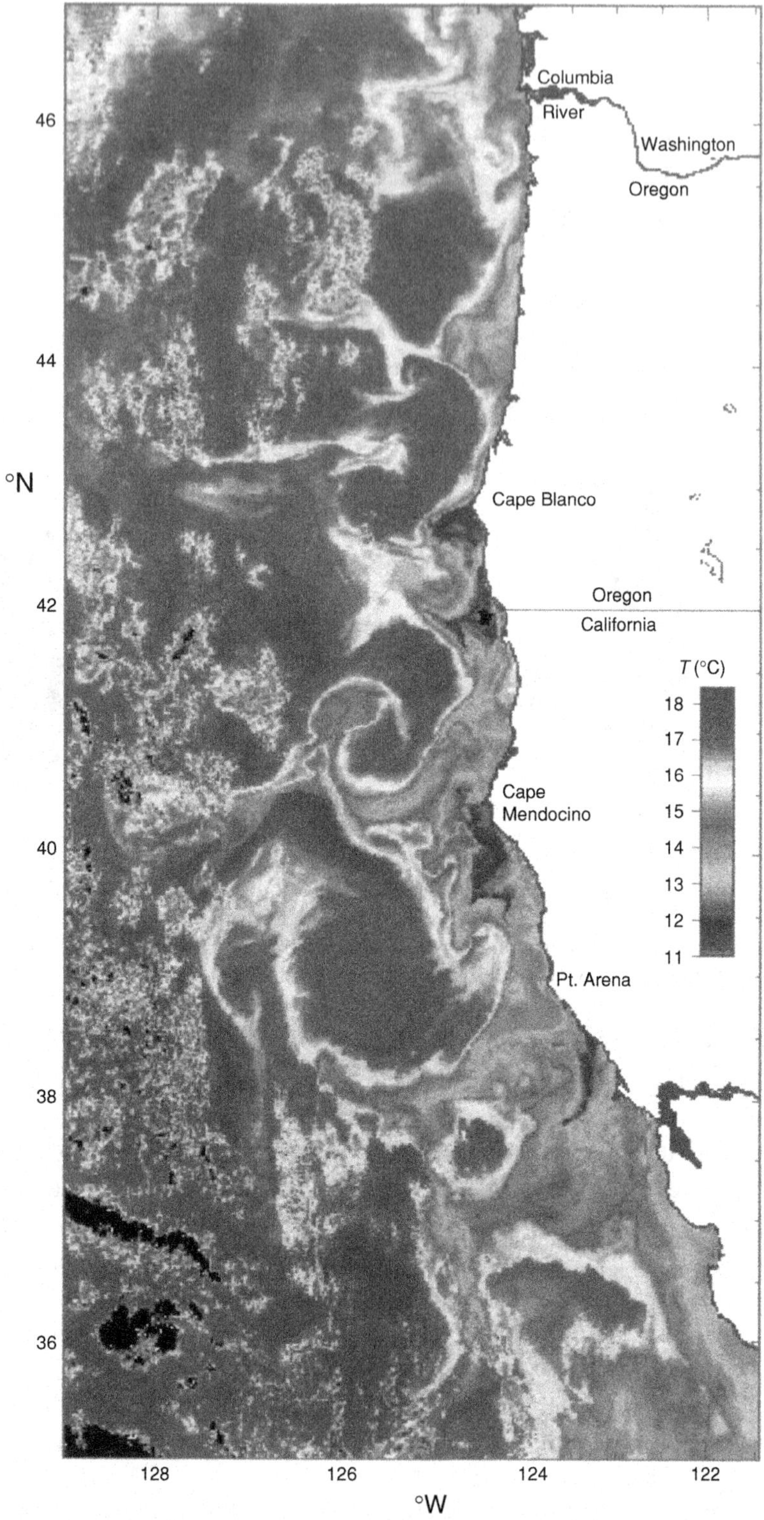

Figure 5.10. A satellite image of sea-surface temperature (SST) in the current system off Oregon and California. The lighter shades denote water of temperature

became detached after a few weeks, leading to some uncertainty in the measurements of dispersion.) In recent years many floats have been released, allowing the study of dispersion at large time and space scales. On average about 2500 ARGOS-tracked drifter observations in the North Atlantic are available for each month in the period from 1993 to 1997, allowing study of the effects of regional and latitudinal variations.

Estimates of the dispersion coefficient, $K_{H\infty}$ in (5.6), have been obtained from observations of floats released in the Pacific during the period 1979–1999 and in the Atlantic between 1989 and 1999. The distributions of the mean currents and of $K_{H\infty}$, with corrections to remove the increases in float separation caused by the mean shear flows, are shown in Fig. 5.11. In general the pattern of the greatest values of $K_{H\infty}$ follows that of the major currents, the Kuroshio in the Northwest Pacific, the Gulf Stream in the Northwest Atlantic, the East Australian Current, the Agulhas Retroflection Zone south of South Africa and the Equatorial currents; the most energetic mesoscale eddies leading to dispersion are generally found, and are often generated, where the mean flows are greatest. The values of $K_{H\infty}$ range from about 2×10^3 to 28×10^3 m^2 s^{-1} and there is a general decrease from the Equator towards the poles, although this trend may be reversed, at least locally, by energetic eddies in the Antarctic Circumpolar Current. Values of the Lagrangian time scale, T_L in (5.2), are 2–6 days and the Lagrangian length scale, L_L in (5.3), is typically 20–50 km, scales which are consistent with the source of dispersion being the large-scale and slowly varying mesoscale eddies.

• It is evident from Figs. 5.9 and 5.10 that, even at the large scales of mesoscale eddies, the ocean is in a state of eddying motion that broadly resembles the turbulence observed at smaller scales, e.g., the structure shown in Fig. 2.3. It does, however, differ in that the motions are largely horizontal, of scale far exceeding the Ozmidov length scale, the largest scale of overturning eddies in the pycnocline, and greater even than the ocean depth. The Lagrangian integral length scale, L_L, providing a measure of eddy scale, is comparable to the internal Rossby radius, L_{Ro}, defined in Section 1.8.2 as $L_{Ro} = c/f$, the scale at which the effects of the Earth's rotation become important. (Here c is the speed on long internal waves of the first mode and f is the Coriolis frequency. Both f and c vary with position and therefore so does L_{Ro}: it generally decreases polewards.)

A theory of the large-scale, or 'geostrophic', turbulence which takes into account the effects of the Earth's rotation and curvature has been devised. Because the mesoscale eddies depend on different dimensional quantities and have different dynamics, the form of their energy spectra, for example, differs from that in non-rotating homogeneous turbulence defined by (2.15). Generally it appears that, at these relatively large scales, the ocean turbulence is non-homogeneous and generally

Figure 5.10. (*cont.*) about 16 °C and water is colder nearer shore. Cloud (black) obscures the sea surface at bottom left. The image shows a remarkable range of features, including eddies, jets and meanders in the flow, some originating near promontories and leading to filaments extending 100 km or more off-shore, for example off Cape Mendocino and southwest of Point Arena. (Data obtained as part of the US GLOBEC Northeast Pacific Program. From Hickey and Royer, 2001.)

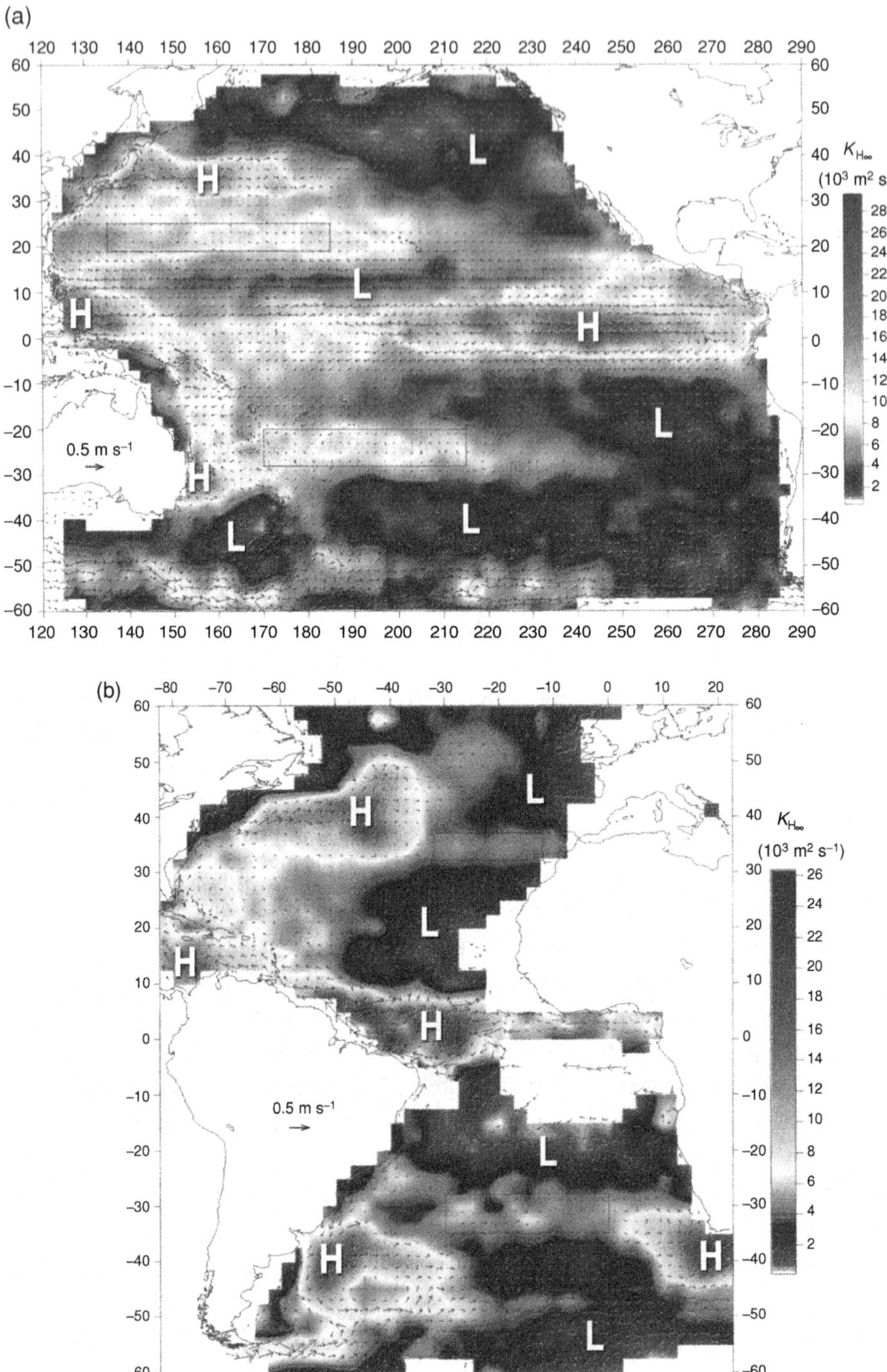

Figure 5.11. The eddy dispersion coefficient, $K_{H\infty}$, of surface drifters in (a) the Pacific and (b) the Atlantic Ocean. Arrows indicate the mean currents; H and L indicate, respectively, the areas of high $K_{H\infty}$ ($> 20 \times 10^3$ m^2 s^{-1}) and low $K_{H\infty}$ ($<4 \times 10^3$ m^2 s^{-1}). (From Zhurbas and Oh, 2004. Care has to be taken in making the estimates at specific locations from drifter tracks, not all of which may pass through a fixed point, and it is recommended that interested readers refer to the original paper for details.)

anisotropic in all three directions, the vertical, meridional (or north–south) and zonal (east–west).

The rate of horizontal dispersion at the sea surface within the Atlantic and Pacific Oceans at the scale of the mesoscale eddies is fairly well determined, although seasonal variations have yet to be assessed. Dispersion is, however, relatively poorly quantified at the lateral boundaries of the ocean basins, for example in regions where enhanced along-slope currents are found over continental slopes.

5.3.2 Subsurface floats

Two methods are available to follow the motions of subsurface floats. The first relies on acoustic tracking using SOFAR (SOund Fixing And Ranging) that was developed by Rossby and Webb in about 1970. Long-range acoustic propagation is possible in the sound channel at the level of minimum sound speed. The channel is at a depth of about 800 m in mid-latitudes but close to the surface in the polar oceans. A measure of the travel time of pulses of sound with the known speed of sound in the ocean, about 1500 m s^{-1}, provides the means of establishing range, and triangulation gives position. The second method uses satellite location. Autonomous Lagrangian Circulation Explorer (ALACE) floats spend most of their time drifting at a set depth, only occasionally returning to the surface so that their new location can be fixed and their drift determined, and to transmit internally recorded data back to shore.

Observations of the dispersion of pairs of floats tracked acoustically at a depth of 700 m in the North Atlantic are consistent with Richardson's four-thirds power law, dispersion following $D \propto t^{3/2}$ as in (5.8) when D, the rms distance between floats, is between the internal Rossby radius and a scale of about 300 m. At greater scales, or typically a few months after pairs of floats are released, dispersion tends to be consistent with a Fickian relationship (5.9), but with the zonal relative dispersion generally greater than the meridional.[9] (At scales smaller than about 300 m there is, however, some evidence that $D \propto \exp(qt)$, where q is a constant, corresponding to exponential growth.)

Analysis of the tracks of subsurface floats at depths between 90 and 4100 m covering a total of 588 float-years between 1973 and 1998 in the North Atlantic finds that the tracks of relatively shallow subsurface floats have Lagrangian time and length scales that are consistent with those of the surface drifters, suggesting that over long periods of time, $t \gg T_L$ (typically 4 days), the dispersive effects of motions at the sea surface, such as wave Stokes drift or those of Langmuir circulation experienced by surface drifters, are insignificant. Outside the Gulf Stream, T_L is relatively constant at about

9 An explanation of the difference between the zonal and meridional dispersivities is yet to be found. It may possibly be related to a reverse (up-scale) cascade of energy proposed by Rhines, that leads to barotropic zonal jets of width of order $(U/\beta)^{1/2}$, where U is the typical speed of current, about 0.05–0.1 m s^{-1}, and $\beta = 2\Omega \cos\phi/R$, where Ω is the angular frequency of rotation of the Earth, ϕ is the latitude and R is the Earth's radius. In the ocean, the Rhines scale, $(U/\beta)^{1/2}$, is typically about 70 km, rather less than the scale at which the difference between zonal and meridional dispersion is observed.

5–7 days for floats at depths between 700 and 2000 m, but the characteristic speeds, u', and Lagrangian integral length scales, L_L, both decrease with increasing float depth below 700 m, the latter from 20–40 km at depth 700 m to 5–20 km at 2000 m. In the Gulf Stream, $T_L \sim 4$ days at 700 m and increases by a factor of about 1.3 between 700 and 2150 m.

5.4 The dispersion of solutes: methods and observations

5.4.1 Dispersion (or horizontal diffusion) of a solute

A different approach has to be taken to obtain a measure of the spread of a patch of solute, rather than the spread of particles. Methods are inevitably based on the growth of a length scale determined from the distribution of the concentration of the solute.

Okubo, for example, analysed measurements of dye released in patches into the upper ocean. He used the measured area, $A(C)$, enclosed by a horizontal curve of constant dye concentration, C, which is set equal to the area, πR^2, of a circle with the same area (i.e., $R = (A/\pi)^{1/2}$), and he then determined a value, σ_r, from

$$\sigma_r(t) = \left(\int R^2 C(R) 2\pi R \, \mathrm{d}R \Big/ \int C(R) 2\pi R \, \mathrm{d}R \right)^{1/2}. \tag{5.10}$$

This is a measure of the second moment of the radial distribution of dye. Plots of $\log \sigma_r^2$ versus $\log t$ are obtained, to which a linear relationship is fitted. The estimates of the horizontal dispersion coefficient, K_H, at time t described in the following section are based on the relation

$$K_H = \sigma_r^2/(4t), \tag{5.11}$$

a definition chosen to be consistent with that in two-dimensional Fickian diffusion with a constant dispersion coefficient.

5.4.2 Dye releases in the surface boundary layer

The horizontal spread of patches of dye released into the surface layer in 20 studies in the North Sea, off Cape Kennedy and Southern California, in New York Bight and in the Banana and Manokin Rivers was analysed by Okubo. The results are often used as a basis for estimating near-surface dispersion at scales ranging from 30 m to 100 km over times of 2 h to 1 month.

The data are in accord with an approximate empirical relationship,

$$\sigma_r^2 = 1.08 \times 10^{-6} t^{2.34}, \tag{5.12}$$

determined using (5.10), with σ_r in metres and t in seconds. The variation of diffusivity

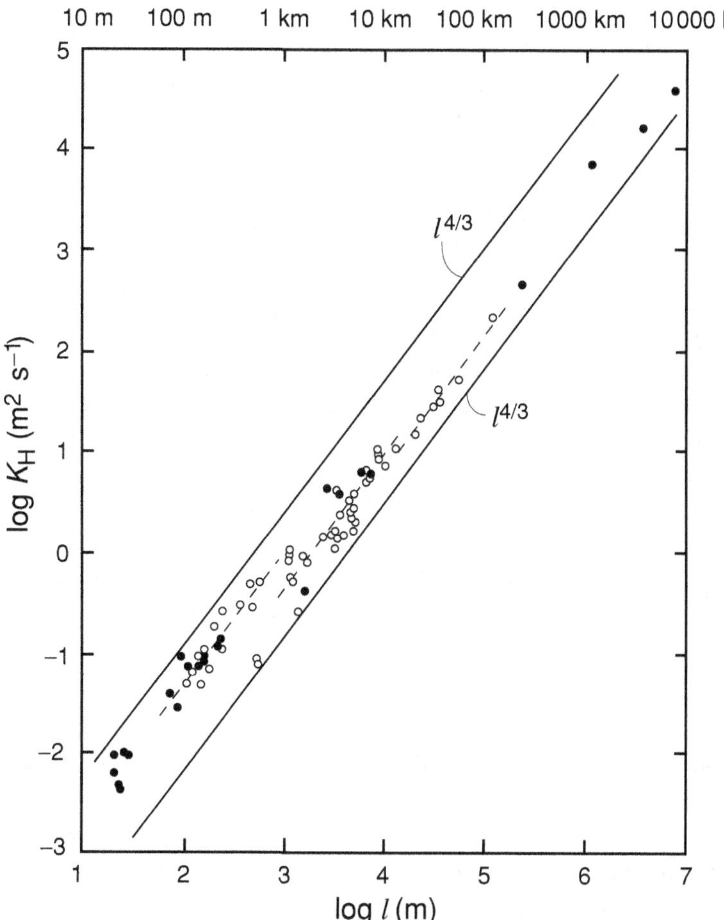

Figure 5.12. Size-related dispersion. The variation of horizontal eddy dispersion coefficient, K_H, with the horizontal scale, l, of patches of dye at the sea surface derived using (5.10) and (5.11). There is a break at a scale of about 1 km (where log $l = 3$) in the dashed $l^{4/3}$ lines that are closely fitted to the data, and possibly another at 10 km. (From Okubo, 1971.)

with length scale l, taken to be equal to $3\sigma_r(t)$,[10] is approximately given by

$$K_H = 1.03 \times 10^{-4} l^{1.15}, \tag{5.13}$$

with K_H in m^2 s^{-1} and l in metres (Fig. 5.12). Although this differs from Richardson's four-thirds power law, the values of K_H are (remarkably) consistent with dispersion coefficients determined from particle tracking. [**P5.3**]

• Closer fits to the data can, however, be found by fitting K_H with $l^{4/3}$ power laws over segments of the data shown in Fig. 5.12, those with 30 m $< l <$ 1 km and 1 km $< l <$ 100 km. A major break in these linear fits occurs at about 1 km and corresponds

10 This is selected on the grounds that, if the concentration distribution is radially symmetric and Gaussian, 95% of the dye is within a circle of diameter $3\sigma_r$.

to a time of about 10 h. A dimensionally correct form for K_H is $c_1\varepsilon^{1/3}l^{4/3}$, with a dimensionless constant c_1 and where ε is a quantity having the dimensions of a rate of energy dissipation. If this form is correct, the two segments in which the four-thirds law seems to apply would correspond to two ranges with different rates of dissipation or transfer of energy by the energy transfer or cascade through the energy spectrum. The break in the spectrum is plausibly a scale at which energy enters the spectrum from some physical process driving energetic dispersive motions with scales of about 1 km and 10 h. Changes in the energy spectrum of turbulent motion where there is an energy input (and, correspondingly, changes in the relationship between K_H and l) are envisaged in a generalized theory of turbulence devised by Ozmidov in 1965. It is possible that tidal flows (the M_2 tide has a period of 12.43 h) or other processes such as Langmuir circulation and eddies (Section 5.4.3) provide such sources of energy. [**P5.4**] The nature of dispersion also changes at large scales, for which the effects of the Earth's rotation and of mesoscale eddies (Section 5.3.1) become appreciable. New dimensional parameters then become significant in scaling the dependence of the rates to the scale of a diffusing patch or plume of solute.

Those performing experiments in the ocean find that patches of dye or other tracers rarely have the nearly circular form implicit in (5.10), since patches often become noticeably longer in the downwind direction than across wind.[11] Even in flows that are regarded as turbulent, anisotropy of the stirring motions, notably of shear, may lead to directional dispersion characteristics. (The higher downwind currents found in windrows imply the presence of differential currents at the 5–100-m across-wind scales of the turbulent Langmuir circulation.)

Measures have been devised to express such directional differences in diffusivity. They commonly depend on measures of the second moments of concentration in each of the dispersion directions (e.g., x, downwind, and y, across) to determine $\sigma_x(t)$ and $\sigma_y(t)$, and on using equations similar to (5.11) to determine separate diffusion coefficients, K_x and K_y, to characterize dispersion. The downwind dispersion coefficients of patches of dye are generally found to exceed those in the across-wind direction. [**P5.5**]

• The significance of such studies of dispersion is emphasized by the results, (5.12) and (5.13): dispersion depends not only on spatial scales (the size of a dispersing patch) but also on the time over which dispersion has occurred. The unsteadiness of external forcing may be important, as remarked in Section 5.1.3. During the period of a day or more during which oil or a toxic chemical may be released following an accident at sea, the effects of changes in wind speed, for example, may alter the nature of turbulence in the upper ocean and even change the processes that are dominant in causing dispersion.

5.4.3 Tracer releases in the pycnocline

As well as giving estimates of diapycnal diffusion, the spread of SF_6 in the NATRE diffusion experiment described in Section 4.7 provides information about isopycnal

11 Okubo (1971) acknowledged this: '*The real diffusion pattern never shows radial symmetry*'.

dispersion. The SF_6 tracer was injected at a depth of about 300 m in bands less than 100 m wide and about 5 km apart, and its general location was subsequently found with the use of acoustically tracked neutrally buoyant floats. Two weeks after injection the bands were found to have blended into what appeared to be a single continuous patch. In the following months the mean motion of this patch was to the west at a rate of about 1 cm s^{-1}, but, as it grew over this period of time, the patch became distorted by the shear in the large mesoscale eddy field (i.e., it was stirred by the large turbulent eddies, much as illustrated in Figs. 1.7 and 5.6(b)). When it was next surveyed, 6 months after release, the tracer was found to be present in sinuous bands or 'streaks' of rms width about 3 km, estimated to be some 1800 km in overall length and separated by tracer-free water, as shown in Fig. 5.13(a) (October and November).

The rate of the reduction in the width of these streaks by shear and divergence produced by the mesoscale eddies may be calculated from knowledge of the mesoscale field of motion, much as described in Section 1.6. An estimate for the mean horizontal dispersion at the 3 km scale of the streaks can be derived by assuming that the observed streak width represents a steady state in which the reduction in width by divergence and shear is countered and balanced by small-scale dispersive processes. The dispersion coefficient is found to be $K_H \approx 2.7\,m^2\,s^{-1}$.[12] Although they constitute a significant achievement and an interesting estimate, the observations do not determine whether this estimate or the processes leading to dispersion are representative of those in the ocean pycnocline at scales of 3 km or are particular only to the regions within the mesoscale eddy field where the streaks are produced; neither do they definitively identify the processes contributing to small-scale dispersion. There is a continuing debate about the extent to which internal waves and very-small-scale eddies are involved.

Further insight into the processes of dispersion at scales of 1–10 km is provided by an analysis of investigations of the spread of dye in stratified water over the New England Continental Shelf by Sundermeyer and collaborators. They conclude that relatively mixed patches of water produced, for example, by breaking internal waves, spread much as illustrated in Fig. 1.16 and, under the effect of the Earth's rotation, begin to rotate, forming a field of small pancake-like eddies within the stratified water. Such eddies, or 'vortical modes', interacting with one another (and perhaps with internal waves) disperse material horizontally, leading to a horizontal dispersion coefficient estimated to be

$$K_H \approx (7/2)[h^6\,\Delta N^4\,\varphi/(L^2 f^3 K_\nu)], \tag{5.14}$$

where h and L are typical vertical and horizontal dimensions of the eddies, ΔN is the difference between the buoyancy frequency within and that outside the eddies, φ is the frequency at which mixing events (e.g., breaking waves) occur, f is the Coriolis frequency and K_ν is the vertical component (as defined by (2.2)) of the eddy viscosity in the region outside the mixed patches. (It is not known whether similar eddies resulting, for example, from patchy wind mixing can affect dispersion in the mixed layer in

12 This estimate by Ledwell *et al.* (1998) compares with an estimate, $K_H = 1.0\,m^2\,s^{-1}$, found using (5.13) for the dispersion of a surface dye patch with a length scale, l, of 3 km.

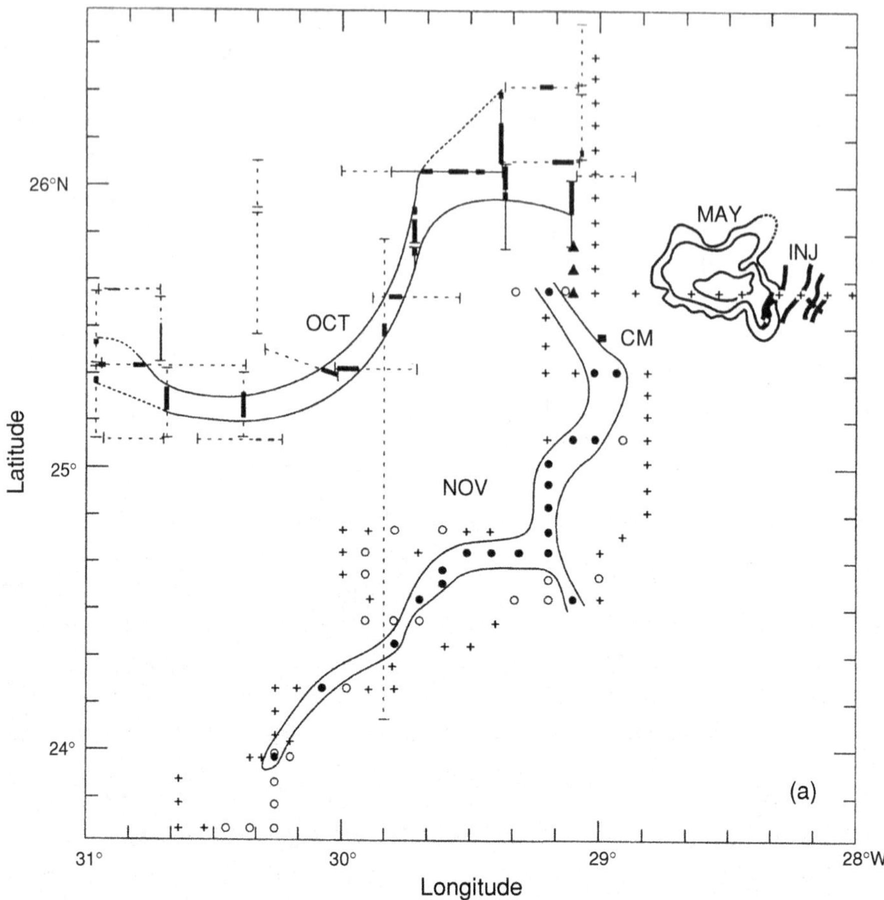

Figure 5.13. Dispersion of a tracer in the NATRE. The horizontal outlines of a patch of SF_6 released at about 300 m depth in the eastern North Atlantic in the North Atlantic Tracer Release Experiment (NATRE). (a) The injection in early May 1992, marked as INJ, and the location of the tracer patch in subsequent surveys in May, October and November 1992. The October and November surveys sampled different parts of the streaky dispersed tracer patch. (b) The tracer's location in surveys about a year after release, in April and May 1993, when (although patchy) the region occupied by the tracer is more uniform and much less streaky than in the previous surveys. 1 N $\approx$ 110 km. (From Ledwell *et al.*, 1998.)

subsequent calm conditions and explain the apparent breaks in the linear fits to data in Fig. 5.12.)

When they were surveyed a year after release in the NATRE study, the streaks of SF_6 tracer had combined to produce a huge patch some 500 km across in the meridional (north–south) direction and of width 700–800 km in the zonal (east–west) direction. The tracer was found almost everywhere within the patch, but at variable concentration, as shown in Fig. 5.13(b). Complete homogenization of tracer concentration had still

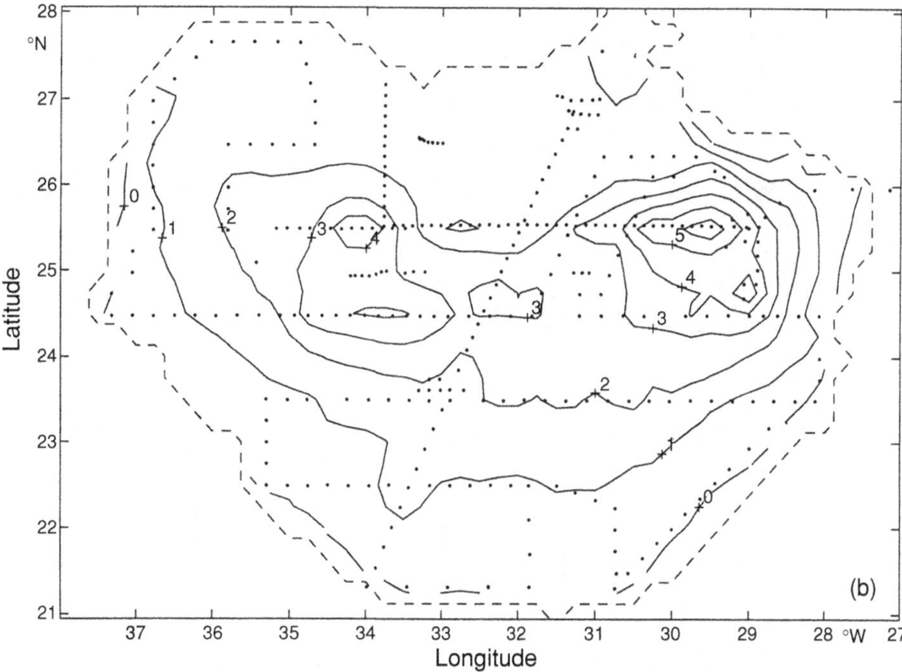

Figure 5.13. *(cont.)*

not been achieved even after a further 18 months, by which time the patch was some 900 km in extent in the meridional direction and 1600 km in the zonal.

Estimates of the isopycnal (approximately horizontal) dispersion coefficient, K_H, based on the evolving size of the streaky SF_6 patch are about 0.07 m^2 s^{-1} at scales of 0.1–1 km, and 2 m^2 s^{-1} at scales of 1–30 km. As expected, these are very much greater than the diapycnal diffusivity of $(1.7 \pm 0.2) \times 10^{-5}$ m^2 s^{-1} found in the same experiment (Section 4.6). After 12–18 months of dispersion and at scales of 300–1000 km, there is a clear difference between the horizontal zonal and meridional diffusivities. The evolving distribution of the tracer indicates horizontal diffusivities, estimated on the basis of Fickian dispersion, of 2300 m^2 s^{-1} in the zonal direction and 650 m^2 s^{-1} in the meridional, values that are consistent with those deduced from subsurface float tracks. [**P5.6**]

The tracer, SF_6, has also been released in other regions. By the time of the first survey, 14 months after release at depth 4000 m in an experiment in the Brazil Basin, about 500 m above the top of east–west-trending fracture-zone valleys and ridges to the west of the Mid-Atlantic Ridge, much of the tracer had spread to the southwest, further from the Ridge (Fig. 5.14). Some had, however, moved eastwards, following isopycnals that extend to greater depths into the valleys of the fracture zone. The lateral scale of the patch was fairly symmetrical after 14 months, with a width of about 300 km. When the tracer was next sampled, 26 months after release, the zonal dimension of the patch had expanded to about 800 km, significantly greater than its meridional extent

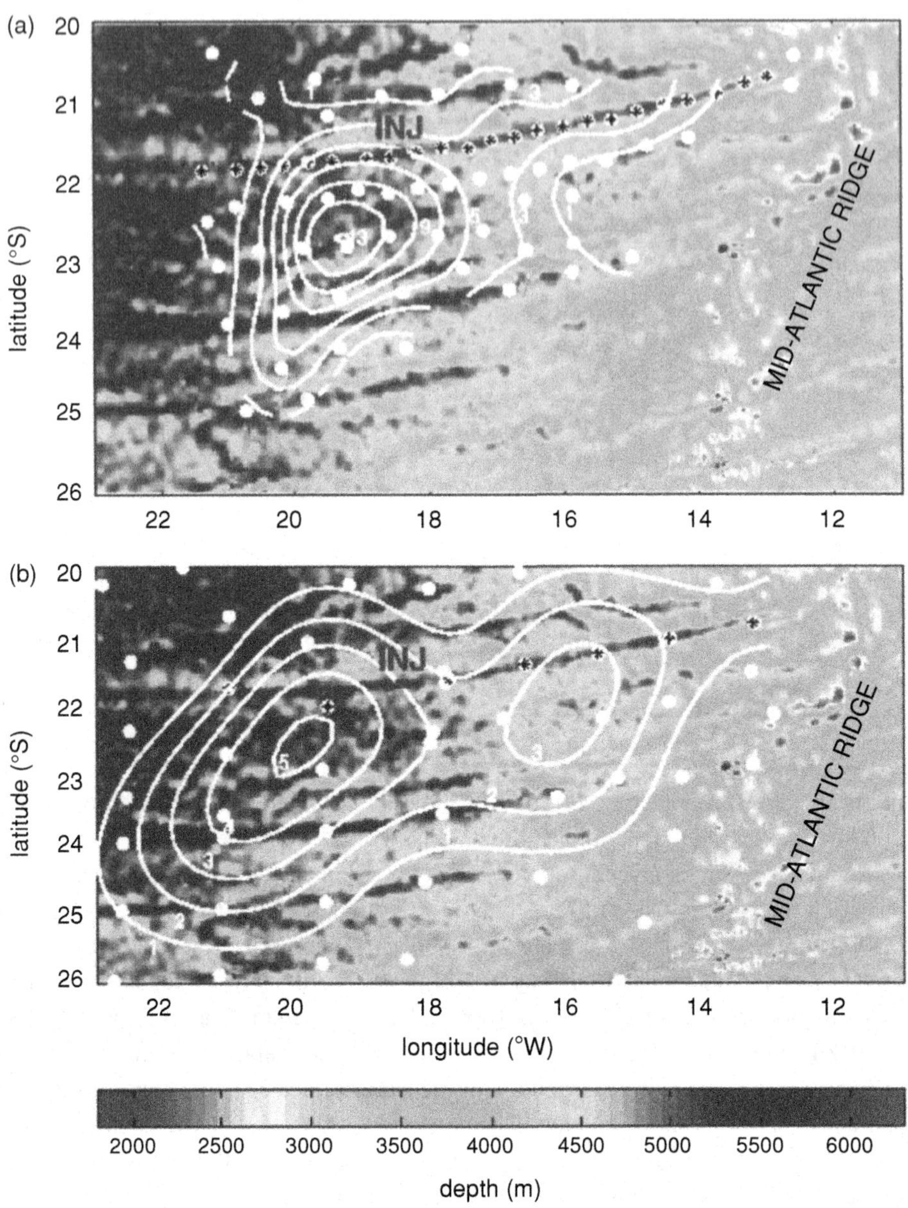

Figure 5.14. Dispersion of a tracer released at great depth. The horizontal spread of a tracer, SF_6, released at the position marked INJ, at about 4000 m depth near 22 °S 18 °W in the Brazil Basin, below and about 650 km west of the top of the Mid-Atlantic Ridge. Surveys are at (top) 14 months and (bottom) 26 months after the release of the tracer. Dots mark sampling points and the grey background indicates the presence and orientation of the underlying rough fractured topography. Depths exceed 4800 m in the dark areas to the west of the Ridge, whilst over the Ridge depths are typically 3500 m or less. The patch has spread along isopycnal surfaces that tilt down towards the Ridge, and is about 300 m thick in the first survey and 500 m thick in the second. (From Schmitt and Ledwell, 2001.)

of 450 km. These scales are about half those of the shallower NATRE tracer study, a consequence of the lower mesoscale eddy activity and stirring in the deep waters of the basin. The dispersion coefficients are significantly lower than those found in NATRE but still have greater values in the zonal than in the meridional direction.

The use of SF_6 as a tracer is becoming a standard means of studying both diapycnal diffusion and isopycnal dispersion.

5.4.4 Natural and anthropogenic tracers

The 'natural tracers' that characterize water masses may be conservative ones such as salt that, except for the small loss through spray to airborne droplets (aerosols), does not readily pass from the ocean, or non-conservative, such as temperature (or heat) that may be passed through the sea surface to and from the atmosphere. Oxygen is another non-conservative tracer that, in addition to exchange with the atmosphere, is subject to biological processes. Natural tracers may be almost neutral, having negligible effect on the density or other properties of water that affect its dynamics (e.g., oxygen concentration), or non-neutral: temperature and salinity change the water density. The study of water masses identified by their particular temperature (T) and salinity (S) has long provided clues to circulation and, aided by the use of T–S diagrams, to where mixing occurs.

Float tracks and satellite observations of sea-surface temperature show that some mesoscale eddies remain coherent for periods of a year or more. These eddies are often associated with water derived from variations or instabilities in major currents, the first to be extensively studied having been the warm and cold core eddies derived from the meandering Gulf Stream and detectable at the surface as illustrated in Fig. 5.15. The best known subsurface eddies are the 'Meddies' (Fig. 5.16), which are composed of water of Mediterranean origin and found in the region of its spread in the eastern North Atlantic (Fig. 5.1). These eddies are isolated from one another and carry water from their source, sometimes over many hundreds of kilometres. They are a discrete, rather than continuous, mechanism of dispersion, rather like that of the bursts of water with high, but localized, Reynolds stress referred to in Section 3.4.4 that are ejected – or detrain – from the benthic boundary layer, although, unlike the bursts or ejections, Meddies may contribute only marginally to a long-term overall dispersion.

Radioactive chemicals, accidentally or deliberately released into the ocean, have also provided tracers of dispersion and are usually neutral in their effect on dynamical processes. During the atmospheric nuclear bomb testing in the 1950s and early 1960s the radionuclide tritium (^3H) was introduced by fallout into the ocean, mainly at latitudes of 40° to 60° in the northern hemisphere.[13] Its radioactive half-life is 12.43 yr, and its concentration decreases as a consequence of radioactive decay as well as diffusion as it is advected from its source. The changes in distribution and concentration

13 By the early 1970s, the total worldwide fallout of tritium was about 110 EBq, with 24 Ebq in the North Atlantic Ocean (1 EBq = 10^{18} Bq).

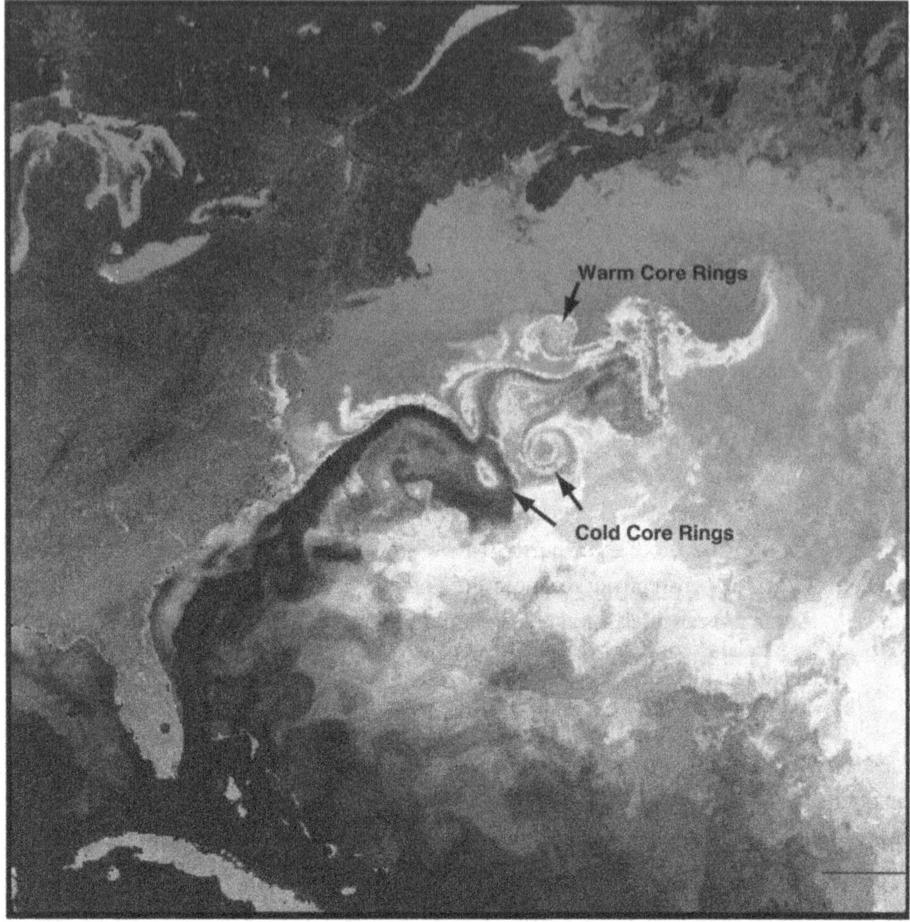

Figure 5.15. Gulf Stream eddies. An image of sea-surface temperature showing the meanders and eddies formed in the Stream after it has separated from the eastern seaboard of the USA. The image covers a north–south scale of about 3000 km, and the eddies are some 200 km across. The darker shades of grey correspond to the warmest water. (Image produced and kindly provided by O. Brown, R. Evans and M. Carle at the Rosenstiel School of Marine and Atmospheric Science, Miami, Florida.)

of ^{3}H since the termination of bomb tests provide information about oceanic transport from the northern to the southern hemispheres, and of its downward dispersion into the ocean by deep convection, as illustrated in Fig. 5.17.

The spread of caesium 137 discharged from the Windscale (renamed Sellafield) Nuclear Plant into the Irish Sea is illustrated in Fig. 5.18. The figure shows the distribution in 1983 after 10 years of continual release, at variable, but documented, rates. (About $2\,\text{PBq} = 2 \times 10^{15}\,\text{Bq}$ was released in 1982.) The concentration of caesium in seawater is reduced by mixing and radioactive decay: caesium has a radioactive half-life of 30.17 yr. Factors such as its absorption by organisms and adsorption onto the surface of particles in suspension and on the seabed cause further reduction in concentration.

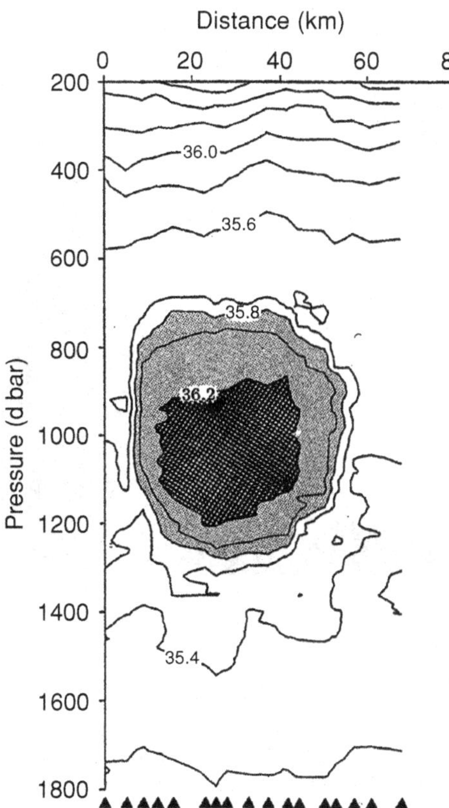

Figure 5.16. The structure of a Meddy. A vertical salinity section through a Meddy at a depth of about 1000 m. The vertical scale is stretched relative to the horizontal; the relatively warm and salty eddy composed of water that has originated from the Mediterranean is flattened, about 500 m thick and 40 km across. This particular eddy was tracked by floats for about 2 years and resampled four times as it decayed, accompanied by the development of horizontal intrusions and layers formed beneath it as a result of double diffusive convection. The black triangles below the figure show CTD sampling locations. (From Armi *et al.*, 1989.)

• As is the case for other dispersants, the dispersion of radioactive material depends on its properties (e.g., half-life) as well as on advection and the turbulent motion of the ocean.

Suggested further reading

Observations of dispersion

Richardson and Stommel's (1948) brief account of their very short, preliminary but illuminating experiment on particle dispersion is worth reading, if only to see how many points of doubt can be found in their conclusions! Stommel (1949a) describes his subsequent experiments.

Okubo's (1971) analysis of dispersion of dye released in the upper ocean is still very useful and thought-provoking.

If not already read after Chapter 4, perhaps it should be now: Ledwell *et al.*'s (1998) account of the spread of SF_6 observed in NATRE is amongst the best available of dispersion and isopycnal mixing.

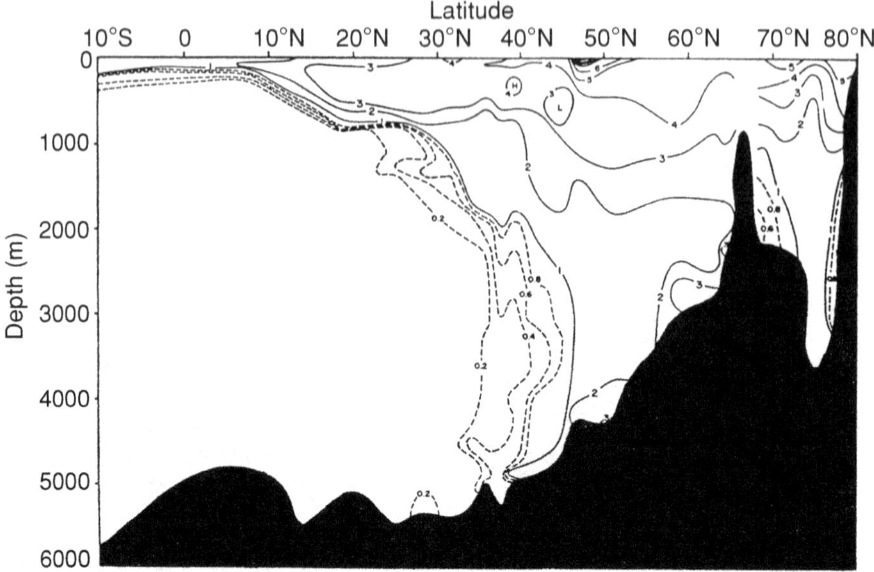

Figure 5.17. Tritium (^{3}H) concentration in the North Atlantic in 1981. Contours give the ratio of tritium-to-hydrogen in so-called tritium units. One tritium unit represents a tritium to hydrogen ratio of 10^{-18}. Releases from the atmospheric nuclear bomb testing ended in the 1970s, and the concentrations at depth in the North Atlantic reflect the vertical transfers by convection, diapycnal diffusion and flows from the Arctic Ocean, and the gradual, radioactive decay of ^{3}H (half-life 12.4 yr). Unlike in Antarctica, where Antarctic Bottom Water (AABW) is formed that flows northwards into the Atlantic, Pacific and Indian Oceans, there is no comparable formation of bottom water from the Arctic. (Data collected and figure produced by Göte Ostlund of the University of Miami as part of the GEOSECS programme of observations.)

Further study

Lagrangian float measurements in the ocean

Rossby and Webb (1970) describe the development of SOFAR. Rossby *et al.* (1983) and Davis (1991a) provide valuable reviews of the methods used to follow motions in the ocean by floats and of their application. Much discussion has been directed towards how mean shears should be accounted for, particularly by Davis (1987, 1991a, 1991b). A method to obtain a long-term dispersion coefficient, $K_{H\infty}$, of floats that accounts for and removes the effects of mean shears has been devised, and this is applied by Zhurbas and Oh (2004) in the analysis leading to Fig. 5.11 (although they refer to $K_{H\infty}$ as a 'lateral diffusivity').

Observations of dispersion

There are numerous papers describing studies using floats, for example see Lumkin *et al.* (2002) and McClean *et al.* (2002). Ollitrault *et al.* (2005) provide a good account

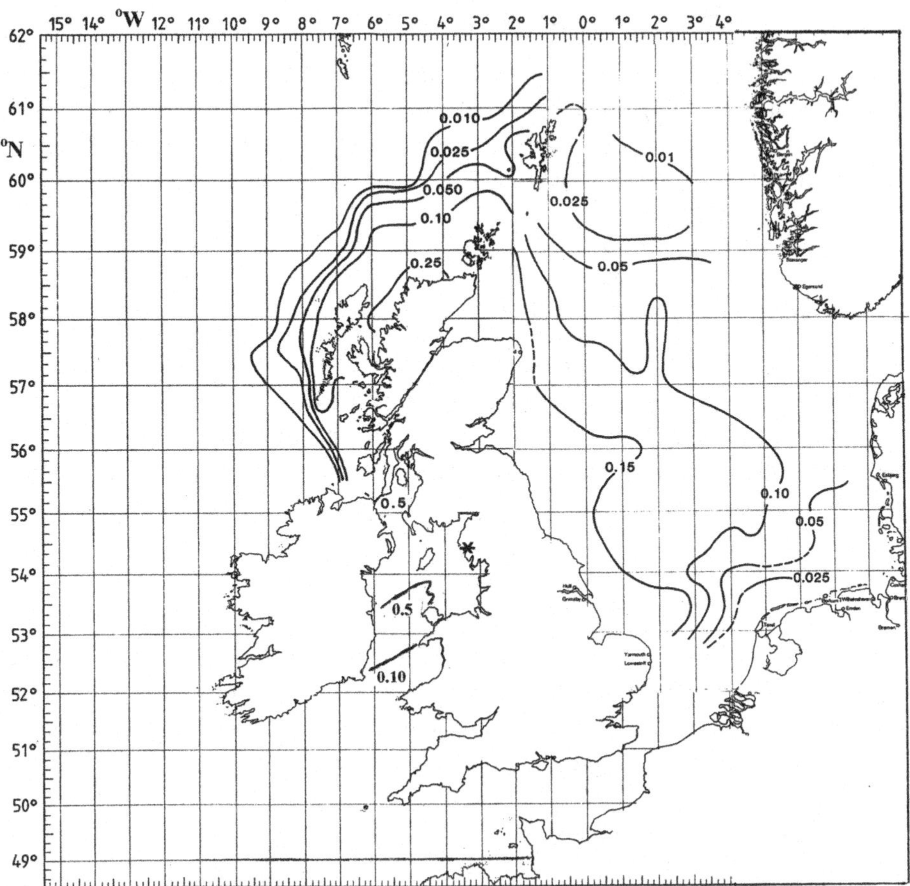

Figure 5.18. The distribution of caesium 137 around the British Isles observed in 1983. The source of caesium, the nuclear reactor site at Sellafield on the coast of the Irish Sea, is marked with a star. Much of the spreading results from advection in the mean flow from the Irish Sea, around the north of Scotland and eastwards through the northern North Sea towards the coast of Norway. The units are Bq kg^{-1}. (From Hunt, 1985; British Crown copyright, 1985, reproduced by permission of CEFAS, Lowestoft, UK.)

of earlier studies and a description of estimates of dispersion from pairs of floats released in the North Atlantic.

Although apparently providing useful estimates of dispersion, much more investigation is needed in order to verify Obuko's empirical relationships (5.12) and (5.13), and to examine their implications about the nature of turbulence dispersion. As Okubo himself makes plain, more careful study is needed of the anisotropy of dispersion and to quantify the 'environmental' effects of wind, waves, mixed-layer depth and stability (or Monin–Obukov length scale, L_{MO}), and their transience, on the dispersion of solutes over time scales of 1 h to about a week, a time period within which, for example, accidentally released substances may present the greatest hazard.

Turbulence, particles and bubbles

There is a rapidly growing literature on the interactions between turbulence and particles or bubbles, relating mainly to the mean fall speeds of the particles, e.g., Maxey (1987), or rise speeds of bubbles (Spelt and Biesheuvel, 1997), but little that can yet be applied with confidence because of the transient and anisotropic state of turbulence in ocean boundary layers.

Dispersion by vortical mode

Kunze (2001) provides an informative introduction to this mode of motion that can coexist with and has similar dimensions to internal waves.

Sundermeyer *et al.* (2005) give an account of mixing by vortical modes in shallow stratified water. Numerical simulations of dispersion are reported by Sundermeyer and LeLong (2005).

Geostrophic turbulence

At scales exceeding the internal Rossby radius (and beyond the scale generally regarded as part of the vortical mode of motion, the field of variable motion is dominated by mesoscale eddies in nearly geostrophic balance. A good introduction to this 'geostrophic turbulence' is given by Rhines (1979).

The deposition of wastes in the ocean

Park *et al.* (1983), in a volume of a series entitled 'Wastes in the Ocean', provided several informative articles on radioactive wastes, including inventories of discharges of radioactive wastes into the ocean and reference to their effect and the strategies adopted for their oceanic disposal.

Dispersion is discussed in several parts of TTO including Sections 9.6 and 10.5 and Chapter 13.

Problems for Chapter 5

(E = easy, M = mild, D = difficult, F = fiendish)

P5.1 (M) Richardson's four-thirds power law. Show that, if the scale-dependent diffusivity coefficient in Richardson's dispersion equation (5.7) is $F(l) = c_l l^q$, where q is a non-dimensional constant, then the dimensions of c_l must be $L^{(2-q)}T^{-1}$. If the variations of time and length depend only on viscosity, ν, and on ε (the dissipation rate, a measure of the flux of energy through the spectrum of the dispersing motions) so that time can be non-dimensionalized by $(\nu/\varepsilon)^{1/2}$ and length by the Kolmogorov scale, $l_K = (\nu^3/\varepsilon)^{1/4}$, show that the only power q that will make c_l

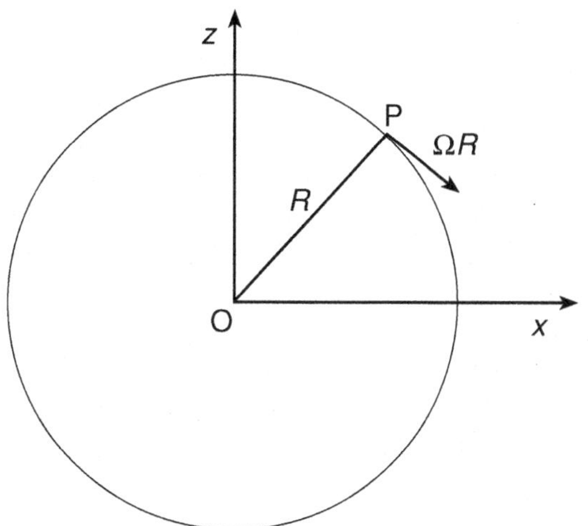

Figure 5.19. An idealized eddy rotating like a solid body about a horizontal axis at constant angular velocity, Ω. The motion is two-dimensional. The axes are x (horizontal, normal to the axis of rotation) and z (vertical). (See P5.2.)

independent of viscosity (and so leave dispersion to be effectively determined by ε alone) is $q = 4/3$.

• This interesting argument, justifying the four-thirds power law, is reported by Pacquill (1962), and comes from Taylor (1959). It differs from the more conventional dimensional argument, which might be to argue that, since the dimensions of the dispersion coefficient are L^2T^{-1}, if the dispersion coefficient is one in which molecular processes are insignificant and the dispersive process is one in which the turbulence and its dispersive effects are characterized completely by ε, the coefficient must have a form, $F(l) = \varepsilon^p l^r$, related to the dissipation rate but not to ν. From this it follows, by comparing dimensions, that the powers, p and r, must be 1/3 and 4/3, respectively. The agreement with Richardson's power law is doubtful because neither argument appears to relate a measure of the range of eddy sizes to the size of the patch.

P5.2 (D) The suspension of falling or rising particles in eddies. For simplicity, we take an idealized eddy motion that is represented by an infinite fluid that rotates as a solid body about a horizontal axis at constant angular velocity, Ω, as in Fig. 5.19. The motion is two-dimensional. The axes, x (horizontal, normal to the axis of rotation) and z (vertical), are taken at a point O on the axis of rotation, and lie in a plane normal to the axis of rotation. Fluid at a point P, at a distance R from the axis, moves in the plane at speed ΩR in a direction normal to OP. A particle at the point P is sinking vertically through the fluid at speed w.

(a) What are the equations that describe its horizontal and vertical components of velocity in terms of x and z? You may suppose that the inertia of the particle is negligible and that its diameter is very small relative to R.
(b) Solve these equations for x and z in terms of t, supposing that the particle is at position $(d, 0)$ at time $t = 0$.
(c) Show that the particle will follow a circular path of radius $|d - w/\Omega|$ about a centre at position $(w/\Omega, 0)$ with a period equal to that of the rotating fluid, and therefore remains suspended in the fluid.

Estimate the fall speed of dust particles in static water, supposing that they are approximately spherical, of density $1500\,\mathrm{kg\,m^{-3}}$ and of radius 20 μm (1 μm $= 10^{-6}$ m) and hence find the smallest upward vertical speed of flow required to maintain them in suspension. (For the fall speeds, see footnote 16 in Chapter 1.)

The centre of the circular path of the particle is the only point in the flow at which the particle can remain at rest, where there is no horizontal component of flow speed and where the particle's speed of exactly sinking matches the upward vertical component of the flow. (This is the limit of the circular paths as their radius tends to zero.) The horizontal distance, d, of the centre of the particle's circular path from O, the centre of flow rotation, increases with the particle's fall speed, w. Larger particles of the same density that sink more rapidly (e.g., $w \propto$ (particle radius)^2 for spherical particles) will follow circular paths with centres at a greater distance from O.

The problem assumes that the fluid in solid-body rotation is of infinite extent. Upward motion of the particle can occur only at a radius $R > w/\Omega$, and, if the extent of the region of solid-body rotation is of smaller radius (as it may be in a real eddy), particles will sink everywhere and cannot therefore be supported in suspension by the fluid motion in the eddy. A field of small eddying motions may cause very slowly sinking (or rising) particles to remain in suspension.

• In practice, turbulent motions, although at any instant possibly resembling closed eddies with speed characterized by (the dimensionally correct) $\varepsilon^{1/4}\nu^{1/4}$ – but not necessarily in solid-body rotation – and able to 'suspend' particles if $\varepsilon^{1/4}\nu^{1/4} > w$ (i.e., cause them to reverse their sinking or rising motion relative to the seabed), generally do not persist as discrete eddies for times long enough for water particles to rotate more than once around a complete closed curve, and are not very effective in maintaining suspensions. Moreover, particles have inertia, and hence do not faithfully follow the accelerating motions to which they are exposed as they move through the water.

P5.3 (M) Okubo's formula. Show that, if $\sigma_r(t)$ is given by (5.12) and K_H by (5.11), then $K_H \approx 2 \times 10^{-4} l^{1.45}$, and not by (5.13) as found by Okubo.

It should be noted that Okubo fitted the curves (5.12) and (5.13) by eye rather than deducing K_H (i.e., (5.13)) from σ_r (5.12) as he might have done. The revised relation between K_H and l is still a reasonably good fit – as it should be – to the data in Fig. 5.12, and the difference between the two formulae for K_H gives a measure of the uncertainty in using (5.13).

P5.4 (M) Dispersion and dissipation at different length scales. Use Fig. 5.12 to compare the mean rates of dissipation of energy, ε, in the linear ranges surrounding scales of 100 m and 10 km, supposing that dispersion is consistent with Richardson's four-thirds power law and depends only on the length scale, l, and on ε.

P5.5 (M) Dispersion in Langmuir circulation. A square patch of floating particles with sides of dimension d, aligned across wind and downwind, is released into a field of motion on the water surface dominated by Langmuir circulation. The distance between windrows aligned in the wind direction is $h \ll d$. Supposing, for simplicity, that the windrows are steady, that the cross-wind flow of the circulation carrying the particles into windrows from locations halfway between windrows is u and that, once

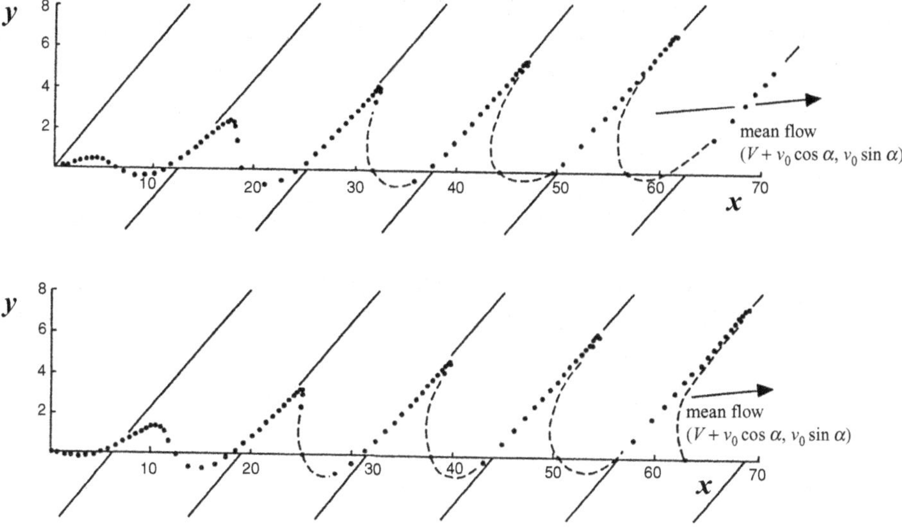

Figure 5.20. Dispersion caused by Langmuir circulation and a mean flow. A numerical model showing the location of floating particles, marked by dots, released from a source on the left (the origin of the x and y axes) into a steady flow, including the idealized effects of Langmuir circulation. The flow, V, is in the x direction to the right. Wind causes the formation of a pattern of Langmuir cells that are advected by the flow and drive surface currents, v_0, along windrows (the circulation converges at the surface along the positions marked by tilted lines) and towards the windrows. The figure shows locations of particles, released at equal intervals of time, at two times separated by half that required to advect a Langmuir cell (with width half the distance between windrows) past the source. The particle pattern resembles that of the oil film in Fig. 5.4(b). Near the source the particles meander, but with a meander length that corresponds to the scale of the windrows rather than (as in Fig. 5.4(b)) to larger eddies or variations in the (possibly tidal) flow. (From Thorpe, 1995.)

in a windrow, the downwind speed is an amount v greater than the flow, supposed to be uniform, between windrows, find

(a) how long it is before all particles reach the windrows,
(b) how much further particles initially in the windrows will have been carried downwind than will the particles that are last to reach the windrows, and
(c) the maximum across-wind and downwind extents of the patch?

What if $h = d$?

A model has been devised to describe the dispersion of floating markers caused by the presence of a uniform current and by a uniform Langmuir circulation that is driven by a wind aligned in a direction different from the direction of the current. The markers, continuously released from a fixed location, are drawn into a set of lines as shown in Fig. 5.20.

Later stages of dispersion depend on the unstable nature of Langmuir circulation.
• Csanady (1973) pointed out that, in conditions under which Langmuir circulation is active, cross-wind dispersion of floating material captured in windrows is determined

largely by the duration, T_{Lc}, before the windrows are disrupted by amalgamating with neighbouring windrows or otherwise dispersing their captured material. This time is typically about 5 min to 1 h for windrow spacing of 2–20 m, but may depend on the depth of the mixed layer or, in well-mixed shallow lakes, on water depth. The crosswind dispersion coefficient, K_y, is then of order $L_{Lc}^2 T_{Lc}^{-1}$, where L_{Lc} is the distance between windrows. Values of K_y obtained using Csanady's estimate in winds of 5–10 m s^{-1} are about 0.01–00.5 m^2 s^{-1}. A more sophisticated numerical model has been devised by Faller and Auer (1988).

P5.6 (E) Dispersion in NATRE and Okubo's relation. Compare the estimates of dispersion in the pycnocline obtained during NATRE with those determined from the near-surface estimates (5.13).

Chapter 6

The energetics of ocean mixing

6.1 Introduction

The study of ocean turbulence may be viewed as a key component in the investigation of the ocean's processes and their energetics: how the energy supplied from external sources is distributed and eventually dissipated by the external and internal processes of mixing referred to in Section 3.1. The ocean is driven mainly by forcing from the atmosphere at the sea surface and by the tidal body forces imposed by the gravitational attraction of the Moon and the Sun. Relatively insignificant are the localized, but spectacular, inputs of energy from hydrothermal vents in the deep ocean ridges, the fortunately infrequent seismic movements of the seabed that may generate devastating tsunamis, the flux of geothermal heat through the floor of the abyssal plains and the energy inputs from rivers and the break-up or melting of ice sheets. The tidal forces and atmospheric inputs are the dominant sources of energy responsible for the overall circulation of the ocean (the kinetic energy of the mean flow) and its density structure (containing potential energy), and are the principal cause of the waves and the turbulence within the ocean.

The discussion in this chapter focuses on how turbulent mixing in the deep ocean is maintained. Much of the energy provided by the atmosphere is used in driving surface waves and the processes that sustain the structure of the upper ocean boundary layer. The mean depth of the ocean is 3795 m and most of the water mass is remote from the atmosphere or the near-surface mixed layer, being separated by thermoclines from direct contact with the atmosphere and therefore from the energy it provides to the ocean. How much of the energy supplied by the atmosphere can be transported from the surface to depth and then made available for mixing?

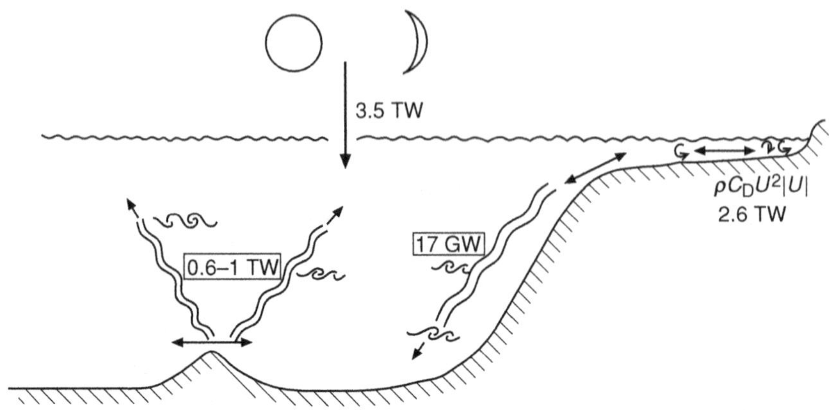

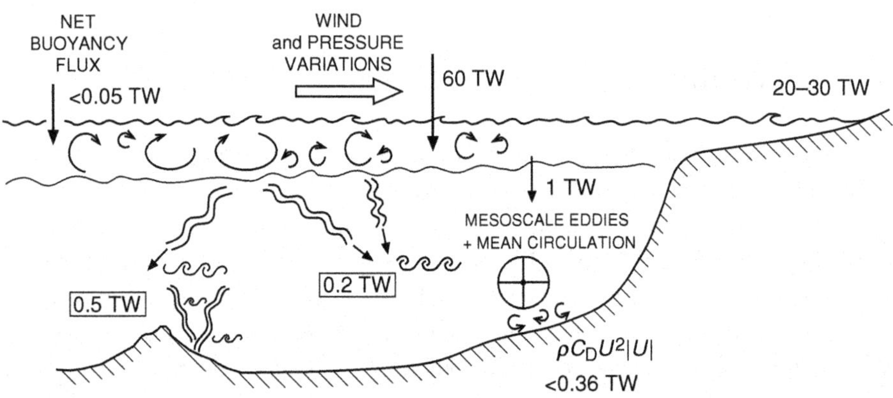

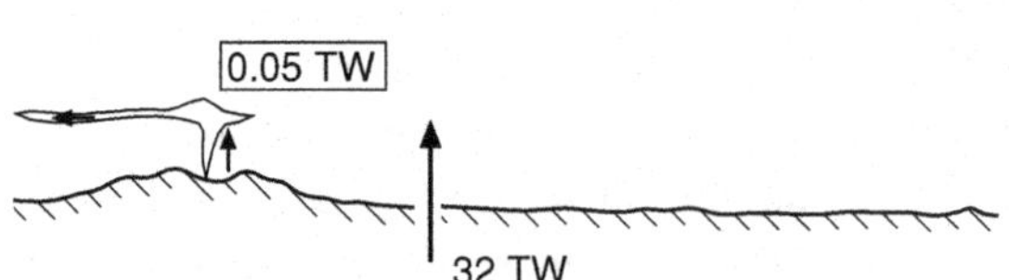

Figure 6.1. The fluxes of energy leading to ocean mixing. The fluxes entering the ocean and dissipation in shallow water or at the seabed are shown, together with, in rectangular boxes, the estimates of the rates available to mix the abyssal ocean. The latter are generally very approximate estimates. (a) Tidal energy. The rate at which the energy of the tides is dissipated in the ocean is the only rate yet known with any degree

As we have seen in earlier chapters, turbulence leading to mixing takes a variety of forms and is derived from numerous sources, such as breaking waves and shear stress on the seabed. The routes of energy transfer are also varied and complex. The mesoscale eddies described in Chapter 5 are themselves a form of turbulent motion derived largely from the baroclinic instability of ocean currents such as the Gulf Stream, but their interactions with each other or with the lateral boundaries of the ocean can generate internal waves and, through their subsequent breaking, relatively small-scale turbulence that leads to mixing. The routes of energy transfer and the contributions that the various sources of energy make to ocean mixing are explained in this chapter.

The magnitudes of the energy fluxes that are described in the following sections are summarized in Fig. 6.1. Not all, indeed relatively few, of the fluxes can be quantified with a high or even moderate degree of accuracy, and discovering the relative importance of the sources of energy that drive diapycnal mixing is a subject of active investigation within the broad field of the science of oceanography.

6.2 How much energy is required to mix the abyssal ocean?

The relation $K_\rho = \Gamma\varepsilon/N^2$ (4.9), with $\Gamma = 0.2$ and the estimates of the mean values of K_ρ and N^2, can be used to derive an approximate estimate of the mean rate of turbulent dissipation per unit mass, ε. Integration over the approximately 85% of the ocean volume that lies below 1500 m then leads to an estimate of the total rate of energy input required to maintain the mixing of the abyssal ocean.

Figure 6.1. *(cont.)*
of accuracy, but where and how dissipation occurs is still uncertain. Pathways for energy transport leading to mixing via the generation of internal tides are as described in Section 6.3.2. (b) The energy flux from the atmosphere. This is known only approximately (Section 6.4). Most is derived from wind action. The net buoyancy flux is probably negligible, although, being distributed unevenly across the ocean surface, it may result in cooling and convective flux into the abyssal ocean. The range, 20–30 TW, represents a rough estimate of the energy lost from surface waves in breaking as they approach the shallow water – the beaches or rocky shores – surrounding the ocean. It will be noticed that the energy flux into and from the mesoscale eddies and general circulation is not balanced; the work done against the stress on the seabed accounts only for a small fraction of the flux from the atmosphere (Sections 6.5 and 6.7). It seems likely that the interaction between eddies and the adjustments in the flow field as eddies are generated leads to a flux into the internal wave field, but this has not yet been quantified. Included in this figure are the approximate values of 0.2 and 0.5 TW in mid-water and near sloping boundaries, respectively, the energy fluxes resulting from breaking internal waves generated largely by atmospheric processes as explained in Section 6.6. (c) The geothermal heat flux. Although the energy transport into the ocean in the form of heat, about 32 TW, is relatively large, the contribution to mixing of the geothermal heat flux is small in relation to others (Section 6.9), except locally, e.g., in rising plumes over hydrothermal vents. The net contribution to mixing is about 0.05 TW.

The available estimates of K_ρ (or K_T) are Munk's canonical value of 10^{-4} m^2 s^{-1} for the abyssal ocean below about 1500 m and the value of about 10^{-5} m^2 s^{-1} typical of the values of ε derived from microstructure measurements (see Section 4.7). Integration leads to values of the required total energy flux of about 2 TW and 0.2 TW,[1] respectively, the difference being accounted for by the relatively intense mixing in regions of rough topography included by Munk's method of estimation but largely undersampled or unaccounted for by the microstructure estimates. [**P6.1**, **P6.2**]

The greater value appears to provide a closer approximation to the total flux of energy required to mix the abyssal ocean. The flux values are subject to considerable uncertainty and provide, at best, approximate estimates of the energy required to support diapycnal mixing in the deep ocean. For example, the value of Γ is in question (see Section 4.4.2) and there is uncertainty about the assessment and integration over the body of the deep ocean of the product $K_\rho N^2$ in the equation $\varepsilon = K_\rho N^2/\Gamma$ derived from (4.9).

•A recent review of available information[2] finds that the flux of energy required to maintain mixing in the abyssal ocean is 1.7 TW but with a possible uncertainty of about $\pm$50%. (This energy flux is of the same order of magnitude as the global electricity-generating capacity.[3])

Where does the energy required to mix the ocean come from?

6.3 The tides

6.3.1 The surface or barotropic tides

Of all the rates of energy flux relating to ocean mixing, the best known is that of the tides.

Present estimates of the total dissipation of tidal energy derived from the times of ancient eclipses, modern measurements of the change in the length of day and lunar ranging give a rate of dissipation equal to 2.5 ± 0.1 TW for the M_2 tide and 3.7 TW for all tides, including M_2 and solar. About 0.2 TW of tidal energy is dissipated within the solid Earth and a relatively small amount, 0.02 TW, is used in driving the atmospheric tides, leaving about 3.5 TW to be dissipated in the ocean, the majority, 2.4 ± 0.1 TW, in the M_2 tides.

The rate of dissipation of energy by the flow of tidal currents, U, over the seabed is proportional to $U^2|U|$, implying that dissipation is very much higher in regions of strong tidal flows; the mean value is dominated by contributions from relatively small areas of the ocean in which tidal currents are largest, the continental shelves. The tides lose relatively little energy at the floor of the abyssal ocean.

1 1 PW = 10^3 TW = 10^6 GW = 10^{15} W.
2 By Wunsch and Ferrari (2004).
3 For further comparison, the generating capacity of the Three Gorges power station on the Yangtze River in China is 18.2 GW.

As mentioned in Section 2.5.3, the earliest calculation of turbulent energy dissipation in the ocean was that made by Taylor in 1919 in seeking to estimate the rate of energy loss by the tides. He found that about 50 GW of tidal energy is lost in the Irish Sea. In 1920, Jeffreys extrapolated Taylor's value to derive a value, 2.2 TW, for the dissipation of the lunar M_2 tide in the shelf seas of the whole ocean. This value comes close to the total M_2 tidal energy dissipated within the ocean, apparently leaving relatively little that might be transferred to turbulence in mid-water in the abyssal ocean and dissipated by turbulent mixing in the water column.

6.3.2 The internal or baroclinic tides

Information about tidal flows obtained since 1920 leaves the accuracy of Jeffreys' estimate in some doubt, and it is probable that a significant fraction of the total flux of 3.5 TW of tidal energy dissipated within the ocean remains available to support mixing after the fluxes lost by working against the bottom stress in shallow seas have been subtracted. (A value of about 2.6 TW is ascribed to mixing in shallow seas in Fig. 6.1(a).)

Although the internal tides receive negligible energy from the direct action of the tidal forces of the Moon and the Sun, by the 1960s it was known that large internal fluctuations in temperature occur with tidal frequency over and around the continental slopes, particularly those where the surface (or barotropic) tides are large. The motion caused by the surface tides generates internal (or baroclinic) tides by carrying the stratified water against and over the continental slope and shelf break surrounding the relatively shallow (depth <200 m) shelf seas. (Turbulent dissipation in an internal tidal ray caused by this process and propagating from the continental slope towards deeper water is illustrated in Fig. 4.10.) A substantial fraction of the energy in the surface tides may be transferred to internal tides, and then lost as they dissipate, sometimes after propagating through the depth of the ocean and horizontally over many hundreds of kilometres.

From calculations carried out in the early 1980s it was found that the rate of total transfer of energy from the surface tides to internal tides around the continental slopes of the ocean is about 14.5 GW (i.e., 0.0145 TW) for the M_2 tide and 2.73 GW for the solar S_2 tide, values that are very much smaller than the known rates of energy loss from the tides and than the rate required to mix the abyssal ocean. It was concluded that, whilst the energy transferred to internal tides and then lost through their dissipation or breaking might be significant in mixing regions close to the continental slopes, it was too small to make a substantial contribution to the mixing of the whole abyssal ocean.

The calculations had not taken into account, however, the tidal motion over oceanic ridges. Much of the tidal motion near the continental slope is along the slopes and is relatively inefficient in moving the water over the sloping topography and in generating internal tides, but the barotropic tidal flows of the deep ocean frequently cross mid-ocean ridges and are there relatively effective at generating baroclinic tides. Persuasive

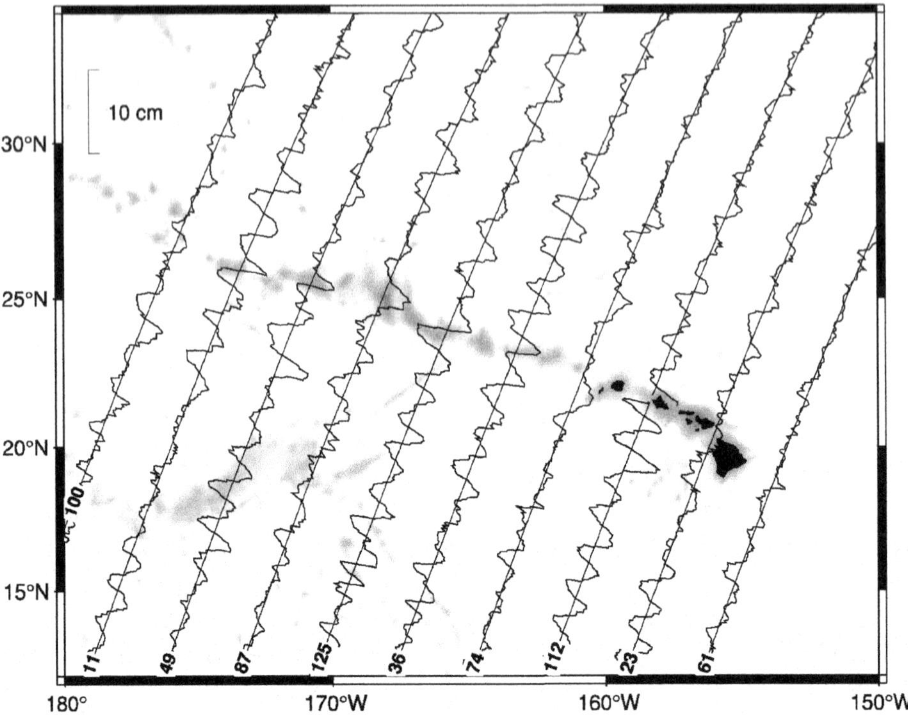

Figure 6.2. The surface manifestation of the radiation of internal tides from the Hawaiian Ridge. Anomalies in mean sea-surface height are detected along ten tracks of the TOPEX-POSEIDON satellite altimeter, with changes of about 5 cm in surface height over distances of about 160 km. The variations in sea-surface height are caused by internal waves generated by the flow of the barotropic tides between the islands lying along the Ridge. The shaded area marks shallow topography. The internal tides radiate to distances of at least 1000 km from the Ridge. (From Ray and Mitchem, 1997.)

evidence for this is found in observations of tidally coherent patterns of small (typically 0.05 m) changes in surface elevation measured by satellite altimetry around the Hawaiian Ridge, the internal tides that cause the changes in surface elevation radiating out to at least 1000 km from the Ridge (Fig. 6.2).

Evidence that tides are involved in mixing in the abyssal ocean has been found in the Brazil Basin. Figure 4.16 shows the high dissipation over the rough topography of the mid-Atlantic Ridge at the eastern edge of the Brazil Basin. The spring–neap cycle in dissipation (Fig. 6.3) observed in this area appears to be an effect of mixing induced by the presence of internal tides caused by the surface (or barotropic) tidal flow over the irregular topography of the Ridge. (The dissipation in the water column in this area is greater by a factor of about 100 than the energy lost from the barotropic tide through stress on the seabed.)

• About 0.6–1 TW is now estimated as the possible size of the flux of energy from the surface tides to the internal tides in regions of ridges and relatively rough bottom topography, and this is the range of values shown in Fig. 6.1(a). The estimated rate of energy transfer in the North Pacific alone is 0.27 TW.

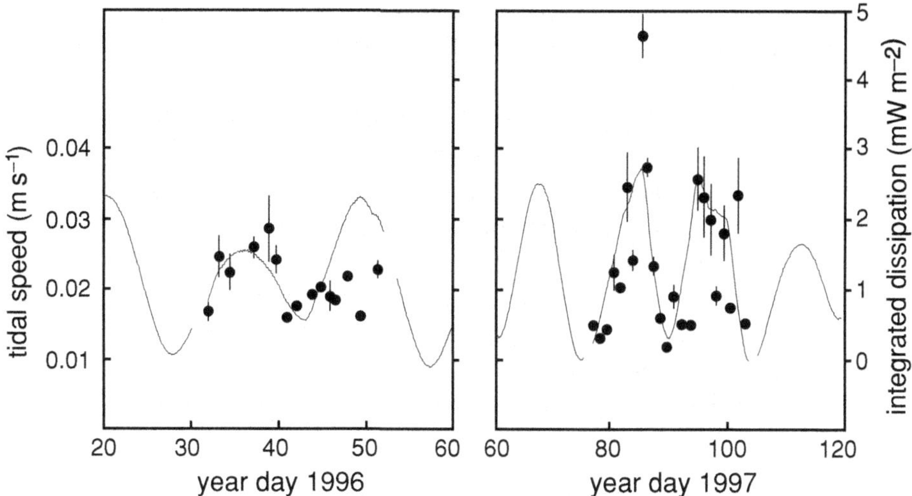

Figure 6.3. Spring–neap variation in mixing in the Brazil Basin. The tidal speeds, varying over the approximately 14-day spring–neap cycle, are shown as full lines and are estimated from a numerical model of the barotropic tides. The points are the observed values of dissipation from observations in 1996 (left) and 1997 (right) averaged over 24-h periods and integrated over the lower 2000 m of the water column. (From St. Laurent *et al.*, 2001)

An observational study called the Hawaiian Ocean Mixing Experiment (HOME) was conducted between 2000 and 2005 to examine the transfer of energy from the surface to internal tides and the consequent mixing at the Hawaiian Ridge. It is estimated that 20 ± 6 GW of energy is lost from the surface M_2 tide at the Ridge and that 10 ± 5 GW radiates away in the form of internal tides, leaving about 10 GW to be dissipated within 50 km (mostly at smaller distances) from the Ridge. The 20 GW extracted from the surface M_2 tide at the Ridge is about 1/30–1/50 of the total loss from the surface tides in deep water around the globe, and the transfer to internal tides from similar ridges appears to be a substantial fraction of the total 0.6−1-TW flux of energy from the surface to the internal tides. But how much of the energy in the internal tides goes into diapycnal mixing?

• The value of K_T in the most energetic region around the Ridge is less than 3×10^{-4} m^2 s^{-1} and, given the relatively small volume of ocean that lies above such topography in the ocean, it seems unlikely at present that the oceanic ridges – or seamounts – can provide an explanation of the canonical abyssal ocean value, $K_T \approx 1 \times 10^{-4}$ m^2 s^{-1}.

Perhaps it is not near their generation sites that internal tides contribute most to ocean mixing, but where they break or are dissipated after propagating some distance away? It is now known from sets of satellite altimetry data like that shown in Fig. 6.2 that internal tides can propagate over great distances, radiating energy from mid-ocean ridges across ocean basins, sometimes to reach as far as the surrounding continental slopes.

A thorough investigation of where radiating internal tides lose their energy has yet to be performed, but it is estimated that about 12 GW of the energy from the internal tides is dissipated in the submarine canyons that incise the continental slopes surrounding the ocean. The internal tidal energy that is dissipated in these canyons may be supplied by local sources – recall that about 17.2 GW is available in the internal tides generated by surface tides on the continental slopes – as well as by the tides radiating and reaching the canyons from the mid-ocean ridges, but in any case the rate of loss in canyons appears to be very small in comparison with the 1.7 TW required to support ocean mixing. The relatively small volume of canyons suggests that, even though enhanced values of K_T are found within them, their overall contribution to ocean mixing is also insubstantial.

We need to examine other sources of energy in the ocean.

6.4 The atmospheric input of energy through the sea surface

6.4.1 The wind stress

The rate of transport of kinetic energy from the wind into the ocean is not known with great certainty. Estimates are generally derived from a flux per unit area given by τu_s, where τ is the wind stress on the sea surface and u_s is a measure of the drift speed of the water surface. Several factors contribute to the uncertainty. There is uncertainty in estimates of the drift speed, u_s, which is composed of two components, the wave Stokes drift, U_S, and a drift generated by the wind stress, the wind drift. The value of U_S is commonly taken as 2% of the wind speed, but in any case is imprecise, and the further contribution of wind drift is even more uncertain. The net surface drift, u_S, is generally estimated to be 2%–7% of the wind speed, W_{10}, at a height of 10 m. The wind stress, $\tau = \rho_a C_{Da} W_{10}^2$, is dependent on the magnitude of the drag coefficient, C_{Da}, and uncertainty regarding its size contributes further to the possible errors in estimates of kinetic energy flux. The flux varies as W_{10}^3 and is therefore very sensitive to wind speed: its mean value depends on correctly finding the average cube of the wind speed, W_{10}, and consequently on correctly accounting for the contributions of rare, but high, winds.

The present rate of energy flux from the atmosphere is thought to be about 60 TW, the value shown in Fig. 6.1(b). The flux appears to have increased from about 50 TW in 1950 to 58 TW in 2000, as a consequence of increasing wind speeds and wave heights. The flux has also considerable spatial and seasonal variation. About two-thirds of the mean energy flux takes place in the southern hemisphere, partly because about 63% of the global ocean area is south of the Equator. The energy flux within the southern hemisphere is dominated by the contribution arising from the high winds in the Southern Ocean surrounding the Antarctic Continent. [**P6.3**]

Where does this flux of energy from the atmosphere go? Much of it goes into driving surface gravity waves as described in the following section. These act as a catalyst in the passage of energy from the wind into the mean currents; wind-waves that are generated in deep water, and subsequently break, transfer some of their momentum to the mean circulation of the ocean: about 1 TW of the flux from the atmosphere goes into driving the mean circulation and fluctuations of frequency less than the inertial frequency, f. The mean circulation also passes on its energy: the instability of the mean currents contributes to driving the mesoscale eddies as described in Section 6.5.

A significant part of the energy passing from the atmosphere into the ocean drives the turbulent motions within the mixed layer that maintain its uniformity or are dissipated in deepening the mixed layer during storms. Some energy transferred from the atmosphere to the interior of the mixed layer is used in driving motions in the underlying thermocline: for example, vertically overturning eddies in the mixed layer collide with and buffet the underlying stratified water, generating internal waves that radiate into the seasonal thermocline and the abyssal ocean, where their energy may be available for its mixing. Moving atmospheric fronts generate internal inertial waves that also transport energy downwards. These energy fluxes to the deep ocean are described further in Section 6.6.

6.4.2 Surface waves

Energy can be carried by surface waves over great distances; the propagation of wave energy has been tracked across the Pacific Ocean from storms in the Southern Ocean to the beaches of Alaska.[4] Most of the wave energy that reaches the shoreline surrounding the ocean is dissipated by wave breaking in the surf zones over sandy beaches or on its rocky shores. Little is reflected. The estimated dissipation of 25 ± 5 TW of wave energy near the shoreline shown in Fig. 6.1(b) is, at best, a rough estimate. [**P6.4**]

Surface waves convey energy and momentum from the wind to the ocean; wind generates waves, some of which break, transferring momentum to the underlying water. However, neither the breaking of waves at the seashore nor the breaking that occurs in deep water contributes directly to the turbulent mixing of the abyssal ocean.

Although some of the quantified information that is required in order to assess the fraction of wind energy that is dissipated in wave breaking in deep water is available, the details are not yet known with sufficient precision for accurate estimation. Results

4 Individual (dispersive) waves of wavelength much less than the water depth do not carry their energy over great distances. Waves travel in groups and interact with one another. Because individual waves propagate at greater speeds than the groups or packets containing them – see P1.3 for the relevant speeds – any particular wave approaches the leading edge of its more slowly moving group, and eventually disappears. (New waves are continually formed at the back of wave groups.) But whilst individual waves cannot therefore be tracked far across the ocean, groups or packets of swell containing waves with periods of 12–20 s are found to carry energy across ocean basins at speeds corresponding to the group velocity of their component waves; wave energy is carried at the speed of the groups, the group velocity, $\boldsymbol{c}_g$.

In contrast, very long (non-dispersive) waves such as tides and tsunamis can, and do, propagate for thousands of kilometres across ocean basins.

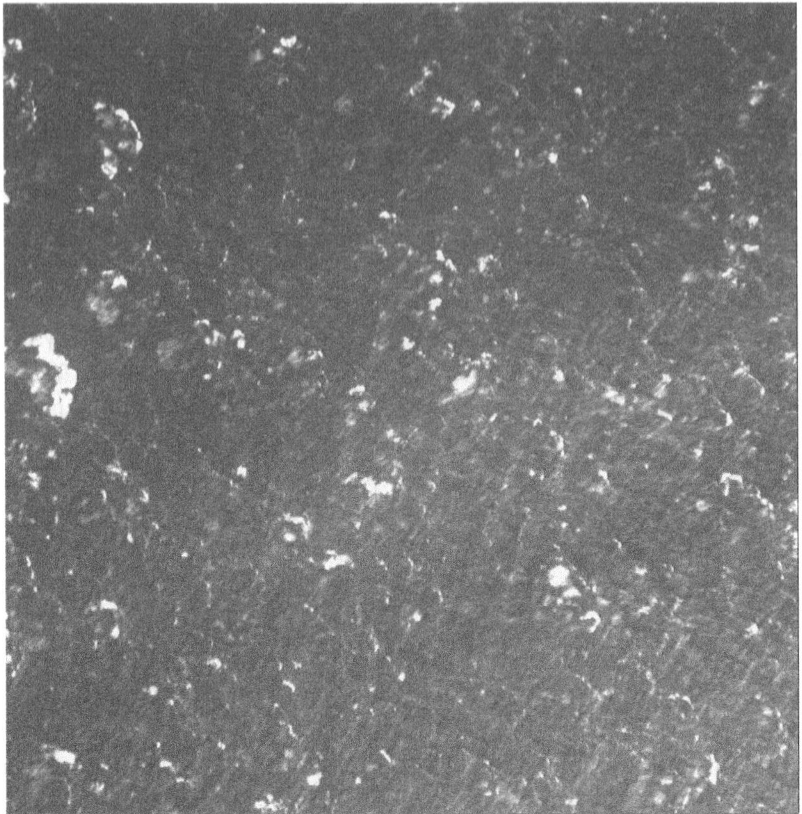

Figure 6.4. Breaking waves and foam patches. This photograph of waves breaking on the sea surface is taken from an aircraft flying at a height of about 1500 m. The image width is about 950 m and the wind speed is 20–25 m s^{-1} directed at about 25° clockwise from the vertical edge of the photograph. The breakers are patchy and variable in size. Wave breaking rarely extends far along a wave crest, and the distance is generally less than the wavelength. The time between occurrences of wave breaking at a single position is very irregular but typically of the order of 1 min. (Photograph kindly provided by Professor W. K. Melville, Scripps Institution of Oceanography, USA.)

of laboratory experiments are consistent with a rate of dissipation of energy per unit length of the crest of a breaking surface gravity wave given by

$$E_t = b\rho_0 c_b^5 / g, \tag{6.1}$$

where b is a constant in the range 10^{-3}–10^{-2}, c_b is the speed of advance of the wave crest and g is the acceleration due to gravity. [**P6.5**] Waves break intermittently, and rarely is breaking found along a wave crest over a length much exceeding the wavelength of the wave. Whitecaps are consequently patchy, as shown in Fig. 6.4. Multiple video photographs of the sea surface taken from an aircraft allow estimation of the speed of breaking wave crests and the length over which breaking occurs along wave crests per unit area of the sea surface at various wind speeds, W_{10}, and an assessment of energy dissipation can be made by applying (6.1). The breaker length per unit sea-surface area is found to vary as W_{10}^3. The mean speed of the waves that are breaking is generally significantly less than the speed of waves with frequency equal to that at the peak of the wave spectrum. Available data are at present insufficient to extrapolate to obtain an estimate of the global flux of energy from the breakers into the mixed layer.

Some of the energy lost from breakers provides the potential energy of subsurface bubble clouds, and so contributes energy to a route for the transfer of gases from the air to the sea as the bubbles lose their constituent gases into solution in the surrounding seawater. Much of the breakers' energy is dissipated in turbulence and some passes into the mean circulation as explained in Section 6.4.1.

6.4.3 Buoyancy flux

There appear to be no reliable estimates of the flux of energy to the ocean from the atmosphere from the buoyancy fluxes associated either with heat transfer or with precipitation and evaporation. Estimates of the latter integrated over the ocean surface give values that are small and uncertain, but of order 0.01 TW. It appears that the buoyancy contribution to abyssal mixing is less than 0.05 TW, which is negligibly small in comparison with that of the wind.

The effect of the buoyancy fluxes is, however, subtle. On average the oceans gain heat in the tropics and transport heat towards the arctic regions, where heat is lost. The poleward heat flux carried by the ocean varies with latitude, but is typically 1–3 PW, and amounts to about a third to a half of the heat carried polewards by the atmosphere.[5] The heating in the tropics and subtropics serves partly to balance the upward advection of cold water in the abyssal ocean, the steady-state balance assumed by Munk and used in the estimation of K_T (see Section 4.7 and P6.10).[6]

Cooling at high latitudes in the northern hemisphere leads to deep convection, particularly in the Labrador and East Greenland Sea, although convection does not penetrate to the 5-km depths of abyssal plains and so supply water that is dense enough to reach and replenish that in the deepest parts of the ocean. In the southern hemisphere, however, there is an advective pathway of energy transport from the surface to the abyss. Cooling beneath coastal polynyas,[7] particularly in the Weddell and Ross Seas, and the melting of ice shelves leads to dense water of temperature less than -1 °C that sinks down the Antarctic Continental Slope, forming Antarctic Bottom Water (AABW). This spreads northwards along the ocean floor of the Pacific,

5 For example, the heat flux from the Equator towards the Arctic at 30° N is about 5.5 PW, about 2 PW of which is transported within the Pacific and Atlantic Oceans, the remainder being carried by the atmosphere.

6 Munk assumed a steady state below a depth of 1500 m, which is a good approximation. In reality, however, changes in the mean ocean temperature and in its total heat content are observed. Levitus *et al.* (2005) report an increase in the heat content of the upper 3000 m of the ocean of 1.45×10^{23} J between the years 1955 and 1998, mostly in the upper 700 m. This is equivalent to a rise in the mean temperature of 37 mK, and corresponds to a rate of increase of heat of 107 TW, or of 200 mW m^{-2}, exceeding the heat fluxes from geothermal sources described in Section 6.9.

7 A polynya is an area of open water in arctic regions that is free of ice, where air–sea interaction can result in ocean cooling much more rapidly than it does in ice-covered areas.

Atlantic and Indian Oceans. The northward flux of AABW with potential temperature less than 0 °C is estimated to be about 10–15 Sv. The mixing that the flow of this water may produce in the abyssal ocean is discussed in Section 6.8.

6.5 The mean circulation and mesoscale eddies

The general circulation of the ocean is comprised of its mean currents, and these include the largely horizontal flows of the major ocean gyres (of which the Gulf Stream and the Kuroshio are the best-known components) and the east-going flow of the Antarctic Circumpolar Current in the Southern Ocean surrounding the Antarctic continent, as well as the vertical circulation, a component of which is the flow of the AABW. The general circulation contains energy amounting to some 150 EJ (1 EJ $= 10^{18}$ J), much of it associated with potential energy. This huge store of energy is sustained mainly through atmospheric forcing; the tides and geothermal heat flux make insignificant contributions. Some of it might be made available for transfer to mixing if the forcing of the circulation were reduced and if the tilted isopycnal surfaces became horizontal.

A relatively small amount of energy is lost from the general circulation in the abyssal ocean through processes involved in sustaining flow and stress at the seabed (Section 6.7) and by the generation of internal lee waves or eddies, particularly where flows pass over or around rough topography or through passages connecting major ocean basins where stationary hydraulic jumps may occur (Section 6.8). A much greater flux of energy from the general circulation occurs through the generation of mesoscale eddies. These have horizontal dimensions of 30–200 km and corresponding periods of about 100 days, and dominate the kinetic energy of the ocean, exceeding that of the mean circulation. The transfer of energy to the mesoscale eddies from the general circulation is largely a consequence of barotropic and baroclinic instabilities with a flux estimated at about 1 TW. The mesoscale eddies are further supported by an energy flux from the wind of about 0.2 TW and these sustain a total energy in the mesoscale field of about 13 EJ.

• Just how energy is dissipated from the mesoscale eddies is at present unknown and unquantified. Like the general circulation, some must be dissipated, perhaps by shear-flow instabilities or in convection following the generation of static instability, as eddies are modified or adjust in approaching or moving over topography or as eddies collide and interact with islands, seamounts and continental slopes. Some energy is almost certainly transferred from the mesoscale eddies to internal waves as eddies interact with one another or as they adjust to a balanced state after their formation or interaction with topography.

Except for the loss of energy near the seabed and in hydraulic jumps, there appears to be little direct flux of energy from these large-scale flows into diapycnal mixing in the deep ocean; the flux from internal waves appears to be more important.

6.6 Internal waves

In addition to the internal waves of tidal periods generated over topography described in Section 6.3.2, internal waves are generated at the sea surface by atmospheric forcing and through turbulent motion within the near-surface mixed layer. Several mechanisms are involved, including resonance interactions between pairs of surface waves and the buffeting of the upper pycnocline by turbulent motions (e.g., convective plumes or Langmuir turbulence) in the mixed layer.

The movement of atmospheric pressure gradients or fronts across the sea surface is known to lead to the generation of waves with near-inertial periods. The associated energy flux is estimated as 0.6 TW, but of this only about 0.1 TW appears to be radiated into the abyssal ocean by internal inertial waves, and there is at present no reliable assessment of how much of the energy provided by the atmosphere reaches the abyssal ocean through internal waves of higher frequencies. Although the total energy contained within the field of internal waves in the ocean, excluding the internal tides, is known to be about 1.4×10^{18} J (i.e., 1.4 EJ, much less than that of the mesoscale eddy field), there is some uncertainty about how rapidly this energy is supplied and dissipated, and, therefore, about the flux of energy that may be available to drive ocean mixing through the breaking of internal waves. [**P6.6**]

• There is, however, general agreement that the energy loss from internal waves sustains mixing at a level leading to $K_T \approx 1 \times 10^{-5}$ m^2 s^{-1}, a value consistent with microstructure measurements and providing an energy flux for mixing of about 0.2 TW, the value shown in Fig. 6.1(b).

Internal waves are modified by reflection from topography, particularly so for incident waves of frequency close to a 'critical frequency', σ_c, equal to $N \sin \alpha$, where α is the mean angle of inclination of the topography to the horizontal and N is the buoyancy frequency in its vicinity. Rays of waves with critical frequency have an inclination to the horizontal (β in Fig. 1.14(b)) that is equal to the topographic slope angle, α.[8] Peaks in wave energy at frequencies close to σ_c have been observed in wave spectra at locations within a few hundred metres of sloping topography, implying that reflected waves with frequency near σ_c are amplified in their reflection, and there is evidence of a loss in *energy flux* from the internal wave field at these frequencies, probably through breaking as a result of the enhanced shear and consequent Kelvin–Helmholtz instability. The estimated energy flux used in mixing near sloping boundaries, 0.5 TW in Fig. 6.1(b), is, at best, very approximate.

8 The direction of propagation of waves of critical frequency that are reflected from a sloping boundary, and the particle motions in the reflected waves, are parallel to the line of greatest slope of the boundary. This is independent of the orientation of the approaching beam, i.e., whether or not it lies in a plane normal to the sloping boundary.

6.7 Dissipation produced by bottom stress

The rate of energy loss by flows per unit area of the seabed is $\rho_0 C_D U^2|U|$ and this, integrated over the area of the deep seabed, provides an estimate of the rate of dissipation by turbulence in the benthic boundary layer. The mean value, $\langle U^2|U|\rangle$, is poorly known but, since the currents are typically less than 0.1 m s^{-1} near the bed, it is unlikely to exceed about 5×10^{-4} m^3 s^{-3}.[9] Taking C_D to be 2.5×10^{-3} and ρ_0 as 1.03×10^3 kg m^{-3}, the mean dissipation of kinetic energy per unit area of the seabed must be less than about 1.3 mW m^{-2}.

The area of the seabed at depths below 3000 m, which includes the abyssal plains and mid-ocean ridges, is about 2.8×10^{14} m^2, so (multiplying by 1.3 mW m^{-2}) 0.36 TW is an upper bound for the energy dissipated at the abyssal seabed. Much of the energy lost at the seabed from the flow produced by the mean circulation and by flows induced by mesoscale eddies and internal waves may be dissipated in turbulent motion close to the bed and used in maintaining the 5–60-m-thick uniform benthic boundary layer. Only a small fraction of the (less than 0.36 TW) energy flux may be radiated as internal waves or made available for mixing well above the boundary layer. Where the bottom topography intersects isopycnal surfaces, such as the continental slopes, turbulence in the boundary layer[10] will result, however, in diapycnal mixing across density surfaces, thus contributing to a diapycnal density flux that may affect the water well above the level of the abyssal plains. [**P6.7**]

6.8 Flow through and around abyssal topography

Evidence that mixing occurs near rough topography through the breaking of internal waves is mentioned in Section 6.6. Results from recent studies, however, suggest that some of the mixing in the abyssal ocean occurs not through the breaking of internal tidal waves or internal waves originating in and radiating downwards from the upper ocean, but in association with relatively steady flows near topography. Mixing may occur in flows through or within the median valleys in the centre of mid-ocean ridges and in the canyons on the sides of the ridges – as suggested in Fig. 4.16.

One possible way in which mixing may occur in such regions is through stationary internal hydraulic jumps. These form in flows over sills, for example in channels cutting through a ridge separating neighbouring deep ocean basins, where dense water overflows from one basin into another. Stationary internal hydraulic jumps are similar in nature to the jumps, the sudden rises in water level with turbulence and bubble entrainment, seen downstream of weirs in rivers. Examples of such 'free-surface' hydraulic jumps are shown in Fig. 6.5, together with a photograph of a similar stationary internal hydraulic jump in the atmosphere.

9 This estimate follows on assuming that current, U, is of amplitude 0.1 m s^{-1} and varies approximately sinusoidally. The mean value of $|\sin^3 z|$ is $4/(3\pi) \approx 0.42$.

10 Turbulence possibly caused or supplemented by the amplification of internal waves undergoing reflection as described in Section 6.6.

(a)

(b)

Figure 6.5. Hydraulic jumps. (a) A free-surface undular jump over a shallow ridge in a river flowing from right to left. The undulations – the waves – are stationary and those on the far side of the river are breaking along part of their crests, losing energy to turbulence. (b) Hydraulic jumps in the lee of a weir. The flow is from right to left, and the water in the sloping channel on the lower right of the photograph is shallow, about 0.2 m deep upstream of the breaking jump that creates foam and throws up spray. (c) An apparently stationary internal hydraulic jump in the stratified flow of air (from right to left) over the Sierra Nevada mountain range, visible at the left, in the USA. The jump is made visible by dust raised by the wind flow over the ground and by the formation of clouds. (Photograph by Robert Symons. Reproduced with kind permission of Professor R. S. Scorer.)

Figure 6.5. (*cont.*)

Overflows between basins are caused, for example, as Antarctic Bottom Water (AABW) flows northwards into the major oceans. The paths of the bottom water are best known in the Atlantic Ocean (Fig. 6.6). The flow is constrained by its density to flow between deep ocean basins as a submarine river through relatively narrow (typically 30 km wide) connecting channels. Speeds reach 0.5 m s^{-1} in some of these channels, and, whilst there have so far been no definitive observations demonstrating that hydraulic jumps occur, there is evidence from the observed reduction in density of the AABW that substantial mixing occurs as it passes through the channels. Direct observations of mixing have been made in the Romanche Fracture Zone (RFZ, marked A in Fig. 6.6), a channel through the Mid-Atlantic Ridge near the Equator that allows AABW to pass from the Brazil Basin on the western side of the Ridge into the Guinea and Sierra Leone Basins on the eastern side. Figure 6.7 shows sections of potential temperature, velocity and turbulent dissipation, ε, through the channel. Estimates of the Ozmidov scales are used for comparison with the displacement or Thorpe scale in Fig. 4.8. Both have values that often exceed 10 m, implying the presence of large eddies. Detailed measurements of dissipation in other channels through which AABW passes have yet to be reported, and, even in the survey of the RFZ shown in Fig. 6.7, the horizontal resolution is insufficient to establish definitely whether or not hydraulic jumps occur, or whether the observed conditions (particularly that of low Richardson number) will definitely allow hydraulic jumps. [**P6.8**]

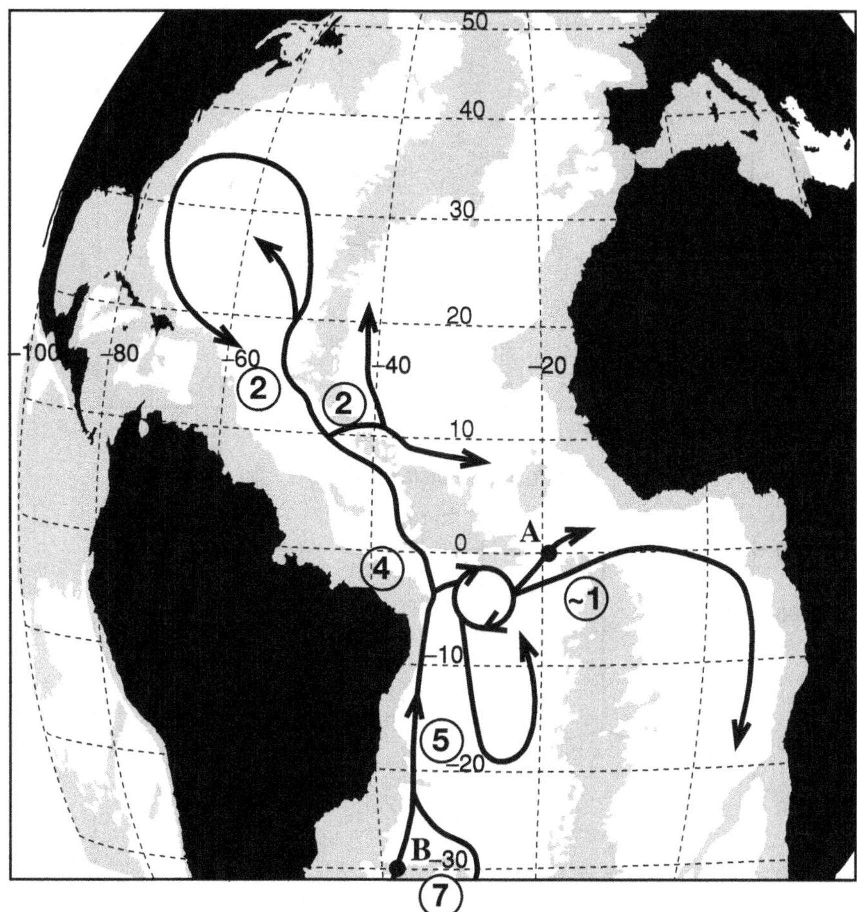

Figure 6.6. The path of the Antarctic Bottom Water spreading on the floor of the Atlantic Ocean. The circled numbers represent estimates of the fluxes in Sv (units of $10^6\ m^3\ s^{-1}$). The shaded areas are where the water depth is less than 4000 m. A flux of 5 Sv is estimated in the Brazil Basin (lying between points A and B). Two channels lead into the Basin from the south. The Vema Channel, marked 'B', carries a greater flux of AABW than does the Hunter Channel, further east. AABW passes from the Basin eastwards through the Mid-Atlantic Ridge in two passages, the Romanche Fracture Zone (marked 'A'), carrying most of the flux, and the more southerly Chain Fracture Zone. (From Stephens and Marshall, 2000.)

The estimated rate of loss of energy flux in an internal hydraulic jump is of order $2\rho h(g'h)^{3/2}$ per unit channel width, where ρ is the density, h is the thickness of the flowing layer and g' is the reduced gravity based on the density difference between the flowing water and that above. Its size suggests that, even though there is evidence of intense mixing, the total contribution from the mixing produced by AABW in flowing through a hydraulic jump at a single sill in the RFZ is less than 60 MW. The total contribution to mixing from flows over sills in the deep ocean channels appears sufficient to contribute no more than 0.1 TW to the total abyssal energy dissipation rate.

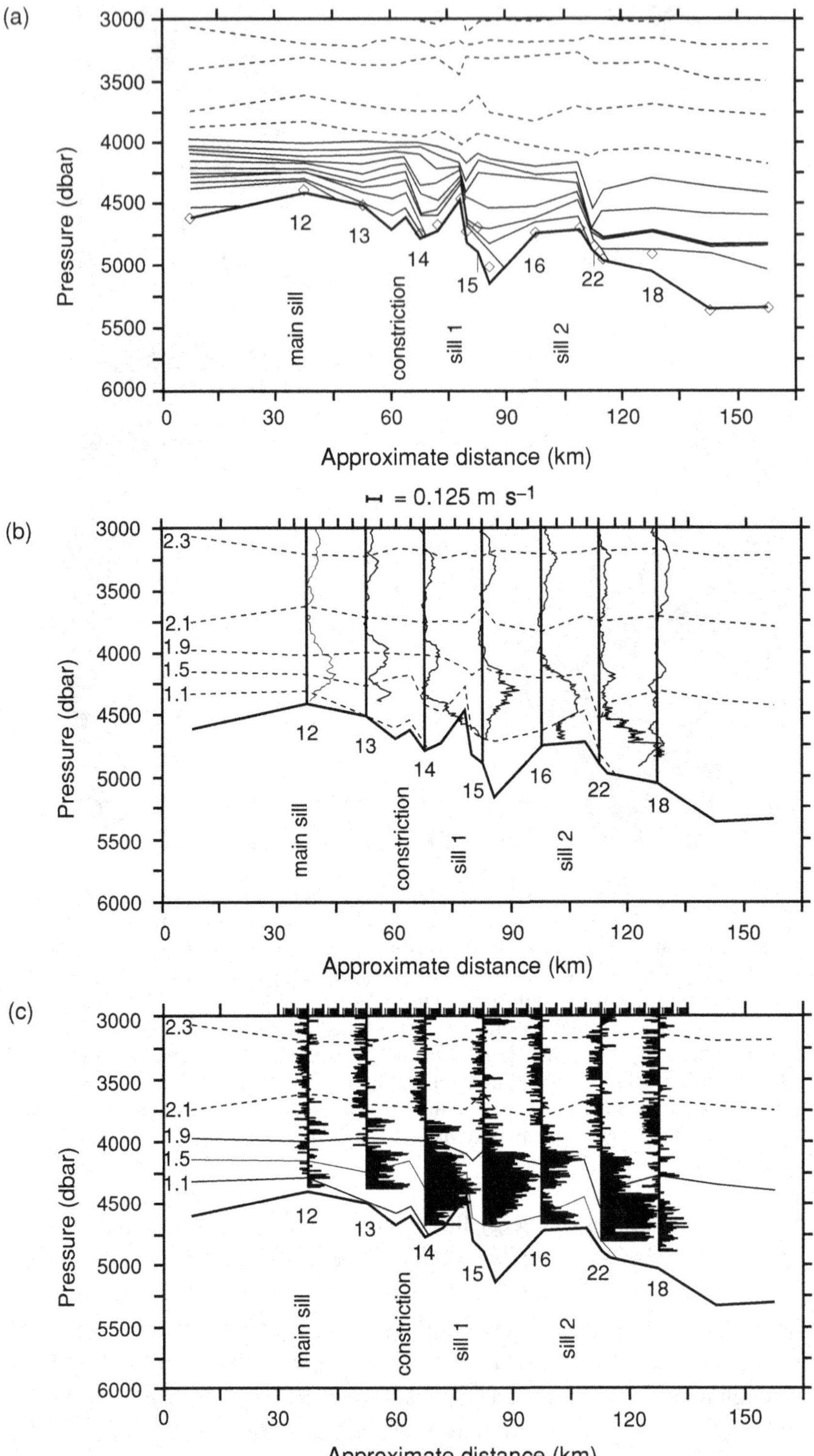
(a)
Pressure (dbar)
3000
3500
4000
4500
5000
5500
6000
12
13
14
15
16
22
18
main sill
constriction
sill 1
sill 2
0
30
60
90
120
150
Approximate distance (km)
= 0.125 m s−1
(b)
2.3
2.1
1.9
1.5
1.1
Pressure (dbar)
Approximate distance (km)
(c)
Pressure (dbar)
Approximate distance (km)

[**P6.9**] Comparison may be made with flows through relatively shallow straits, such as the Bosphorus and the Strait of Gibraltar, that have been studied more thoroughly. Mixing is most intense near the sills and where, because of the changes caused by tidal flows, internal bores are formed. These propagating internal hydraulic jumps are similar to the tidal bores known in some river estuaries, such as the UK River Severn and the Canadian Bay of Fundy, and to the walls of foaming water advancing in the surf zone (Section 1.4).[11] The rate of energy dissipation over the Camarinal Sill in the Straits of Gibraltar is about 0.34 GW.

The mixing in the channels in the deep ocean may, however, lead to relatively substantial diapycnal heat transfer. Estimated values of K_T in the RFZ are high, reaching $0.1\ m^2\ s^{-1}$. The total heat flux within the Zone, although of relatively small area, about $10^3\ km^2$, is comparable to that within the whole of the much larger adjacent basins, of area about $10^6\ km^2$, into which the AABW flows.

The source of the flux of energy driving the AABW differs fundamentally from that driving baroclinic tides or the inertial waves of the upper ocean. The barotropic tides are generated by lunar and solar gravitational forces and, interacting with topography, cause baroclinic tides. Inertial waves are also produced as a result of the energy flux to the ocean from the atmosphere. In contrast, the source of energy of the AABW is the increase in potential energy in the Antarctic resulting from a flux of energy in the form of heat *from* the ocean, strictly a *loss* of energy from the ocean. A positive flux of energy in the form of heat from the ocean leads to a gain in potential energy and results in convective motions; the potential energy is released as kinetic energy in the water cascading from the Antarctic shelves to form the AABW. The dynamical effect of a flux of heat energy depends on whether it increases or decreases the potential energy, and so on whether or not it results in buoyancy forces that drive motion (i.e., whether

---

Figure 6.7. The potential temperature, velocity and turbulent dissipation, ε, at seven stations in the Romanche Fracture Zone (marked 'A' in Fig. 6.6). The section runs from west (12; left) to east (18; right) across three sills in the channel. The lower curve indicates the bottom of the channel. (a) Contours of equal potential temperature at a contour interval of 0.1 °C. The full lines are temperatures $\leq$1.9 °C within the Antarctic Bottom Water. Temperatures increase as the AABW passes through the channel, indicating that there is mixing with the overlying water. (b) Eastward current speeds, reaching $0.5\ m\ s^{-1}$ at station 22. The spacing of stations and probable variations of flow across the channel (about 30 km wide) are, however, insufficient to determine the maximum speeds or to establish the continuity of the volume, density or momentum fluxes. (c) Rates of dissipation, ε, measured using the free-fall HRP (Fig. 2.13(b)). These are relatively high in the AABW, particularly downstream of the sills within the channel. The dissipation is shown on a logarithmic scale, the vertical lines representing $1 \times 10^{-10}\ W\ kg^{-1}$. Values at stations 15 and 22 reach $1 \times 10^{-6}\ W\ kg^{-1}$. The units of pressure, decibars, correspond approximately to depth in metres. (From Polzin *et al.*, 1996.)

11 Propagating and stationary jumps are equivalent in principle, one transforming into the other in a moving frame of reference. There may, however, be differences in the levels of turbulence within the flow approaching or ahead of such hydraulic transitions.

or not, in the terminology of Section 3.4.1, it is destabilizing) and therefore on where in the ocean – at its surface or bed – the heat flux occurs. This should be compared with the situation described in the following section.

How much of the energy lost by mesoscale eddies as they encounter and cross the mid-ocean ridges is made available to drive diapycnal mixing has yet to be quantified (see also Section 6.5).

6.9 Geothermal heat flux

As mentioned earlier, the geothermal flux of heat entering the ocean through the seabed, despite its having the potential to drive convective motion, is a relatively small contributor to ocean mixing. The heat entering the ocean from the Earth's core is transported into the ocean above the sea floor in two ways, by molecular processes through the porous or solid seabed and by a hydrothermal flux of heat carried by fluid circulating through and exiting from the seabed. The latter accounts for about a third of the total geothermal heat flux into the ocean.

The mean heat flux through the floor of the abyssal plains is about 46 mW m^{-2}, but fluxes of heat reaching 1 Wm^{-2} are found in geophysically and often hydrophysically active regions such as the median valleys in the centres of the mid-ocean ridges. The global mean flux through the seabed is estimated to be about 87.8 mW m^{-2}, giving a total flux of heat through the seabed into the ocean of about 32 TW, a flux that is substantial in comparison with the energy flux of about 1.7 TW required for mixing. However, little geothermal heat flux is used in this way.

Although the mean heat flux exceeds the mean dissipation of kinetic energy at the seabed, estimated in Section 6.7 to be less than 1.3 mW m^{-2}, its effect is relatively slight. Using (2.8), the mean buoyancy flux, B_0, resulting from the mean heat flux of 87.8 mW m^{-2} is about 3.6×10^{-11} m^2 s^{-3}.[12] The resulting rate of change of potential energy per unit mass (Section 2.4.2) is therefore about -3.6×10^{-11} W kg^{-1}. The friction velocity, u_*, is approximately $0.05U$ if $C_D \approx 2.5 \times 10^{-3}$ (Section 3.4.1), and, taking a typical value of $U = 0.05$ m s^{-1}, a representative value of the Monin–Obukov length scale, $L_{MO} = -u_*^3/(kB_0)$, at the seabed is about -1060 m. (The negative sign is consistent with the convention for an upward flux of heat and the contribution of the flux towards unstable convective conditions.) Free convection driven by the mean heat flux may occur only at heights that exceed the magnitude of the Monin–Obukov length, $|L_{MO}|$. This is much more than the greatest observed thickness, about 100 m, of deep-ocean benthic boundary layers. Consequently, in these average conditions, turbulence is dominated by stress on the bottom, and the buoyancy flux is too small to produce any substantial convective motion; convective mixing produced by the geothermal heat flux is not important compared with the mixing caused by turbulence in the benthic boundary layer produced by work done against the bottom stress. **[P6.10]**

12 This is estimated by taking $g = 9.81$ m s^{-2}, the expansion coefficient, $\alpha = 1.71 \times 10^{-4}$ K^{-1} (see Section 1.7.1), the density $\rho_0 = 1030$ kg m^{-3} and $c_p = 3.99 \times 10^3$ J kg^{-1} K^{-1}.

• Geothermally produced convection does not contribute significantly to mixing in the stratified water overlying the benthic boundary layer. Mixing within the layer is dominated by turbulence produced by the work done against the bottom stress. (See also P3.10.)

Plumes are observed, however, above hydrothermal vents, and in such locations the geothermal heat flux will contribute substantially to the local mixing. The heat flux from the hydrothermal vents in the Juan de Fuca Ridge in the northeast Pacific Ocean, for example, is about 7 GW. Summed over the whole ocean, about 10 TW of the total geothermal heat flux is carried by such hydrothermal fluid flows, and the total volume flux entering the ocean through the sea floor is estimated to be about 12 Sv. [**P6.11**] The overall contribution to mixing is small, however; the mean rate of increase of potential energy from the geothermal heat flux into the ocean is only about 0.05 TW.

• Although the total geothermal energy flux through the seabed into the ocean, about 32 TW, greatly exceeds the total rate of tidal energy dissipation, it is not in a form that can be used efficiently in mixing. The tides and atmospheric forcing generate kinetic energy and, in particular, internal waves that can radiate through the ocean and may break, contributing to mixing. The geothermal energy is in the form of heat and, overall, its transfer through buoyancy into potential energy and thence into motion and mixing is relatively small.

6.10 Discussion

We now return to Fig. 6.1, with which we began in Section 6.1. It summarizes the uncertain conclusions of this chapter about the sources of the energy flux that drives mixing in the abyssal ocean. There is (just) enough energy flux to account for the 1.7 ± 0.8 TW required for mixing, 0.6–1.0 TW possibly coming from the internal tides, about 0.2 TW from the breaking of near-inertial internal waves and those of higher frequency in mid-water, and perhaps 0.5 TW from the waves undergoing reflection from sloping topography. The possible additional contributions resulting from interactions between mesoscale eddies and internal waves are at present unquantified.

It is customary in dynamical physical problems to solve a set of equations in some specific volume within which body forces may act. The equations may involve both motion and state variables, e.g., velocities, accelerations, pressures and fluid density. The equations must satisfy given conditions on the boundary of the specified volume where external forcing may be applied. Problems often involve finding the mean flow and its variability, an objective being to determine the circulation of fluid within the volume and to identify critical features, in this case, for example, where and at what rates dissipation occurs and how the rates depend on the forcing.

The ocean tides and the circulation of the ocean can now be reasonably well replicated using numerical models driven by external forcing. Dissipation is commonly represented parametrically in these models, the formulation implicitly representing the processes that occur at scales that are too small to be resolved or included explicitly in the models. These 'small-scale' processes may include internal waves, thought to be a major process in the transport of energy from the sea surface to the ocean deeps, and their breaking. As we have seen, however, it is not yet known exactly how, where or when internal waves are generated by the atmospheric forcing or how and where their energy is lost through breaking. The rate of exchange of energy between mesoscale eddies and internal waves is at present unquantified. It is therefore not surprising that parametric representations available at the moment are uncertain, if not known to be poor. Further development and critical testing of suitable parametric formulations remain to be done. The effects on the circulation of the highly non-uniform or patchy distribution of mixing that occurs within the ocean have yet to be resolved.

The problem posed by the deep-ocean mixing described in the sections above and summarized in Fig. 6.1 is unconventional. It is presented as one of the class of 'inverse problems' in which statistical inferences are made by using observations in combination with dynamical or kinematic models. In this case the total dissipation has been estimated (although only to within a factor of about 50%) from observations within a body of water – the abyssal ocean – and the fluxes across the boundaries, the sea surface and the sea bed, are known, but again sometimes only with great uncertainty and often much less precisely than is the total dissipation. The objective is to explain the inferred rate of dissipation and to obtain an understanding of the dominant physics, an understanding that is constrained by the existing fragmentary and often inadequate observations.

The problem, as stated, seems well posed, but the constraints, such as the flux of energy from the atmosphere into the deep ocean and the processes leading to mixing in particular regions, are not known to levels of accuracy sufficient to constrain severely or to quantify the internal processes leading to dissipation.

• The value of the present analysis of the energy fluxes is consequently not to explain the imprecise rates or to identify the dominant processes – these cannot yet be regarded as known – but rather to reveal where information is lacking and where more or better observations are needed in order to obtain more accurate measures.

It is an embarrassment to physical oceanographers that, whilst the energy fluxes to the ocean by solar and lunar forces and the geothermal heat flux from the solid Earth are known to within 10% and sometimes better, other fluxes of importance are known far less precisely. The processes leading to mixing, and the geographical distribution and seasonal variation of the mean rates of dissipation of turbulence kinetic energy within the body of the ocean, are still largely unknown.

Suggested further reading

The balance of energy flux

Wunsch and Ferrari's (2004) review of the fluxes of energy that may support mixing in the deep ocean provides a valuable guide to which reference should be made, if only to appreciate better the uncertainty of most of the estimates and to follow the routes of energy transfer in their figure 5. (The energy within the internal wave field should be 1.4 EJ, rather than 14 EJ as shown.)

Tidal energy

If Taylor's (1919) paper on *tidal friction* in the Irish Sea has not been read following Chapter 2, and Munk and Wunsch's (1998) discussion of *deep ocean mixing* after Chapter 4, perhaps they should be now.

Munk (1997) gives an elegant review of knowledge of the dissipation of tidal energy in the ocean.

Rudnick *et al.* (2003) describe the main findings of the Hawaiian Ocean Mixing Experiment (HOME) devised to study the conversion of barotropic tidal energy into baroclinic tides around the Hawaiian Ridge.

Further study

The uncertainty in the estimates of energy and energy fluxes given above is highly unsatisfactory. Better estimates of almost all rates, except those of the tides and geothermal heating, and better understanding of the processes that lead to these rates, are required if the dynamical machinery of the ocean is to be regarded as known and its response to climatic changes or variations in solar heating predicted with confidence.

Energetics and thermohaline circulation

Huang (1999) makes quantitative estimates of the fluxes involved and expresses some interesting, if debatable, ideas about the energetics of thermohaline circulation, including 'tidal mixing may be the most important energy source driving the thermohaline circulation' and that 'a large basin without midocean ridges' (to generate internal tides) 'would have a dramatically different thermohaline circulation'. Wunsch and Ferrari (2004) correct Huang's estimate of the energy input to mixing from geothermal heating by a factor of 10. Wang and Huang (2004) estimate the energy transferred from the wind into surface waves.

Breaking surface waves

Melville (1996) has reviewed the knowledge of breaking waves. Recent observations using aerial photographs like that shown in Fig. 6.4 are described by Melville and Matusov (2002).

Internal tidal energy

Egbert and Ray (2001) and Nycander (2005) estimate the total rate of transfer of energy from the barotropic tides to the baroclinic tides. Although the methods of calculation differ, the conclusions are in fair agreement. The baroclinic tidal energy radiating from the Hawaiian Ridge is quantified in observations described by Lee *et al.* (2006) and agrees with model predictions to within a factor of 2. The distance to which the internal tide may radiate – a dissipative length scale – exceeds 1000 km (see P.6.6).

Mixing over deep ocean sills and in canyons

Ferron *et al.* (1998) discuss observations and estimates of mixing in the Romanche Fracture Zone. Bryden and Nurser (2003) present a simple argument for the apparent importance of the density flux and mixing within passages between deep ocean basins (see P.6.8). Although Wunsch and Ferrari (2004) ascribe a total flux of energy to ocean mixing of only 0.1 TW to mixing in the flow through deep-ocean passages, the subject is one of continued interest and debate. Mixing occurs also over the many (possibly 10^4) sills in canyons on the flanks of the mid-ocean ridges, as pointed out by Thurnherr *et al.* (2005).

Gregg *et al.* (2005) estimate that 12 GW of energy is dissipated by the internal tides in canyons on the continental slopes.[13]

Inverse problems

A very helpful introduction to inverse models, formal procedures in which observations are combined with equations, usually those describing the dynamics or kinematics of fluids in general and the ocean in particular, so as to derive information about the ocean's properties, is given by Wunsch (2001).

Problems for Chapter 6

(E = easy, M = mild, D = difficult, F = fiendish)

P6.1 (M) Dissipation in the abyssal ocean. The total mass of the ocean is about 1.4×10^{21} kg. If 85% of the ocean is deeper than 1500 m and the mean buoyancy frequency in this deep water is $1.7 \times 10^{-3}\ \mathrm{s}^{-1}$, estimate the flux of energy required in order to sustain the rate of dissipation of turbulent kinetic energy at depths below 1500 m, assuming that the mixing it produces results in a diapycnal eddy diffusion coefficient of temperature equal to Munk's canonical value of $1 \times 10^{-4}\ \mathrm{m}^2\ \mathrm{s}^{-1}$.

P6.2 (D) The heat produced by turbulent dissipation and the advection–diffusion balance equation. Construct an equation relating the rate of increase in temperature, $\mathrm{d}T/\mathrm{d}t$, resulting from dissipation of turbulent kinetic energy per unit mass at a rate ε. Relating this dissipation rate to the eddy diffusion coefficient, and supposing that the contribution of salinity to the vertical gradient of density is negligible, express $\mathrm{d}T/\mathrm{d}t$ in terms of the local vertical gradient in temperature, $\mathrm{d}T/\mathrm{d}z$.

13 This value is a correction of the earlier estimate of 58 GW given by Carter and Gregg (2002) that is quoted in TTO (Thorpe, 2005).

From this, show that the rate of increase in temperature caused by dissipation is much less than that caused by the upward advection of the ambient temperature gradient at a speed, $w \approx 4\ \mathrm{m\ yr^{-1}}$, typical of the deep ocean, and can therefore be neglected in the advection–diffusion balance equation used by Munk to estimate K_T. You should make appropriate choices of any required parameters.

P6.3 (D) The energy flux from the wind. Supposing that the mean wind over the ocean is $7.5\ \mathrm{m\ s^{-1}}$ (the value may be slightly higher since the gales of the Southern Ocean are not well represented in the data used in determining this value), make an estimate of the mean flux of energy from the wind into the ocean. The area of the ocean surface is $3.61 \times 10^{14}\ \mathrm{m^2}$, and you may assume that the drag coefficient, C_{Da}, is 1.1×10^{-3} and that the density of the air at the sea surface is $1.25\ \mathrm{kg\ m^{-3}}$.

To estimate a mean value of W_{10}^3, you could suppose that the probability distribution function (or pdf), $y(x)$, of wind speed, $x = W_{10}$, is given by the positive skew function, $y = (x/a^2)\exp[-x^2/(2a^2)]$, where $a = \langle W_{10}\rangle (2/\pi)^{1/2}$ and $\langle W_{10}\rangle$ represents the mean wind speed (which you might prove is true!), or that it is equal to the exponential function $y = [\exp(-x/a)]/a$, where $a = \langle W_{10}\rangle$. (The integral $\int_0^\infty z^4 \exp(-z^2)\mathrm{d}z$ equals $3\sqrt{\pi}/8$, as can be found by integration by parts and noting that $(2/\sqrt{\pi})\int_0^z \exp(-t^2)\mathrm{d}t = \mathrm{erf}(z)$, which tends to 1 as z tends to infinity.)

P6.4 (M) Dissipation around the ocean coastline. The total length of the coastlines of the ocean is approximately 3.1×10^6 km. Use the information in problem P1.3 to estimate the mean rate of loss of energy from waves breaking at the coasts if the root mean square (rms) height of waves approaching shore is 1.0 m and their period is 7 s, supposing that no energy is reflected. Compare the dissipation rate with the estimates of the total flux of kinetic energy from the atmosphere to the sea.

(The mean wave heights and periods are very rough estimates. The total dissipation of wave energy around the ocean coastlines is not known with any certainty, but in some locations the energy flux of waves, when suitably harnessed, provides a very useful form of renewable power.)

P6.5 (E) Dissipation by a breaking wave. Show that the expression (6.1) is dimensionally correct.

• If it is assumed that the rate of loss of energy from a wave through breaking depends on density, the acceleration due to gravity, g, and some measure of the speed of the wave that relates to breaking (the speed of the breaking crest appears a suitable choice), but does not depend on surface tension, viscosity and wave amplitude, the form of (6.1) may be predicted on dimensional grounds. Surface tension and viscosity may affect only the breaking of very short waves, e.g., capillary waves (which have a dispersion relation that includes the effect of surface tension), and can be disregarded for the longer surface gravity waves described here. But what about the wave amplitude?

The amplitude of these longer waves may be involved in a relation or condition for wave breaking – e.g., a condition that the wave slope is sufficiently great or that the acceleration of the water produced by the wave motion as breaking is approached is some significant fraction of g. Both of these conditions indirectly, if not explicitly,

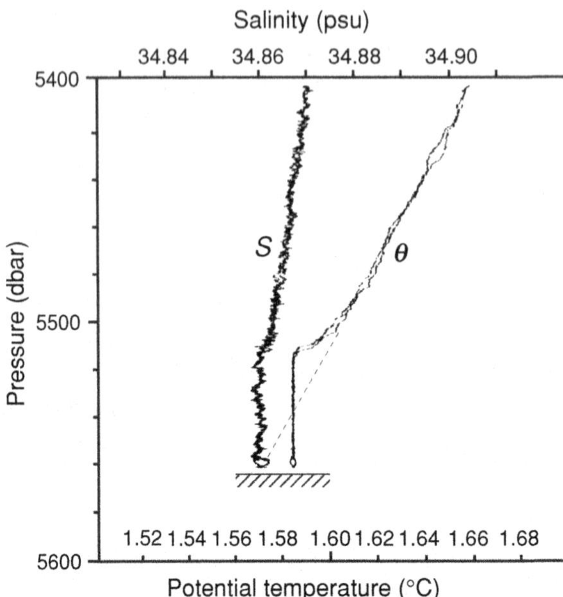

Figure 6.8. The benthic boundary layer. Profiles of potential temperature, θ, and salinity, S, in the Hatteras Abyssal Plain. The units of pressure, decibars, correspond approximately to depth in metres. The dashed line is an interpolation to the seabed of the uniform gradient in potential temperature above the mixed benthic boundary layer. (From Armi and Millard, 1976.)

involve the wave amplitude. The condition that a wave is breaking (together with the wave dispersion relation) may therefore provide an implicit relation involving amplitude, g and a measure of wave speed. The amplitude of a breaking wave is therefore a dimensional variable that is not independent of g and the speed characterizing the wave. The inclusion of g and the speed characterizing the wave in (6.1), but not wave amplitude, does not therefore imply a complete absence of dependence on wave amplitude. Indeed, if the empirical relation (6.1) based on laboratory studies is generally valid, it appears likely that there is also a relation between the amplitude of a breaking wave and the other dimensional variables.

P6.6 (M) Propagation of internal tides across ocean basins. About 20 GW of barotropic M_2 tidal energy is lost at the 3000-km-long Hawaiian Ridge and some 6 GW of baroclinic tidal energy is radiated away in a beam towards the northeast (the remainder of the energy being radiated to the southwest or dissipated locally). Estimate how far the baroclinic tide can propagate, supposing that the beam does not diverge, that *all* the energy lost by the internal tide is lost through interactions that sustain the ambient internal wave field and that this is dissipated through internal wave breaking at a vertically integrated rate (energy loss per unit area of ocean surface) of about 1×10^{-3} W m^{-2} that is independent of distance from the Ridge.

P6.7 (M) Mixing near the seabed. Temperature profiles through the benthic boundary layer is shown in Fig. 6.8. It appears that mixing near the bed has homogenized the relatively uniform temperature gradient in the overlying water.

Supposing that the water overlying the seabed has a uniform density gradient with corresponding buoyancy frequency, N, equal to 3×10^{-3} s^{-1}, estimate the change in potential energy per unit horizontal area required in order to homogenize this uniform

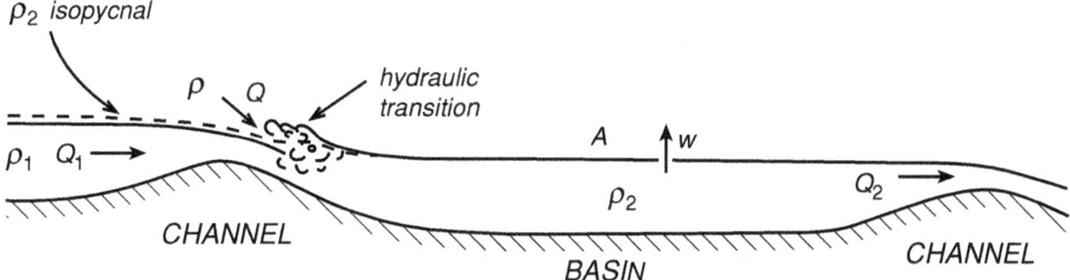

Figure 6.9. Mixing in flows over sills. The sketch represents mixing in the northward flow of AABW over the sill in the Vema Channel ('CHANNEL', on the left, and marked 'B' in Fig. 6.6) into the Brazil Basin ('BASIN'). The volume flux, Q, of water of density $\rho(<\rho_1)$, is entrained by turbulent mixing into the flow, Q_1, of density ρ_1 passing through the Vema Channel. The entrained water is mixed across the dashed ρ_2 isopycnal, to produce water of supposedly uniform density ρ_2 in the Brazil Basin. This water (i.e., that of density ρ_2, with $\rho < \rho_2 < \rho_1$, in the Basin) upwells at speed w across the area, A, of the upper surface of the ρ_2 water, and overflows from the Basin through the Romanche Fracture Zone (and other channels) with net flux Q_2. (This illustration has been modified from that of Bryden and Nurser (2003) by specifically including the entrained flux, Q.)

density gradient and to form a bottom layer of thickness $2H = 20$ m. Assuming that 20% of the energy available from the working of the shear stress on the seabed when $\langle U^2|U|\rangle = 1.25 \times 10^{-4}$ m^3 s^{-3} (chosen to correspond to $\langle U^2|U|\rangle^{1/3} = 0.05$ m s^{-1}, a speed typical of the deep-ocean benthic boundary layer) is used in mixing, estimate how long it will take to complete the mixing of the near-bed uniform density gradient to the thickness of 20 m.

P6.8 (D) Mixing in flows through channels. (This problem should be compared with P4.9.) Figure 6.9 represents the northward flow of Antarctic Bottom Water (AABW; see Section 6.4.3) through the Vema Channel (marked 'B' in Fig. 6.6) into the deep Brazil Basin. The flux of AABW of density ρ_1 entering the Channel is Q_1 and there is a hydraulic jump (or perhaps several jumps) within the channel that entrains overlying water of density $\rho < \rho_1$ at a rate Q. The mixing at the jump results in water of density $\rho_2 < \rho_1$ (but $\rho_2 > \rho$) that fills the deep Brazil Basin. (The flux, Q, is across the ρ_2 isopycnal surface.) The flux of water of density ρ_2 from the Basin is Q_2. The area of the Basin is A and the imbalance of the fluxes into and out of the Basin results in upwelling at speed w. Bryden and Nurser (2003) determined the flux of density anomaly $Q_1(\rho_1 - \rho_2)$ associated with the flow of AABW into the Brazil Basin.

Write down the equation for conservation of volume in the basin.

Supposing that the upward advective transport of density at speed w across the horizontal surface of area A of the basin is exactly balanced by a diffusive density flux $(K_\rho A|\mathrm{d}\rho/\mathrm{d}z|)$, so there is no net vertical density flux through the area A, and neglecting the effects of geothermal heat flux, write down an equation for the conservation of density and the maintenance of a steady state within the Brazil Basin. Use the equation of volume conservation to derive an equation relating the flux identified by Bryden and

Nurser to that associated with the entrainment flux, $Q(\rho_2 - \rho)$, across the ρ_2 isopycnal surface and to the advective flux, $\rho_2 w A$.

If the vertical density gradient in the water at the top of the layer of density ρ_2 is $(0.6 \pm 0.3) \times 10^{-4}$ kg m^{-4} and the area of the Brazil Basin is 3×10^6 km^2, use Munk's canonical value of eddy diffusivity and the diffusion–advection balance (P4.8 and P4.9) to estimate $\rho_2 w A$. Supposing that the flux of AABW, Q_1, is 6.9 Sv and that $(\rho_1 - \rho_2) = 0.05$ kg m^{-3}, determine the entrainment flux and compare it with the total diapycnal diffusive flux over the area of the Brazil Basin.

• A similar calculation led Bryden and Nurser to conclude that the diapycnal flux of density, a measure of 'mixing', as water passes through channels separating neighbouring basins in the deep ocean may exceed that within the basins themselves, even though the area of the basins greatly exceeds that of the narrow channels.

P6.9 (M) Energy loss in channels. The Vema Channel (marked 'B' in Fig. 6.6) is about 400 km long. Supposing that the thickness of the AABW in the Channel is less than 1 km, show that the energy lost through the bottom stress may exceed the flux of kinetic energy in the Channel. (You may approximate the cross section of the Channel by a rectangle and ignore mixing at the sides of the Channel.)

Further energy may be lost in mixing and in work done against the Reynolds stress at the upper boundary of the AABW. The flux of kinetic energy is sustained by work done by the pressure driving the flow through the Channel.

P6.10 (E) The geothermal heat flux and the diffusive heat flux in the water column. Show that the mean upward geothermal flux of heat through the seabed is small in comparison with the downward diffusive flux of heat required to maintain the steady-state advection–diffusion balance in the deep water at moderate latitudes in the Pacific (e.g., see Section 4.7 and P4.8). You should assume a mean temperature gradient of 1.3 °C km^{-1} in the deep stratified water and Munk's canonical value for the eddy diffusion coefficient of heat.

Estimate the total mean downward diffusive heat flux in the deep water of the tropical regions of the Pacific and Atlantic Oceans, supposing that their total area is about 7.6×10^7 km^2 and that the flux per unit area in the Atlantic is about the same as in the Pacific. Show that this diapycnal flux of heat is much less than the heat flux of about 2 PW transported northwards at 30 °N within these Oceans (see footnote 5).

P6.11 (M) Vent temperatures and mixing. Use the values of the fluxes given in Section 6.9 to show that the mean temperature difference between ambient seawater and the fluid discharged as a hydrothermal flux of water carrying geothermal heat into the ocean is generally very small in comparison with the values of over 300 °C reported for some vents (e.g., see Section 3.2.2.)

• This implies that hydrothermal vents will not, globally, contribute very substantially to mixing within the deep ocean.

References

The page numbers at which references are to be found are given in square brackets after each entry.

Alford, M. H. and M. C. Gregg 2001. Near-inertial mixing: modulation of shear, strain and microstructure at low latitude. *J. Geophys. Res.*, **106**, 16 947–16 968. [137]

Alford, M. and R. Pinkel 2000. Observations of overturning in the thermocline: the context of ocean mixing. *J. Phys. Oceanogr.*, **30**, 805–832. [138]

Andreas, E. L., K. J. Claffey, R. E. Jordan *et al.* 2006. Evaluations of the von Kármán constant in the atmospheric surface layer. *J. Fluid Mech.*, **559**, 117–149. [107]

Armi, L. and E. D'Asaro 1980. Flow structures in the benthic ocean. *J. Geophys. Res.*, **85**, 469–484. [103, 106]

Armi, L. and D. M. Farmer 1988. The flow of Mediterranean water through the Strait of Gibraltar, *Prog. Oceanogr.*, **21**, 1–105. [150]

Armi, L. and R. C. Millard 1976. The bottom boundary layer of the deep ocean. *J. Geophys. Res.*, **81**, 4983–4990. [222]

Armi, L., D. Hebert, N. Oakey *et al.* 1989. Two years in the life of a Mediterranean salt lens. *J. Phys. Oceanogr.*, **19**, 354–370. [189]

Ashford, O. M. 1985. *Prophet or Professor? The Life And Work of Lewis Fry Richardson.* Bristol: Adam Hilger Ltd. [140]

Batchelor, G. K. 1967. *An Introduction to Fluid Dynamics*. Cambridge: Cambridge University Press. [15]

Batchelor, G. 1996. *The Life and Legacy of G. I. Taylor*. Cambridge: Cambridge University Press. [51]

Bowden, K. F. and L. A. Fairbairn 1956. Measurements of turbulent fluctuations and Reynolds stresses in a tidal current. *Proc. Roy. Soc. Lond. A*, **237**, 422–438. [67, 69, 73]

Brown, G. L. and A. Roshko 1974. On the density effects and large structure in turbulent mixing layers, *J. Fluid Mech.*, **64**, 775–816. [38]

Brügge, B. 1995. Near-surface mean circulation and kinetic energy in the central North Atlantic from drifter data. *J. Geophys. Res.*, **100**, 20 543–20 554. [175]

Bryden, H. and A. J. G. Nurser 2003. Effects of strait mixing on ocean stratification. *J. Phys. Oceanogr.*, **33**, 1870–1872. [220, 223]

Caldwell, D. R. and T. M. Chriss 1979. The viscous boundary layer at the sea floor. *Science*, **205**, 1131–1132. [114, 115]

Cardwell, D. S. L. 1989. *James Joule, a Biography*. Manchester: Manchester University Press. [6]

Carter, G. S. and M. C. Gregg 2002. Intense, variable mixing near the head of the Monterey submarine canyon. *J. Phys. Oceanogr.*, **32**, 3145–3165. [220]

Chriss, T. M. and D. R. Caldwell 1982. Evidence for the influence of form drag on bottom boundary layer flow. *J. Geophys. Res.*, **87**, 4148–4154. [87, 106]

Csanady, G. T. 1973. *Turbulent Diffusion in the Environment*. Dordrecht: Reidel. [195]

D'Asaro, E. A. 2001. Turbulent vertical kinetic energy in the ocean mixed layer. *J. Phys. Oceanogr.*, **31**, 3530–3537. [55]

D'Asaro, E. A. and R. C. Lien 2000. Lagrangian measurements of waves and turbulence in stratified flows. *J. Phys. Oceanogr.*, **30**, 641–655. [55]

Davis, R. E. 1987. Modelling eddy transport of passive tracers. *J. Mar. Res.*, **45**, 635–666. [190]

1991a. Lagrangian ocean studies. *Annu. Rev. Fluid Mech.*, **23**, 43–64. [190]

1991b. Observing the general circulation with floats. *Deep-Sea Res.*, **38**, S531–S571. [190]

Davis, R. E., R. de Szoeke, D. Halpern and P. Niiler 1981. Variability in the upper ocean during MILE. Part 1: the heat and momentum budgets. *Deep-Sea Res.*, **28**, 1427–1451. [127]

Drazin, P. G. and W. H. Reid 1981. *Hydrodynamic Stability*. Cambridge: Cambridge University Press. [150]

Durst, F. and B. Ünsal 2006. Forced laminar-to-turbulent transition of pipe flows. *J. Fluid Mech.*, **560**, 449–464. [32]

Eckart, C. 1948. An analysis of the stirring and mixing in incompressible fluids. *J. Mar. Res.*, **7**, 265–275. [13, 32]

Egbert, G. D. and R. D. Ray 2001. Estimates of M2 tidal energy dissipation from TOPEX/Poseidon altimeter data. *J. Geophys. Res.*, **106** (C10), 22 475–22 502. [219]

Ellison, T. H. and J. S. Turner 1959. Turbulent entrainment in stratified flows. *J. Fluid Mech.*, **6**, 423–448. [133]

Eriksen, C. C. 1978. Measurements and models of fine structure, internal gravity waves, and wave breaking in the deep ocean. *J. Geophys. Res.*, **83**, 2989–3009. [127]

Faller. A. J. and S. J. Auer 1988. The role of Langmuir circulation in the dispersion of surface tracers. *J. Phys. Oceanogr.*, **18**, 1108–1123. [196]

Ferron, B., H. Mercier, K. Speer, A. Gargett and K. Polzin 1998. Mixing in the Romanche Fracture Zone. *J. Phys.*

Oceanogr., **28**, 1929–1945. [130, 220]

Fringer, O. B. and R. L. Street 2003. The dynamics of breaking progressive interfacial waves. *J. Fluid Mech.*, **494**, 319–353. [126, 153]

Ganachaud, A. and C. Wunsch 2000. Improved estimates of global ocean circulation, heat transport and mixing from hydrographic data-transport. *Nature*, **408**, 453–457. [141]

Gargett, A. E. 1999. Velcro measurements of turbulent kinetic energy dissipation rate, ε. *J. Atmos. Oceanic Technol.*, **16**, 1973–1993. [74]

Gargett, A. E., T. R. Osborn and P. W. Naysmth 1984. Local isotropy and the decay of turbulence in a stratified fluid. *J. Fluid Mech.*, **144**, 231–280. [75]

Gill, A. E. 1981. Homogeneous intrusions in a rotating stratified fluid. *J. Fluid Mech.*, **103**, 275–295. [32]

1982. *Atmosphere–Ocean Dynamics*. London: Academic Press. [6, 21, 22, 32, 157]

Graf, G. 1989. Benthic–pelagic coupling in a deep-sea benthic community. *Nature*, **341**, 439–441. [113]

Grant, H. L., A. Moilliet and W. M. Vogel 1968. Some observations of turbulence in and above the thermocline. *J. Fluid Mech.*, **34**, 443–448. [74, 117]

Grant, H. L., R. W. Stewart and A. Moilliet 1962. Turbulence spectra from a tidal channel. *J. Fluid Mech.*, **12**, 241–268. [60, 61, 74]

Gregg, M. C. 1980. Microstructure patches in the thermocline. *J. Phys. Oceanogr.*, **10**, 915–943. [58]

1987. Diapycnal mixing in a thermocline: a review. *J. Geophys. Res.*, **92**, 5249–5286. [150]

1989. Scaling turbulent dissipation in the thermocline. *J. Geophys. Res.*, **94**, 9686–9698. [152]

1999. Uncertainties and limitations in measuring ε and χ_T. *J. Atmos. Oceanogr. Technol.*, **16**, 1483–1490. [62, 74]

2003. Reduced mixing from the breaking of internal waves in equatorial waters. *Nature*, **422**, 513–515. [153]

Gregg, M. C., G. S. Carter and E. Kunze 2005. Corrigendum. *J. Phys. Oceanogr.*, **35**, 1712–1715. [220]

Griffiths, R. W. and P. F. Linden 1981. The stability of vortices in a rotating stratified fluid. *J. Fluid Mech.*, **105**, 283–316. [32]

Heathershaw, A. D. 1979. The turbulent structure of the bottom boundary layer in a tidal current. *Geophys. J. Roy. Astron. Soc.*, **58**, 395–430. [71]

Hickey, B. M. and T. C. Royer 2001. California and Alaska Currents. In *Encyclopedia of Ocean Sciences*, ed. J. H. Steele, S. A. Thorpe and K. K. Turekian, vol. 1. London: Academic Press, pp. 368–379. [177]

Hinze, J. O. 1959. *Turbulence*. New York: McGraw-Hill. [74]

Hogg, N., P. Biscaye, W. Gardner and W. J. Schmitz 1982. On the transport and modification of Antarctic Bottom Water in the Vema Channel. *J. Mar. Res.*, **40** (suppl), 231–263. [155]

Howard, L. N. 1961. Note on a paper by John W. Miles. *J. Fluid Mech.*, **10**, 509–512. [119, 150]

Huang, R. X. 1999. Mixing and energetics of the oceanic thermohaline circulation. *J. Phys. Oceanogr.*, **29**, 727–746. [219]

Huppert, H. E. and J. S. Turner 1981. Double-diffusive convection. *J. Fluid Mech.*, **106**, 299–329. [147]

Hunt, G. N. 1985. *Radioactivity in Coastal and Surface Waters of the UK*. Aquatic Environment Monitoring Report, MAFF Directorate of Fisheries Research, Lowestoft. [191]

Hunt, J. C. R., J. R. Pacheco, A. Mahalov and H. J. S. Fernando 2005. Effects of rotation and sloping terrain on the fronts of density currents. *J. Fluid Mech.*, **537**, 285–315. [32]

Joule, J. P. 1850. On the mechanical equivalent of heat. *Phil. Trans. Roy. Soc. Lond.*, **140** (1), 61–82 (also in *Scientific Papers*, published by Taylor and Francis for the Physical Society of London, 1884, pp. 298–328). [6, 32]

Karpen, V., L. Thomsen and E. Suess 2004. A new 'schlieren' technique application for fluid flow visualisation at cold seep sites. *Mar. Geol.*, **204**, 145–159. [74]

Kunze, E. 2001. Vortical modes. In *Encyclopedia of Ocean Sciences*, ed. J. H. Steele, S. A. Thorpe and K. K. Turekian, vol. 6. London: Academic Press, pp. 3174–3178. [192]

Kunze, E., A. J. Williams III and R. W. Schmitt 1987. Optical microstructure in the thermohaline staircase east of Barbados. *Deep-Sea Res.*, **34**, 1697–1704. [74, 150]

Kunze, E., M. G. Briscoe and A. J. Williams III 1990. Interpreting shear and strain from a neutrally buoyant float. *J. Geophys. Res.*, **95**, 18 111–18 125. [152]

Kunze, E., E. Firing, J. M. Hummon, T. K. Chereskin and A. M. Thurnherr 2006. Global abyssal mixing inferred from lowered ADCP shear and CTD strain profiles. *J. Phys. Oceanogr.*, **36**, 1553–1576. [152]

Lamarre, E. and W. K. Melville 1994. Void-fraction measurements and sound-speed fields in bubble plumes generated by breaking waves. *J. Acoust. Soc. Amer.*, **95**, 1317–1329. [33]

Langmuir, I. 1938. Surface motion of water induced by wind. *Science*, **87**, 119–123. [95, 107]

Ledwell, J. R., A. J. Watson and C. S. Laws 1998. Mixing of a tracer in the pycnocline. *J. Geophys. Res.*, **108**, 21 499–21 529. [143, 150, 183, 184, 189]

Lee, C. M., E. Kunze, T. B. Sanford *et al.* 2006. Internal tides and turbulence along the 3000-m isobath of the Hawaiian Ridge. *J. Phys. Oceanogr.*, **36**, 1165–1183. [220]

Leibovich, S. 1983. The form and dynamics of Langmuir circulation. *Annu. Rev. Fluid Dyn.*, **15**, 391–427. [107]

Levitus, S., J. Antonov and D. Boyer 2005. The warming of the world ocean, 1955–2003. *Geophys. Res. Lett.*, **32**, L02604, doi:10.1029/2004GL021592. [207]

Lien, R.-C. and M. C. Gregg 2001. Observations of turbulence in a tidal beam and across a coastal ridge. *J. Geophys. Res.*, **106**, 4575–4591. [136]

Lu, Y. and R. G. Lueck 1999. Using broadband ADCP in a tidal channel. Part II: turbulence. *J. Atmos. Oceanic Technol.*, **14**, 1568–1579. [74]

Lumkin, R., A.-M. Treguier and K. Speer 2002. Lagrangian eddy scales in the North Atlantic Ocean. *J. Phys. Oceanogr.*, **32**, 2426–2440. [190]

Lupton, J. E. 1995. Hydrothermal plumes: near and far field. In *Seafloor*

Hydrothermal Systems. Physical, Chemical, Biological and Geological Interactions. Washington, DC: American Geophysical Union, pp. 317–346. [159]

McClean, J. L., P.-M. Poulain and J. W. Pelton 2002. Eulerian and Lagrangian statistics from surface drifters and high-resolution POP simulation in the North Atlantic. *J. Geophys. Res.*, **22**, 2472–2491. [190]

MacKinnon, J. A. and M. C. Gregg 2003. Mixing on the late-summer New England Shelf – solibores, shear and stratification. *J. Phys. Oceanogr.*, **33**, 1476–1492. [152]

McWilliams J. C., P. P. Sullivan and C.-H. Moeng 1997. Langmuir turbulence in the ocean. *J. Fluid Mech.*, **334**, 31–58. [107]

Maxey, M. R. 1987. The gravitational settling of aerosol particles in homogeneous turbulence and random flow fields. *J. Fluid Mech.*, **174**, 441–465. [192]

Melville, W. K. 1996. The role of surface-wave breaking in air–sea interaction. *Annu. Rev. Fluid Mech.*, **28**, 279–321. [109, 219]

Melville, W. K. and P. Matusov 2002. Distribution of breaking waves at the ocean surface. *Nature*, **417**, 58–63. [219]

Miles, J. 1961. On the stability of heterogeneous shear flows. *J. Fluid Mech.*, **10**, 496–508. [119, 150]

Miles, J. W. and L. N. Howard 1964. Note on a heterogeneous shear layer. *J. Fluid Mech.*, **20**, 331–336. [120]

Morris, M. Y., M. M. Hall, L. C. St. Laurent and N. G. Hogg 2001. Abyssal mixing in the Brazil Basin. *J. Phys. Oceanogr.*, **31**, 3331–3348. [155]

Morton, B. R., G. I. Taylor and J. S. Turner 1956. Turbulent gravitational convection from maintained and instantaneous sources. *Proc. Roy. Soc. Lond. A*, **234**, 1–13. [83, 106, 110]

Moum, J. N. and W. D. Smyth 2001. Upper ocean mixing processes. In *Encyclopedia of Ocean Sciences*, ed. J. H. Steele, S. A. Thorpe and K. K. Turekian, vol. 6. London: Academic Press, pp. 3093–3100. [107]

Moum, J. N., M. C. Gregg, R.-C. Lien and M.-E. Carr 1995. Comparison of turbulent kinetic energy dissipation rates from two microstructure profilers. *J. Atmos. Oceanic Technol.*, **12**, 346–366. [46, 64, 74]

Moum, J. M., D. M. Farmer, W. D. Smyth, L. Armi and S. Vagle 2003. Structure and generation of turbulence at interfaces strained by internal solitary waves propagating shoreward over the continental shelf. *J. Phys. Oceanogr.*, **33**, 2093–2112. [26]

Mowbray, D. E. and B. S. H. Rarity 1967. A theoretical and experimental investigation of the phase configuration of internal waves of small amplitude in a density stratified liquid. *J. Fluid Mech.*, **28**, 1–16. [27]

Munk W. 1966. Abyssal recipes. *Deep-Sea Res.*, **13**, 207–230. [140, 141, 150]

1997. Once again: once again – tidal friction. *Prog. Oceanogr.*, **40**, 7–35. [219]

Munk, W. and C. Wunsch 1998. Abyssal recipes II: energetics of tidal and wind mixing. *Deep-Sea Res.*, **45**, 1976–2009. [150, 219]

Nimmo Smith, W. A. M., S. A. Thorpe and A. Graham 1999. Surface effects of bottom-generated turbulence in a shallow sea. *Nature*, **400**, 251–254. [12]

Nimmo Smith, W. A. M., J. Katz and T. R. Osborn 2005. On the structure of turbulence in the bottom boundary layer of the coastal ocean. *J. Phys. Oceanogr.*, **35**, 72–93. [44]

Nycander, J. 2005. Generation of internal waves in the deep ocean by tides. *J. Geophys. Res.*, **110** (C10), C10028, doi:10.1029/2004JC002487. [219]

Oakey, N. S. 1982. Determination of the rate of dissipation of turbulent energy from simultaneous temperature and velocity shear microstructure measurements. *J. Phys. Oceanogr.*, **12**, 256–271. [74]

2001. Turbulence sensors. In *Encyclopedia of Ocean Sciences*, ed. J. H. Steele, S. A. Thorpe and K. K. Turekian, vol. 6. London: Academic Press, pp. 3063–3069. [74]

Okubo, A. 1971. Oceanic diffusion diagrams. *Deep-Sea Res.*, **18**, 789–802. [181, 182, 189]

Ollitrault, M., C. Gabillet and A. C. de Verdière 2005. Open ocean regimes of relative dispersion. *J. Fluid Mech.*, **533**, 381–407. [190]

Osborn, T. R. 1974. Vertical profiling of velocity microstructure. *J. Phys. Oceanogr.*, **4**, 109–115. [62]

1980. Estimates of the local rate of vertical diffusion from dissipation measurements. *J. Phys. Oceanogr.*, **10**, 83–89. [132]

Osborn, T. R. and C. S. Cox 1972. Oceanic fine structure. *Geophys. Fluid Dyn.*, **3**, 321–345. [131]

Ozmidov, R. V. 1965. On the turbulent exchange in a stably stratified ocean. *Izvestia Acad. Sci. U.S.S.R., Atmos. & Ocean Phys.*, **1**, 861–871. [129, 182]

Park, P. K., D. R. Kester, I. W. Duedall and B. H. Ketchum 1983. Radioactive wastes and the ocean. In *Wastes in the Ocean*, ed. P. K. Park *et al.*, vol. 3. New York: John Wiley and Sons, pp. 4–46. [192]

Pasquill, F. 1962. *Atmospheric Diffusion*. Toronto: D. van Nostrand Co. Ltd. [193]

Peters, H., M. C. Gregg and J. M. Toole 1988. On the parametrization of equatorial turbulence. *J. Geophys. Res.*, **93**, 1199–1218. [153]

Polzin, K. 1996. Statistics of the Richardson number: mixing models and fine structure. *J. Phys. Oceanogr.*, **26**, 1409–1425. [128, 152]

Polzin, K., K. G. Speer, J. M. Toole and R. W. Schmitt 1996. Intense mixing of Antarctic Bottom Water in the equatorial Atlantic Ocean. *Nature*, **380**, 54–56. [215]

Polzin, K. L., J. M. Toole, J. R. Ledwell and R. W. Schmitt 1997. Spatial variability of turbulent mixing in the abyssal ocean. *Science*, **276**, 93–96. [150]

Proudman, J. 1953. *Dynamical Oceanography*. London: Methuen & Co. Ltd. [149]

Rapp, R. J. and W. K. Melville 1990. Laboratory measurements of deep-water breaking waves. *Phil. Trans. Roy. Soc. Lond. A*, **331**, 735–800. [32]

Ray, R. D. and G. T. Mitchem 1997. Surface manifestations of internal tides in the deep ocean: observations from altimetry and island gauges. *Prog. Oceanogr.*, **40**, 135–162. [202]

Reynolds, O. 1883. An experimental investigation of the circumstances which determine whether the motion of water shall be direct or sinuous, and of the law of resistance in parallel channels. *Phil. Trans. Roy. Soc. Lond. A*, **174**, 935–982 (also in *Scientific Papers* (1901), **2**, 51–105). [3, 32]

1895. On the dynamical theory of incompressible viscous fluids and the determination of the criterion. *Phil. Trans. Roy. Soc. Lond. A*, **186**, 123–164 (also in *Scientific Papers* (1901), **2**, 535–577). [39]

1900. On the action of rain to calm the sea. In *Papers on Mechanical and Physical Subjects*, vol. 1. Cambridge: Cambridge University Press, pp. 86–88. [40]

Rhines, P. B. 1979. Geostrophic turbulence. *Annu. Rev. Fluid Mech.*, **11**, 401–411. [192]

Richardson, L. F. and H. Stommel 1948. Note on eddy diffusion in the sea. *J. Meteorol.*, **5**, 238–240. [189]

Richardson, P. L., A. S. Bowers and W. Zenk 2000. A census of Meddies tracked by floats. *Prog. Oceanogr.*, **45**, 209–250. [159]

Rossby, H. T., S. C. Riser and A. J. Mariano 1983. The western North Atlantic – a Lagrangian viewpoint. In *Eddies in Marine Science*, ed. A. R. Robinson. Berlin: Springer-Verlag, pp. 66–91. [190]

Rossby, T. and D. Webb 1970. Observing abyssal motions by tracking Swallow floats in the SOFAR channel. *Deep-Sea Res.*, **17**, 359–365. [179, 190]

Rudnick. D. L., T. J. Boyd, R. E. Brainard *et al.* 2003. From tides to mixing along the Hawaiian Ridge. *Science*, **301**, 355–357. [219]

St. Laurent, L. and R. W. Schmitt 1999. The contribution of salt fingers to vertical mixing in the North Atlantic Tracer Release Experiment. *J. Phys. Oceanogr.*, **29**, 1404–1424. [149]

St. Laurent, L. C., J. M. Toole and R. W. Schmitt 2001. Buoyancy forcing by turbulence above rough topography in the abyssal Brazil Basin. *J. Phys. Oceanogr.*, **31**, 3476–3495. [144, 203]

Saunders, P. M. 1987. Flow through Discovery Gap. *J. Phys. Oceanogr.*, **17**, 631–643. [155]

Schmitt, R. W. 1981. Form of the temperature–salinity relationship in the Central Water: evidence of double-diffusive mixing. *J. Phys. Oceanogr.*, **11**, 1015–1026. [148]

2001. Double-diffusive convection. In *Encyclopedia of Ocean Sciences*, ed. J. H. Steele, S. A. Thorpe and K. K. Turekian, vol. 2. London: Academic Press, pp. 757–766. [150]

Schmitt, R. W. and J. R. Ledwell 2001. Dispersion and diffusion in the deep ocean. In *Encyclopedia of Ocean Sciences*, ed. J. H. Steele, S. A. Thorpe and K. K. Turekian, vol. 2. London: Academic Press, pp. 726–733. [186]

Schott, F., M. Visbeck, U. Send *et al.* 1996. Observations of deep convection in the Gulf of Lions, northern Mediterranean, during winter of 1991/2. *J. Phys. Oceanogr.*, **26**, 505–524. [107]

Sharples, J. and J. H. Simpson 2001. Shelf-sea and slope fronts. In *Encyclopedia of Ocean Sciences*, ed. J. H. Steele, S. A. Thorpe and K. K. Turekian, vol. 5. London: Academic Press, pp. 2760–2768. [109]

Shay, T. J. and M. C. Gregg 1984a. Turbulence in an oceanic convective layer. *Nature*, **310**, 282–285. [107]

1984b. Turbulence in an oceanic convective layer – corrigendum. *Nature*, **311**, 84. [92, 107]

1986. Convectively driven turbulent mixing in the upper ocean. *J. Phys. Oceanogr.*, **16**, 1777–1798. [93, 107]

Simpson, J. H. 1998. Tidal processes in shelf seas. In *The Sea*, vol. 10, ed. K. H. Brink and R. Robinson. New York: John Wiley and Sons, pp. 113–150. [109]

Simpson, J. H., J. Brown, J. Matthews and G. Allen 1990. Tidal straining, density currents and stirring in the control of estuarine stratification. *Estuaries*, **13**, 125–132. [109]

Simpson, J. H., T. P. Rippeth and A. R. Campbell 2000. The phase lag of turbulent dissipation in tidal flow. In *Interactions between Estuaries, Coastal Seas and Shelf Seas*, ed. T. Yanagi. Tokyo: Terr Scientific Publishing Co. (TERRAPUB), pp. 57–67. [109]

Smyth, W. D. and J. N. Moum 2000. Anisotropy of turbulence in stably stratified mixing layers. *Phys. Fluids*, **12**, 1343–1362. [75, 124]

Smyth, W. D. and K. B. Winters 2003. Turbulence and mixing in Holmboe waves. *J. Phys. Oceanogr.*, **33**, 694–711. [151]

Smyth, W. D., D. Hebert and J. N. Moum 1996. Local ocean response to a multiphase westerly wind burst 2. Thermal and freshwater responses. *J. Geophys. Res.*, **101** (C10), 22 513–22 533. [94]

Smyth, W. D., P. O. Zavialov and J. N. Moum 1997. Decay of turbulence in the upper ocean following sudden isolation from surface forcing. *J. Phys. Oceanogr.*, **27**, 810–822. [112]

Sparrow, E. M., R. B. Husar and R. J. Goldstein 1970. Observations and other characteristics of thermals. *J. Fluid Mech.*, **41**, 793–800. [82]

Spelt, P. D. M. and A. Biesheuvel 1997. On the motion of gas bubbles in homogeneous isotropic turbulence. *J. Fluid Mech.*, **336**, 221–244. [192]

Staquet, C. and J. Sommeria 2002. Internal gravity waves: from instabilities to turbulence. *Annu. Rev. Fluid Mech.*, **34**, 559–593. [150]

Stern, M. E. 1960. The 'salt fountain' and thermohaline convection. *Tellus*, **12**, 172–175. [145]

Stephens, J. C. and D. P. Marshall 2000. Dynamical pathways of Antarctic Bottom Water in the Atlantic. *J. Phys. Oceanogr.*, **30**, 622–640. [213]

Stommel, H. 1949a. Horizontal diffusion due to oceanic turbulence. *J. Mar. Res.*, **8**, 199–225. [189]

1949b. Trajectories of small bodies sinking slowly through convection cells. *J. Mar. Res.*, **8**, 24–29. [173]

Stommel, H., A. B. Arons and D. Blanchard 1956. An oceanographical curiosity: the perpetual salt fountain. *Deep-Sea Res.*, **3**, 152–153. [145, 150]

Strang, E. J. and H. J. S. Fernando 2001. Entrainment and mixing in stratified shear flows. *J. Fluid Mech.*, **428**, 349–386. [150]

Sundermeyer, M. A. and M. P. LeLong 2005. Numerical simulations of lateral dispersion by the relaxation of diapycnal mixing events. *J. Phys. Oceanogr.*, **35**, 2368–2386. [192]

Sundermeyer, M. A., J. R. Ledwell, N. S. Oakey and B. J. W. Greenan 2005. Stirring by small-scale vortices caused by patchy mixing. *J. Phys. Oceanogr.*, **35**, 1245–1262. [36, 192]

Tait, R. I. and M. R. Howe 1971. Thermohaline staircases. *Nature*, **231**, 178–179. [150]

Taylor, G. I. 1919. Tidal friction in the Irish Sea. *Phil. Trans. R. Soc. Lond. A*, **220**, 1–92. [33, 70, 73, 201, 219]

1931. Internal waves and turbulence in a flud of variable density. *Rapp. et Proc.-Verb. des Réunions du Conseil Perm. Int. pour l'Expl. de la Mer*, **76**, 35–42 (also in *Scientific Papers* 1960, ed. G. K. Batchelor, vol. 2, pp. 240–246). [133, 149]

1959. The present position in the theory of turbulent diffusion. In *Atmospheric Diffusion and Air Pollution*, ed. F. N. Frenkiel and P. A. Shephard. London: Academic Press, pp. 101–112. [193]

Tennekes, H. and J. L. Lumley 1982. *A First Course in Turbulence*, 2nd edn. Cambridge, MA: MIT Press. [74]

Thorpe, S. A. 1968. A method of producing a shear flow in a stratified fluid. *J. Fluid Mech.*, **32**, 693–704. [151]

1971. Experiments on the instability of stratified shear flows: miscible fluids. *J. Fluid Mech.*, **46**, 299–319. [122]

1985. Small-scale processes in the upper ocean boundary layer. *Nature*, **318**, 519–522. [97, 101]

1995. On the meandering and dispersion of a plume of floating particles caused by Langmuir circulation and a mean current. *J. Phys. Oceanogr.*, **25**, 685–690. [164, 195]

2004. Langmuir circulation. *Annu. Rev. Fluid Dyn.*, **36**, 55–79. [107]

2005. *The Turbulent Ocean*. Cambridge: Cambridge University Press (referred to as TTO). [i, ix, 6, 21, 33, 75, 109, 152, 192, 220]

Thorpe, S. A. and A. J. Hall 1980. The mixing layer of Loch Ness. *J. Fluid Mech.*, **101**, 687–703. [100]

Thorpe, S. A., T. R. Osborn, J. F. E. Jackson, A. J. Hall and R. G. Lueck 2003. Measurements of turbulence in the upper ocean mixing layer using Autosub. *J. Phys. Oceanogr.*, **33**, 122–145. [66, 68]

Thurnherr, A. M., L. C. St. Laurent, K. G. Speer, J. M. Toole and J. R. Ledwell 2005. Mixing associated with sills in a canyon on the mid-ocean ridge flank. *J. Phys. Oceanogr.*, **35**, 1370–1381. [220]

Toggweiler, J. R. and R. M. Key 2001. Thermohaline circulation. In *Encyclopedia of Ocean Sciences*, ed. J. H. Steele, S. A. Thorpe and K. K. Turekian, vol. 6. London: Academic Press, pp. 2941–2947. [141]

Troy, C. D. and J. R. Koseff 2005. The instability and breaking of long internal waves. *J. Fluid Mech.*, **543**, 107–136. [119]

Turner, J. S. 1973. *Buoyancy Effects in Fluids*. Cambridge: Cambridge University Press. [106, 149, 150]

Veron, F. and W. K. Melville 2001. Experiments on the stability and transition of wind-driven water surfaces. *J. Fluid Mech.*, **446**, 25–65. [107, 108]

Wang, W. and R. X. Huang 2004. Wind energy input to surface waves. *J. Phys. Oceanogr.*, **34**, 1276–1280. [219]

Welander, P. 1955. Studies of the general development of motion in a two-dimensional, ideal fluid. *Tellus*, **7**, 141–156. [14, 32]

Wesson, J. C. and M. C. Gregg 1994. Mixing at the Camarinal Sill in the Strait of Gibraltar. *J. Geophys. Res.*, **99**, 9847–9878. [46, 53, 140, 150]

Wijesekera, H. and T. J. Boyd 2001. Upper ocean heat and freshwater budgets. In *Encyclopedia of Ocean Sciences*, ed. J. H. Steele, S. A. Thorpe and K. K. Turekian, vol. 6. London: Academic Press, pp. 3079–3083. [96]

Williams, A. J. 1975. Images of ocean microstructure. *Deep-Sea Res.*, **22**, 811–829. [74]

Wimbush, M. 1970. Temperature gradient above the deep-sea floor. *Nature*, **227**, 1041–1043. [106]

Wimbush, M. and W. Munk 1971. The benthic boundary layer. In *The Sea*, ed. A. E. Maxwell, vol. 4(1). New York: John Wiley and Sons, pp. 731–758. [106]

Winkel, D. P., M. C. Gregg and T. B. Sanford 1996. Resolving oceanic shear and velocity with the Multi-Scale Profiler. *J. Atmos. Oceanic Technol.*, **13**, 1046–1072. [74]

Winkel, D. P., M. C. Gregg and T. H. Sanford 2002. Patterns of shear and turbulence across the Florida Current. *J. Phys. Oceanogr.*, **32**, 3269–3285. [46, 134]

Woods, J. D. 1968. Wave-induced shear instability in the summer thermocline. *J. Fluid Mech.*, **32**, 791–800. [118, 149]

Wunsch, C. 2001. Inverse models. In *Encyclopedia of Ocean Sciences*, ed. J. H. Steele, S. A. Thorpe and K. K. Turekian, vol. 3. London: Academic Press, pp. 1368–1374. [220]

Wunsch, C. and R. Ferrari 2004. Vertical mixing, energy, and the general circulation of the oceans. *Annu. Rev. Fluid Mech.*, **36**, 281–314. [200, 218, 219, 220]

Yaglom, A. M. 1994. A. N. Kolmogorov as a fluid mechanician and founder of a school of turbulence research. *Annu. Rev. Fluid Mech.*, **26**, 1, 22. [48]

Zhurbas, V. and I. S. Oh 2004. Drifter-derived maps of lateral diffusivity in the Pacific and Atlantic Oceans in relation to surface circulation patterns. *J. Geophys. Res.*, **109** (C5), C05015, doi:10.1029/2003JC002241. [178, 190]

Index

Printed in the United States
By Bookmasters